Open Cities | Open Data

Scott Hawken • Hoon Han
Christopher Pettit
Editors

Open Cities | Open Data

Collaborative Cities in the Information Era

palgrave
macmillan

Editors
Scott Hawken
Urban Development and Design
Faculty of the Built Environment
University of New South Wales
Sydney, NSW, Australia

Hoon Han
City Planning, Faculty of the Built Environment
University of New South Wales
Sydney, NSW, Australia

Christopher Pettit
Urban Science, Faculty of the Built Environment
University of New South Wales
Sydney, NSW, Australia

ISBN 978-981-13-6607-9 ISBN 978-981-13-6605-5 (eBook)
https://doi.org/10.1007/978-981-13-6605-5

Cover image: Anna Klepatckaya
Cover design: Tom Howey

This Palgrave Macmillan imprint is published by the registered company Springer Nature Singapore Pte Ltd.
The registered company address is: 152 Beach Road, #21-01/04 Gateway East, Singapore 189721, Singapore

The original version of this book was revised. The correction to this book can be found at https://doi.org/10.1007/978-981-13-6605-5_18

Foreword

Open Cities Open Data addresses, and indeed links, two of the most transformational processes at work in the contemporary world: urbanisation and digitalisation. While the academic, policy and indeed market discussions about 'Smart Cities' have been underway for over a decade, the chapters and research in this collection reflect a more recent, and I think more interesting, re-framing of the discussion around what Goldsmith and Crawford (2014) have termed the 'data-driven and responsive city'. Some commentators have questioned whether or not 'Smart Cities' itself remains a useful term or whether it has become tarnished by rhetorical overkill, under-delivery by city governments, or over-selling by vendors of their technology. Regardless of these questions it is a fact that digital technologies and platforms have created an exciting potential for improved city management and governance and thus urban outcomes. That is to say we can, by leveraging 'digital' and the fast-morphing reality of the Internet of Things with the right policies and values, manage our cities in a smarter and more inclusive manner.

Given the decisive shift towards an urban future for the majority of the world's population and the strains that the world-historical process is putting on our urban systems and infrastructure, nothing could be more important than to ensure that as our cities get bigger, they get smarter and more responsive to their inhabitants. We have already seen how technologies and platforms such as hand-held smart devices, social media,

WiFi and fibre networks, have transformed service design and delivery for consumers. They have also reduced the gap, blurred the distinction even, between consumers and producers of products and services. We have also begun to see the more digitally savvy and progressive public bodies and city governments redesign their business processes and community engagement using such tools, with the best ensuring that their city has access to the necessary digital infrastructure and the skills required to benefit from it.

While we have been through bouts of technological determinism and indeed optimism before, I think it is fair to conclude that the trend in this direction is irreversible and welcome. We are moving towards a more digitally enabled city government. With enhanced fibre networks, the Internet of Things, and the advent of 5G, more data-driven and responsive infrastructure networks and services will augur the first real instrumentalising of the Smart City and the making of active prosumers out of once passive consumers. We will see digitally empowered communities demanding to shape not just government service design and delivery but also the very strategic planning objectives of their council and indeed their city. We will also—through the proliferation of data from sensors in infrastructure, the streets we walk on, the buildings we pass by, the energy and water systems we deploy and the transport that we use, know as never before about how our core urban systems are performing; both in relation to expectations and in relation to other cities. In a digital era, 'big data' will be ubiquitous, whether or not it is formally released by governments and councils to inform civil society about the development path of its cities. In prospect, our governments and cities will have the opportunity to be better managed but also more accountable through the translation of Big Data into Open Data. With Open Data the true costs and benefits of urban infrastructure—did those new roads really reduce congestion or just make our cities worse?—will be evermore apparent.

This direction of technological development is, I believe, universally applicable and the articles in the *Open Cities Open Data* collection provide insights of international relevance. The collection is of particular importance in Australia. This is because despite being one of the more urban societies on the planet and notwithstanding Australia's strong performance in the wider region as a centre of tech start-ups and fintech innovation, it would be hard to describe any Australian city as the lead in terms of realising Smart Cities' objectives. It is not one of the international city benchmarking exercises in which our cities yet do well. This I suspect, has a lot to do with the governance challenge in Australian cities, as the cities which are doing best in such benchmarking—Singapore, Boston, London—are smart in a governance sense and not just in technology. Specifically, such cities have effective city governance at the metropolitan level or have responsibility at the council level for significant budgets and services. The City of Boston has the usual range of services you would see in an Australian council plus housing, public transport, police, education and health. Singapore, perhaps the leading global Smart City, is effectively a city-state with aligned government services promoting an effective integration of technology, land use and transport. In both, the city government is either big enough or integrated enough for data-driven approaches and technology tools and platforms to be applied across the government, across the whole city or indeed both.

By contrast digital innovation and exemplars may be being developed in Australian cities, but they will be found in say one or more of the 31 separate councils operating in Sydney, or one of 20 or more siloed New South Wales (NSW) Government departments or agencies. Such fractured governance makes it hard for such innovation to be scaled up or spread across government or across the whole of a city at a metro level. Relatedly, without effective governance at the metropolitan level and in the context of a siloed state government, there is as yet little capacity to feed data on urban or infrastructure performance from councils or state departments back into the management of the city. And without accountability for performance at the metropolitan level to electorates, Australian cities have both a democratic and a managerial deficit whose consequences for liveability, productivity and equity have begun to be noted in global city benchmarking exercises.

It has been well said of Australian cities that they are orphans of public policy with their management and governance falling between too many small and under-powered councils and a too mighty and remote state government on the other. This has also left city inhabitants without the institutions or platforms to enable them to shape their cities at a metropolitan level or even have a cross-city discussion about urban ideas that matter at the local level.

However, this situation makes the drive towards more 'data-driven and responsive' cities in Australia even more important, not less. Without conventional governance and urban management arrangements for our metros, data openness and digitally enabled civic engagement become crucial alternative tools of accountability and performance improvement in and for Australian cities.

Consequently, academics, researchers and policy experts, in and out of the university, are essential in promoting an understanding of best practice in urban performance and transformation. By identifying good 'data-driven and responsive' practices and great projects they have a unique opportunity to show the way for public decision-makers and identify strategies to scale up Open Data innovation across government or indeed the city as a whole. By providing, collating and analysing data on the actual performance of a key piece of urban infrastructure in relation to its claimed benefits, when initially justified and prioritised, better policy and investment decisions can flow. By seeking to ensure that such findings are not confined to academic outputs but are disseminated also via broader media channels, a better civic debate about the direction and performance of our cities can be created.

With much controversy now in Australia about these matters, and a strong sense that our cities are neither understood fully nor managed and yet are on a path to double in population just after mid-century, there was never a better moment for 'big data' to shape the critical debate about the future of our now rather 'big cities'. And therefore, no better time for a collection of this title to be published.

Tim Williams is Cities Leader for Arup in Australasia and Chair of Open Cities whose mission is to promote more data-driven, responsive, resilient and inclusive cities in Australia.

Sydney, Australia — Tim Williams

Reference

Goldsmith, S., & Crawford, S. (2014). *The responsive city: Engaging communities through data-smart governance*. Hoboken: Wiley.

Acknowledgement

Each chapter in this book went through a rigorous double to triple blind peer review process. The editors would like to thank the many reviewers for their critical contributions and considerable investment in time. The editors would like to acknowledge the dedication and intelligent support provided by the urban designer and scholar Ashley Bakelmun. Ashley assisted with the production of the book in many valuable ways including communications with the various chapter authors, through the compilation of the chapters and through the facilitation of the rigorous double- and triple-blind peer-review process. This work was carried out with great professionalism and humanity.

The book was financially assisted by the Faculty of the UNSW Built Environment, Sydney, which also supported the 2015 Open Cities Open Data workshop that inspired many of the ideas and connections evident in the book. The workshop was convened by the UNSW Built Environment's Smart Cities Research Cluster. The Smart Cities Research Cluster is a research network that supports collaboration and research on Smart Cities and has annual high profile events. The Smart Cities Research Cluster (SCRC) seeks to promote and advance the design, planning and delivery of urban environments and services through the use of information and communication technologies with a focus on spatial technologies.

Contents

Notes on Contributors

Irina Anastasiu is a PhD candidate at the Queensland University of Technology, Australia. She holds a BSc and an MSc in Media Informatics and Communication Science from Ludwig Maximilian University of Munich, and an Honours degree in Technology Management from the Center for Digital Technology and Management. She is interested in exploring participatory citymaking as an opportunity to build solidarity with those neglected or affected by dominating smart city visions in order to strengthen urban social movements towards systemic change in how cities are produced and governed. In doing so, she seeks to integrate social, political and urban theory into civic technology, further drawing upon her extensive industry experience in the design and implementation of digital technology.

Aristide Athanassiadis is Chair of Circular Economy and Urban Metabolism at the Université Libre de Bruxelles, Belgium. Within this framework, he attempts to build bridges between the academia, public administrations and "circular" actors in order to accelerate Brussels' transition towards a more circular economy and metabolism. During the last years, he has advised and has acted as an external consultant for a number of local, regional and international administrations and organisations on the topics of urban metabolism and circular economy. Finally, Athanassiadis co-created the non-profit organisation and open-source platform Metabolism of Cities that promotes urban metabolism by centralising data and publications and developing free online tools for teaching and research.

Ashley Bakelmun is Founder and Director of Urban Equity Lab, an urban strategy consultancy advocating for the most vulnerable populations in cities. Her research and design projects address complex "urban questions" with a goal of working towards socially and environmentally just cities. Her current work focuses on the intersection of urban design and planning with inequality and segregation. In her past consulting roles, she led design and construction teams to deliver sustainable master plans on education/corporate campuses.

Saswat Bandyopadhyay is a professor at the School of Planning, CEPT (Centre for Environmental Planning and Technology) University, India. Bandyopadhyay is the Area Chair for the infrastructure planning programme at CEPT with over two decades of experience in the urban development sector in South Asia. He has spearheaded several Indian urban missions and capacity development activities. Government of India, The World Bank, Asian Development Bank (ADB), Deutsche Gesellschaft für Internationale Zusammenarbeit (GIZ) and United Nations Development Program (UNDP) are some of the clients that Bandyopadhyay has consulted for.

Sarah Barns is a research fellow based at the Institute for Culture and Society, Western Sydney University, Australia. In August 2013, Barns was awarded a three-year Urban Studies Postdoctoral Research Fellowship by the Urban Studies Foundation for a project titled 'Platform urbanism: The Role of City Labs, Data Infomediaries, and Open Government Experiments in Urban Governance'. The project examines how urban knowledge is being shaped through smart technologies and pervasive data and will address new institutional alignments and governance arrangements emerging in key digital cities. Barns' current research builds on her doctoral thesis, 'The Death & Life of the Real-Time City: Re-imagining the City of Digital Urbanism'.

Sean Beevers helped establish the London Air Quality Network and the Environmental Research Group at King's College London, is a member of the MRC-PHE Centre for Environment & Health and leads the MSc Global Air Pollution and Health (https://www.kcl.ac.uk/study/postgraduate/taught-courses/global-air-pollution-and-health-management-and-science.aspx). He is also Senior Lecturer in Air Pollution Modelling, King's College London, UK. Beevers has written 74 peer-reviewed papers and has worked closely with London policymakers to implement major changes to the city, from the London Congestion Charging Zone to the recent London Ultra Low Emissions Zone. Beevers' future research goals are to investigate associations between air pollution and health by applying King's air quality models to cities globally and to develop models of personal exposure indoors and outdoors. He aims to develop policies that reduce exposure to air pollution and to investigate the interaction between air quality and climate change policy.

Devanjan Bhattacharya holds a PhD in Geomatics Engineering and his research interests are in applications of geoinformatics for societal challenges, geohazard management, smart cities, and spatial technologies. He is a postdoctoral manager of the EU H2020 project GEO-C at NOVA IMS, Universidade Nova de Lisboa, Lisbon, Portugal.

Carole Bodilis is an engineer with a major in Geographic Information Systems (GIS) and Urbanism. She is completing her Joint European Master in Environmental Studies—Cities and Sustainability (Erasmus Mundus programme JEMES CiSu) in Portugal, working on relevant indicators for planning support systems focused on environmental issues.

Paul Burton is Professor of Urban Management and Planning and Director of the Cities Research Institute at Griffith University, Australia.

Bronwyn Cameron originally joined Sydney Water in 2012 as an undergraduate Co-op Scholar. She returned to Sydney Water on the Graduate Program in 2014. During her four years at Sydney Water, she has worked in civil delivery, network operations and strategic analytics.

She works in Service Planning, developing asset strategies. Cameron holds a First Class Honours degree in Civil Engineering from University of New South Wales (UNSW) and a Masters in Water Resources.

Fang Chen is a prominent leader in AI/data science with international reputation and industrial recognition. She is the winner of the "Oscars" of Australian science, 2018 Australian Museum Eureka Prize for excellence in data science. She has created many innovative research and solutions, transforming industries that utilise AI/data science. She has helped industries worldwide advance towards excellence in increasing their productivity, innovation, profitability and customer satisfaction. The transformations to industry with practical impact won her many industrial recognitions including being named as "Water Professional of The Year" in 2016. She has actively led in developing new strategies, which prioritise the organisation's objectives, and capitalise on any growth opportunities. She has built up a career in creating research and business plans, and executing with leadership and passion. In science and engineering, Chen has 300+ refereed publications, including several books. She has filed 30+ patents in Australia, US, Canada, Europe, Japan, Korea, Mexico and China.

Auriol Degbelo is a postdoctoral researcher at the Institute for Geoinformatics, University of Münster, Germany. His current research interests include semantic integration of geospatial information, re-use of open government data, and interaction with geographic information.

Jonathan Doig is working on establishing Urban Analytics Data Infrastructure involving geospatial data services and semantic web techniques. He has nearly 30 years' experience with geospatial information systems, web services, and data management in government and private firms in Australia and the UK, with a focus on environmental data publishing and reporting.

Lex Drennan is an Adjunct Industry Fellow at the Policy Innovation Hub, Griffith University, Australia.

Marcus Foth is Professor of Urban Informatics in the QUT Design Lab, Australia, and an Honorary Professor in the School of Communication and Culture at Aarhus University, Denmark. His transdisciplinary work is at the international forefront of human-computer interaction research and development with a focus on smart cities, community engagement, media architecture and sustainability. Foth founded the Urban Informatics Research Lab in 2006 and the QUT Design Lab in 2016. Ahead of their time and before the term "smart cities" became popular, Foth pioneered a new field of study and practice: Urban informatics examines people creating, applying and using information and communication technology and data in cities and urban environments.

Mark S. Fox is Distinguished Professor of Urban Systems Engineering, Professor of Industrial Engineering and Computer Science and founding director of the Centre for Social Services Engineering at the University of Toronto, Canada. He is a fellow of the Association for the Advancement of Artificial Intelligence.

Carlos Granell holds a five-year Ramón y Cajal postdoctoral fellowship at the Universitat Jaume I (UJI) of Castellón, Spain. Before re-joining The Geospatial Technologies (Geotec) research group in 2014, he worked for three years as a post-doc in the Digital Earth and Reference Data Unit of the European Commissions' Joint Research Centre (JRC), and was a post- and pre-doctoral researcher during the period 2003–2010 at the Universitat Jaume I of Castellón, from which he holds a PhD (2006). His research interests lie in multi-disciplinary GIS, model web, and spatial analysis and visualization.

Ori Gudes is a research fellow at the City Futures Research Centre and a lecturer in the City Analytics Program at the University of New South Wales (UNSW), Australia. His research focuses on GIS, urban informatics, decision supports systems for urban planners, evaluation and usability, spatial analysis and visualisation. He works on developing Planning Support Systems (PSS) in the urban planning settings and evaluates its effectiveness and impact.

Ting Guo is an associate lecturer at the University of Technology Sydney, Australia. He received his PhD degree in Computer Science from University of Technology Sydney. His research mainly focuses on data mining, machine learning, multimedia systems, and bioinformatics. As a team member at Analytics Research Group in Data61, Guo is involved in designing and building machine-learning algorithms and systems for several projects including water pipe failure prediction projects, chock prediction project and data city building project. He has years of work experience in the programming, algorithm design and data analysis. He has also written several papers in top-tier conferences and journals in The Institute of Electrical and Electronics Engineers (IEEE), Association for Computing Machinery (ACM) and Association for the Advancement of Artificial Intelligence (AAAI) societies.

Hoon Han is an associate professor and Director of the City Planning Program at the University of New South Wales (UNSW), Faculty of the Built Environment, Australia. His research focuses on smart cities, urban renewal and human behaviour. Hoon Han is an associate editor of the journal *City, Culture and Society* (Elsevier) and sits on the international editorial board of both *Housing Studies* (Taylor & Francis) and *Spatial Information Research* (Springer).

Jeremy Harkins is the Founding Director of "ineni Realtime", an innovative building technology company focused on the development of the Realtime Visualisation Industry.
Delivering keynote presentations, Harkins is at the leading edge of realtime virtual technologies and has spoken internationally about the visual interaction with smart cities through 3D immersive environments, Virtual Reality (VR), and Augmented Reality (AR).

Harkins is a strong advocate of Building Information Modelling (BIM), believing that the seamless integration of intuitive visual interfaces with robust data is a vital direction for architecture, construction, and infrastructure.

With over 15 years of experience in Architectural Technologies, including professional work, consultancy, and full-time academia, Harkins has been a lecturer at the University of New South Wales (UNSW), helping to educate industry in what can be possible and is currently focused on the rapid growth and positioning of "ineni Realtime" as market leaders in virtual interaction design.

Scott Hawken is a convenor of the Smart Cities Research Cluster at the University of New South Wales (UNSW) and a lecturer in the Urban Development of Design Program, UNSW Built Environment, Australia. His interests in technology and design are evident from his research which brings together innovative methods for data collection and interpretation with urban planning and design approaches. He leads a series of international research collaborations on data augmented design and biophilic cities and urban design. Each project promotes an inclusive and human-centred vision of smart cities.

Christopher Heard is the projects director at "ineni Realtime", with a particular interest in applying emerging technologies to the built environment. His role sees him co-ordinating a team to produce game-changing software solutions for visualisation and buildings.
An early and eager adopter in the Virtual Reality (VR) realm, Heard has been working with headset hardware since the original Oculus Developer Kit 1 arrived. When his head is not buried in a VR headset, you will find him experimenting with a new input device.

Heard's experience has taken him overseas to Shanghai to speak at an international transport symposium about the benefits of realtime technology for visualisation, as well to Singapore to present to companies about applying VR technology to their development process.

He is teaching architecture students at University of Technology Sydney (UTS) how to apply the latest VR and realtime technology solutions to their workflow.

Sonia Hugh is a research analyst in the Centre for Social Services Engineering at the University of Toronto, Canada. She obtained her MSc in Environmental Studies from Australian National University, Australia. Her work primarily involves analysing global and regional open datasets with GIS multi-purpose applications.

Parisa Izadpanahi is a lecturer at the School of Design and Built Environment, Faculty of Humanities, Curtin University, Australia. Her research interests are focused on sustainable urban design, environmental design and the role of data and data analytics on design processes.

Frank Kelly holds the chair in Environmental Health at King's College London, UK, where he is Director of the Analytical & Environmental Sciences Division. His other positions of responsibility are Director of the Environmental Research Group, Director of the NIHR Health Protection Research Unit in Health Impact of Environmental Hazards and Deputy Director of the Medical Research Council and Public Health England (MRC-PHE) Centre for Environment & Health. From these dual positions, he is able to combine his two main research interests, namely free radical/antioxidant biochemistry and the impact of atmospheric pollution on human health. Further details of this activity can be found on the ERG and London Air websites.
In addition to his academic work, Kelly is past President of the European Society for Free Radical Research and past Chairman of the British Association for Lung Research. He is also involved with providing policy support to the World Health Organization (WHO) on air pollution issues and is a member of the Committee on the Medical Effects of Air Pollution (COMEAP).

Simone Z. Leao is a research fellow in Urban Modelling and Simulation at the City Futures Research Centre and a lecturer in the City Analytics Program at the University of New South Wales (UNSW). Interested in the interdependencies between built environment, society and technology, she works on developing knowledge and methodologies to assist urban planning to generate/regenerate urban areas in relation to contemporary challenges.

Lochlan Morrissey is a postdoctoral research fellow at the Policy Innovation Hub, Griffith University, Australia.

Bitange Ndemo is Associate Professor of Entrepreneurship at the University of Nairobi's Business School, Kenya. His research centres on the link between information and communication technologies (ICTs) and small and medium enterprises with emphasis on how ICTs influence economic development in Africa. Ndemo is the Chairman of Kenya's Distributed Ledgers and Artificial Intelligence Taskforce that will develop a roadmap for the country's digital transformation. He is also an adviser and board member to several organisations, including Safaricom (one of the leading telecommunications companies in Africa), the Mpesa Foundation and Research ICT Africa (based in South Africa). He is a former Permanent Secretary of Kenya's Ministry of Information and Communication, where he was credited with facilitating many transformative ICT projects. He is an Open Data/big data evangelist and is dedicated to simplification (i.e. visualisation) of data for ordinary citizens to consume. He writes two columns every week for the *Business Daily* and Nation online.

Pascal Perez is a senior professor and also the Director of the SMART Infrastructure Facility at the University of Wollongong (UOW), Australia. He has a 30-year experience in advanced data analytics and simulation to explore complex interactions within social and technological systems. He is a fellow of the Royal Society of New South Wales (NSW) and of the Modelling and Simulation Society of Australia and New Zealand (MSSANZ). In 2002, he received an ARC-International Linkage Fellowship to develop social modelling research at the Australian National University. Perez has written 196 refereed papers and book chapters. In 2006, he co-edited with his colleague David Batten the book *Complex Science for a Complex World.*

Christopher Pettit is Professor of Urban Science, Faculty of the Built Environment, University of New South Wales; the inaugural Chair of Urban Science at the University of New South Wales; and the Director of the new Master of City Analytics Program, https://www.be.unsw.edu.au/degrees/postgraduate-coursework/master-of-city-analytics. His expertise is in the convergence of the fields of urban planning and GIS, where he has written more than 150 peer-reviewed papers. For the last 20 years, he has been undertaking research and development in the use of spatial information and mapping technologies for envisioning "what if?" scenarios across both urban and regional landscapes. His research expertise also spans into applications, the development and the critique of geographical visualisation tools, including advanced spatial decision-support systems and city dashboards.

Sarbeswar Praharaj is a Sessional Academic Staff at the Faculty of the Built Environment, University of New South Wales (UNSW), Australia. In the Smart Cities Research Cluster UNSW, Praharaj leads research on smart cities policy, business models and urban analytics. He is the coordinator for the Australia-India Smart Cities Knowledge Exchange Network sponsored by the Australia-India Council, Australia Government.

Homa Rahmat has completed a PhD on a data-driven analysis of urban processes using Twitter data. Her research interests include citizen participation and innovations, Web 2.0 applications, social media data, network analysis, and urban data visualisation.

Bill Randolph joined the Faculty of the Built Environment at the University of New South Wales, Australia, in August 2004 as a professor and Director of the City Futures Research Centre. He served as an associate dean research between 2009 and 2013. At City Futures he leads a research team specialising in housing policy, housing markets and affordability, urban renewal, city data and analytics, urban well-being and metropolitan planning policy issues. Randolph has 40 years of experience as a researcher on housing and urban policy issues in the academic, government, non-government and private sectors. He holds a PhD from the London School of Economics.

Jonathan Reades has been Lecturer in Quantitative Human Geography in the Department of Geography at King's College London, UK, since 2013. Previously, Reades had been a research associate for two years at University College London (UCL)'s Centre for Advanced Spatial Analysis, following the completion of his MPhil/PhD at the Bartlett School of Planning. He also holds a BA (1997) in Comparative Literature from Princeton University. In the intervening years, Reades worked for a database mining and marketing start-up based in New York and London in a range of capacities: graphic designer, web application developer and project manager. This work stimulated his interest in 'big data' and its potential as a platform for examining and acting upon 'smart cities'.

Markus Rittenbruch is Senior Lecturer in Interaction Design at the School of Design, Creative Industries Faculty, Queensland University of Technology, Australia. He leads the "Design and Technologies of Tomorrow" program at the QUT Design Lab, where he fosters transdisciplinary research, innovation and experimentation on the design and application of interactive technologies. His research focuses on the design and evaluation of large-scale collaborative systems, collaborative visualisation systems, human-robot collaboration, ambient, ubiquitous and physical computing, natural user interfaces, large-screen multi-touch interfaces and innovative ways of interfacing with and visualising sensors and sensor data.

Roger Rooney is a senior project manager of smart parking, Canberra, ACT Government. He is a public entrepreneur, passionate about generating new ideas, conducting UX design sprints and delivering a human-centred Smart City. He knows from first-hand experience that technology is a force multiplier that can be harnessed to speed up work, create value and deliver better services for end-users. He is collaborating with Data61 to develop predictive parking to decrease congestion and travel times and help deliver the 30-minute city.

Peter Runcie is the New Industry and Platforms Leader for Data61—CSIRO's data science unit. In this role, he is developing mission-based innovation programmes to address Australia's societal, environmental and economic challenges. He is an inventor of over 25 granted patents in voice and video communications, data networking, biometrics with others pending in structural health monitoring and data analytics. He holds an MBA (Exec) from the Australian Graduate School of Management and is a graduate of the Australian Institute of Company Directors. Runcie is also an independent director on the board of the Australian Smart Communities Association and also of the New South Wales Smart Sensing Network.

David Sanderson has over 25 years of experience working across the world in development and emergencies. From 1994 to 1998, Sanderson was a project manager at the Oxford Centre for Disaster Studies. He worked for eight years for the non-governmental organisation (NGO) CARE International UK, as head of policy and subsequently regional manager for southern and west Africa. From 2006 to 2013, Sanderson was Director of Centre for Development and Emergency Practice (CENDEP), a centre at Oxford Brookes University focusing on development and emergencies. Between 2013 and 2014, he was a full-time visiting professor at Harvard University, where he taught a course, 'Design for Urban Disaster', and from 2014 to 2015, he was a professor at the Norwegian University of Science and Technology (NTNU). Sanderson was appointed the Inaugural Judith Neilson Chair of Architecture at University of New South Wales (UNSW) in February 2016.

Ronald Schroeter completed his PhD at the Urban Informatics Research Lab, QUT, in 2012, during which he developed the award-winning "Discussions In Space," a fun, fast-paced, short-text platform for public urban screens and mobile phones that facilitates public civic engagement, collective expression and public discourse among (particularly young) local citizens. He is a senior research fellow at the Centre for Accident Research and Road Safety—Queensland (CARRS-Q), QUT, Brisbane, Australia. His research focus is the design of innovative driving experiences that make transport by car or bike more fun and safe. This work allows him to embrace multi-disciplinary research across Human-Computer Interaction (HCI)/Human-Machine Interface (HMI), psychology and road safety.

Sarah Jane Shoenfeld co-directs the digital public history project "Mapping Segregation in Washington DC," which is documenting the former extent of racially restricted housing in the U.S. capital city along with other mechanisms of segregation and displacement and their legacy. Her company Prologue DC engages in a variety of history projects, including research for exhibitions and films, historic landmark and district nominations, oral histories, and walking tours. Shoenfeld writes and gives presentations for both scholarly and general audiences and received an MA in History and Certificate in Public History from Northeastern University.

Carmela Ticzon finished her Masters in Environmental Management from the University of New South Wales, where she also worked as a technical assistant under the City Futures Research Centre. Her interests lie in science communication and interdisciplinary approaches to building environmental solutions.

Anne Tiernan is Professor of Public Policy and Dean (Engagement) of the Griffith Business School, Australia.

Ian Tilley is an MPhil candidate at the University of New South Wales (UNSW) Faculty of the Built Environment, Australia. Having lived in Far North Queensland for six years, Tilley brings his lived experience of cyclones to his research at UNSW and is keen to improve how communities experience and recover from natural disasters. He brings a breadth of experience to UNSW from the diverse range of professional roles and is technical director at a Sydney-based IT services company.

Sergio Trilles received his PhD in Integration of Geospatial Information from the Universitat Jaume I in 2015. He had the opportunity to work for four months as a researcher in the Digital Earth and Reference Data Unit of the European Commissions' Joint Research Centre (JRC). He is a post-doc researcher at the GEOTEC group.

Kalpana Viswanath is a researcher who has been working on issues of urban and women's safety for over 20 years. She is the co-founder of SafetiPin, a mobile app developed to support community and women's safety. She has worked as a consultant with UN Women and UN-Habitat on issues of gender and urban safety, and spearheaded the 'Safe Delhi for Women' initiative led by Jagori in Delhi. She has also provided technical support to safe city for women programmes in Cambodia, Pakistan, Indonesia, Vietnam, Kerala, Mumbai and Kolkata. She is a member of the Advisory Group on Gender Issues for UN-Habitat, board member of the International Centre for the Prevention of Crime (ICPC) and the chairperson of Jagori. Her work has been published widely and she has co-edited a book on building gender inclusive cities.

Elizabeth Wentz is Dean of Social Sciences in the College of Liberal Arts and Sciences and a professor in the School of Geographical Sciences and Urban Planning at Arizona State University, USA. Her research focuses on the design, implementation and evaluation of geographic technologies with particular emphasis on how such technologies can be used to understand urban environments. Geographic technologies, which include GIS, remote sensing and spatial analysis, offer insight into how human activities and physical space relate in urban systems by using quantitative methods to measure and analyse such activities. Her teaching focuses on geographic technologies and graduate-level research design and proposal writing.

Jonas Wissing is a student assistant at the Institute of Geoinformatics (ifgi) at the University of Münster, Germany. He is studying a BSc in Geoinformatics and a BSc in Information Systems at the University of Münster, Germany.

Malcolm Wolski is Director of eResearch Services at Griffith University, Australia.

Komali Yenneti is a lecturer (New Generation Network Scholar) at UNSW Built Environment and Honorary Fellow at the Australia India Institute. She has extensive evidence-based research and policy experience in climate change adaptation and mitigation, energy policy and smart sustainable cities and communities using data analytics and human communications sciences. She leads or manages a range of international collaborative research projects including 'Urban Heat Mitigation in Low-Income Households in India', 'Urban Heat Mitigation in Western Sydney', 'Climate-Smart Cities in India and Australia' and 'Mapping the Decision-Making Process to Identify Barriers and Drivers to Meet or Exceed BASIX Requirements of New Builds'.

Bang Zhang is a senior research scientist and team leader at Data61, CSIRO, Australia. He received his PhD degree in Computer Science from the University of New South Wales. His research interests focus on machine learning, data mining and data-driven smart urban planning. He has extensive experience in utilising advanced data analytics techniques to help industry improve productivity. He led and contributed to collaborative projects with 30 water utilities globally for providing predictive asset maintenance solutions. He led the collaborative project between Data61 and Telstra for predictive network maintenance. He contributed significantly to the real-time structural health monitoring project for Sydney Harbour Bridge. He also serves as a reviewer for prestigious international academic journals, such as *IEEE Transactions on Knowledge and Data Engineering (TKDE)* and *IEEE Transactions on Image Processing (TIP)*. He won the best patent activity award at NICTA (currently Data61) in 2014. He is the finalist of the young water professional of the year in NSW 2016.

Lelin Zhang is a senior research engineer at University of Technology Sydney, Australia. He received the BEng degree in Computer Science from South China University of Technology, Guangzhou, China, in 2007, and the Master of Information Technology, Master of Information Technology Management and PhD degrees from the University of Sydney, Australia, in 2009, 2010 and 2016, respectively. He is a postdoctoral fellow at Data61, CSIRO, Australia. He has written research papers in the fields of multimedia retrieval, computer vision, distributed computing and social network analysis, and applied his expertise on machine learning to various projects with telecommunication and utility industries and governments.

List of Figures

List of Tables

1

Introduction: Open Data and the Generation of Urban Value

Scott Hawken, Hoon Han, and Christopher Pettit

The World's Most Valuable Resource

The world has made two significant transitions in the new millennium. The first involves the transition to a knowledge economy where the most lucrative industry is now the production and management of information or "data". "Data" is the new "oil" of our age (The Economist, 2017). The second significant transformation is the transition from a rural to an

S. Hawken (✉)
Urban Development and Design, Faculty of the Built Environment, University of New South Wales, Sydney, NSW, Australia
e-mail: s.hawken@unsw.edu.au

H. Han
City Planning, Faculty of the Built Environment, University of New South Wales, Sydney, NSW, Australia
e-mail: h.han@unsw.edu.au

C. Pettit
Urban Science, Faculty of the Built Environment, University of New South Wales, Sydney, NSW, Australia
e-mail: c.pettit@unsw.edu.au

S. Hawken et al. (eds.), *Open Cities | Open Data*,
https://doi.org/10.1007/978-981-13-6605-5_1

urban world, a process known as urbanisation. Today most people live in cities (United Nations, 2014). As the world has crossed these thresholds, there are new opportunities to address wicked policy problems such as environmental degradation, migration and climate change, through data-assisted solutions. This book endeavours to provide sound evidence and case studies to support renegotiating the terms for information exchange, ownership and data use in cities for the benefit of urban citizens. In other words, this book is about the best use of the world's most valuable economic commodity–data—and how it can be used to assist the planetary transition to sustainable, productive, resilient and liveable urban futures.

Where has this incredible wealth come from? The data explosion has resulted from an urban setting where an extraordinary number of people carry around a powerful mobile device which, when linked to the World Wide Web, generates a rich information footprint. In this growing digital landscape, such devices are now also interacting with a growing number of Internet of Things (IoT), which is powered through a growing array of sensors. Such sensors are embedded in products ranging from TVs, watches and toasters, to cars, trains, building, precincts and cities. As Anthony Townsend (2013, 2014) has said, we are now outnumbered by our own digital devices. There are more than three devices for every human on the planet and this will increase to more than six by 2020 (Evans, 2011).

Today the world's largest economies and corporations trade in data and its products to generate value in new disruptive markets. The wealth and power that the information economy generates is dangerously concentrated. Current trends suggest this concentration will continue, leading to a greater divide across society (Manjoo, 2016; Pollock, 2018). The five wealthiest companies on the globe are all information communication and technology companies that monopolise their respective markets through a combination of "platform" economics and the inherent and emergent properties of the data economy (Andersson Schwarz, 2017; Pollock, 2018). In contrast to the fossil fuels that powered the economies of the twentieth century, digital information flows through the world economy in different ways as it can be reused and value added almost infinitely.

Facebook, one of the big US-based five infotech companies, now controls 80% of social media traffic. Even within the companies themselves, the power imbalance is extreme with a few founders and investors controlling the majority of the company's equity. The data economy has had such a big impact on cities because it is different kind of economy. The way it is mined, extracted, collected, collated, bought, sold, refined, processed and enhanced is distinctive from any other resource. For a start in many instances it can be easily duplicated with no or minimal cost.

This is what has made the rise of the big five, Amazon, Alphabet, Apple, Microsoft and Facebook, so rapid. When the zero cost of duplication is combined with a strong market platform and intellectual property rights, the right conditions are set for the formation of monopolies (Pollock, 2018). The economy built on this new information is staggering. Most of it has been gathered relatively cheaply or even for free. The business model of the platform capitalists is to create algorithms that invite participation by other businesses, governments and the general public who then generate the content, networks and interactions that create the value that is then on-sold at a profit. The information or data economy has changed the rules for markets and it is becoming ominously clear that a new set of regulations and approaches are needed for the new data market. In May 2018 the European Union (EU) introduced a significant regulation in EU law on data protection and privacy for all individuals, known as the General Data Protection Regulation (EU 2016/679 ("GDPR")). This was in a significant regulatory response to the opportunities and challenges facing us in a data-rich world. New moves from regulatory organisations in Australia are also rising to challenge such monopolies (Duke & McDuling, 2018).

The natural habitat for the information economy is the city. Platforms are locations or places where participants in an economy connect. Cities are the original platforms. Cities have always been centres for the exchange of information. This information has never been completely open but managed and administered according to complex bureaucratic, commercial and legal codes and interests. This intersection of information, finance and politics has dramatic repercussions for the city and its citizens. Despite the informational city being no new thing, as Castells (1989) makes clear, the infotech company, the data economy and the smart city all make use of radical technologies which change the way cities are used and are changing the way they conduct their business and are planned and governed. Hardware companies such as IBM, CISCO and SIEMENS have complemented the initiatives of information companies such as Alphabet and Microsoft, hardwiring both existing cities and new bespoke "smart cities", such as Korea's Songdo City, to handle the new data economy. Both new and old smart cities don't look so different from not so smart cities. Rather the smart city infrastructure that generates and supports data in all of its forms (open, big, dark, thick, etc.) is not so much a digital overlay, or physical redesign, but a rewiring of the way the city operates.

This chapter sets out initial arguments for the rethinking of today's informational society through the promotion and greater use of Open Data. It provides a critical framework for understanding and reflecting on the important contributions made by researchers in the ensuing chapters of this volume. This introductory chapter describes the nature of the new contested data landscape and then reflects on the call for action for more and better quality Open Data and an overview of the complexity of openness. The arguments for Open Data are situated in a collaborative, networked vision of society in place of a hierarchical, monopolistic structure. The chapter then provides a high-level discussion of three critical themes which organise the structure of the book. These are (1) urban inclusion and social entrepreneurship, (2) knowledge ecosystems and resilience and (3) civic innovation and transparency. These themes address major needs within the new economy of the global data landscape and set the context for the diverse chapters which follow.

Challenging Urban Information Monopolies

The provocative scholar Greenfield (2017) has stated "there is no such thing as raw data" as all data has its patterns. The mining, refinement and processing of data improve the signal to noise ratio so that it is intelligible. Urban big data is a case in point. The patterns of human and machinic movement that occur within cities have until recently only been perceptible through simple arithmetic methods of the census and manual record keeping. Today's web of sensors, information infrastructures and the

Internet of Things make the production and accounting of information exponentially greater in volume, velocity and variety but also in intelligibility. The combination of mobile devices, the Internet of Things (IoT) and rapid urbanisation is resulting in a spatialised or urban intelligence, but this intelligence is not equally shared or accessible amongst those who reside in our cities.

The ability to direct and influence the dataspace by a few powerful players has resulted in a string of corruption and infotech scandals over the last few years. These scandals are commonplace today. Ranging from business-government scandals such as the Cambridge Analytica–Facebook Scandal (Cadwalladr, 2018; Richterich, 2018), to Facebook's Russian and a long list of other scandals (Debatin, Lovejoy, Horn, & Hughes, 2009; We're keeping track of all of Facebook's scandals so you don't have to, 2018), to the 5-billion fine handed out to Alphabet by the EU for monopolistic practices (Cassidy, 2018; Finley, 2017), to government mishandling of data indicated by whistleblowers such as Snowden (Greenwald, 2015) and WikiLeaks (Benkler, 2011; Domscheit-Berg, 2011; Roberts, 2012), these events are powerful signals of the breach of trust between government, business and the citizen.

The present status quo threatens the norms of a free society with democratic access and participation. Such power dynamics exclude choice and coerce the citizen. There has been much written on how information monopolies limit freedoms and influence how we think, behave and act (Bridle, 2018). Less has been written about how urban infotech companies are changing the way people interact with cities. From share-ride companies such as Uber to delivery and digital retail companies such as Amazon, urban patterns and movements are being re-choreographed without our conscious understanding of the consequences. For example, internet giants such as Airbnb and Uber are sitting on top of mountains of geospatial data that includes addresses, boundaries and vehicle and human movements. This information is being used to make decisions as to how the city is to be used without much consideration of public benefit beyond an immediate transactional fulfilment.

Despite being the platform for the infotech giants, cities have been slow to catch onto the idea that it is their data rather than any new tech infrastructure, that is the most valuable asset. The characteristics of the informational society, in Manuel Castells (1989) sense of the phrase, are the increasing importance of software over hardware and the idea that information is both the raw material and the product of the economy.

The internet giants possess "big data" and only make a small amount of this available as "Open Data". Equally, governments are fearful of the consequences of making their data available. Big data is sometimes thought of with trepidation. It's what businesses and governments use to know how you think and trace what you do. In contrast, Open Data is an investment in society. Gurin (2014) makes the important distinction between these two forms of data. Businesses use Big Data to develop customer databases and marketing profiles, for brand building and market intelligence. In contrast Open Data is used by government and business to engage with society and to encourage democratic participation (Gurin, 2014, p. 12). In contrast to Big Data which is usually held privately for business or security reasons, Gurin (2014) argues that Open Data is "public and purposeful". It is a conscious civic act that releases and presents data in a way that anyone can use. Open Data is often the intelligent use of Big Data for public purposes. For example, much of the data that is now open was previously held by private companies or governments before being "aggregated", "developed", "enriched", "supplied" or "enabled" by organisations that present such data as a public asset (Gurin, 2014, p. 15). It is important to highlight that the Open Data movement is maturing and gaining momentum with governments across the planet as reported by the Open Data Barometer which provides each country with a rating based on its Open Data maturity. Now in its fourth iteration, the Open Data Barometer tracks progress in nations around the world and highlights areas for improvement (World Wide Web Foundation, 2018). In other words the initiative emphasises what needs to be done better and what impact Open Data programmes are having. What data to open up and how to open up that data are not straightforward questions. As the Open Data Barometer demonstrates, it is a multifaceted and ongoing challenge that takes strategy and collaboration.

Rethinking Value: The Potential of Open Data for Cities

The types of data that the largest businesses in the world are gathering include geospatial data describing addresses, movements and activities, and could be used for a range of innovative urban applications. As places with spatial dimensions made up of neighbourhoods, districts and complex interlinked physical infrastructures, cities have much to gain from better understanding of how people relate to space. Geospatial data can help government make more transparent decisions and help communities address social disadvantage, environmental problems and other economic externalities (Marsal-Llacuna, Colomer-Llinàs, & Meléndez-Frigola, 2015; Miller & Tolle, 2016). However, significant swathes of urban data are inaccessible to the public. The Open Data Institute (ODI) has argued that such data should form part of a new national infrastructure as key to building the future of the urban world, as the streets, sewers and powerlines were in the nineteenth and twentieth centuries (Yates, Keller, Wilson, & Dodds, 2018). Geospatial data is the reference point for a range of new and disruptive technologies such as drones, driverless cars and digital urban services as well as yet unimagined businesses and social enterprises. The contribution of geospatial data for cities is substantial. Estimates suggest that geospatial data contributed $21 billion to the Canadian economy in 2013 (Zeiss, 2015). Estimates place the value of geospatial data at $75 billion to the US economy in 2012, as well as being the basis for almost 500,000 jobs, which is close to the number employed in residential construction (Henttu, Izaret, & Potere, 2012). The company AlphaBeta (2017) has estimated the value of Geospatial Data to the global economy at 400 billion in global revenue whilst saving people time and fuel to the value of $550 billion. In 2013 McKinsey calculated the "potential" global value of Open Data at $3 trillion (Manyika et al., 2013).

An interesting transition from closed to Open Data is evident in the story of LiDAR. LiDAR is a form of detailed mapping that involves scanning the earth using laser systems. In recent years this data has increasingly been released as Open Data. In 2012 Finland opened its LiDAR for public use with Denmark and the Netherlands following in 2013 and 2014. Many such data assets were originally carefully guarded and sold at high cost, but Open Data advocates demonstrated that the sales revenue of the data was minimal in comparison to the value of the innovation ecosystem that could be stimulated should the data be opened. The fact that government agencies are one of the largest users of such data was a convincing reason to open it up to minimise bureaucratic costs between government agencies. Many governments now realise that they get a better return on their investment if more businesses and citizens freely use government data to create new technologies and enterprises, than if they were to keep the datasets locked up behind a paywall. The City of Helsinki is a leader in geospatial Open Data (Jaakola, Kekkonen, Lahti, & Manninen, 2015) through their 3D Helsinki platform which provides access to over 18,000 buildings as openly available to 3D models (Sulopuisto, 2014). Such data is in turn used for a wide range of commercial applications and businesses.

Governments around the world have been opening data for public benefit. Over 1 million government datasets are globally available as a result of Open Data policies. There is also a keenness to develop a culture of innovation around such data. For example, International Open Data Hackathon Day has been held for little under a decade and in 2019 will be celebrated in over 400 events in cities around the world. Open Data and hackathons are emerging public research areas which are held to enhance and promote the use of Open Data to benefit the public. The events inspire innovation and can be focused on specific challenges like a recent Australian event which focused on the problem of "transport congestion".

Such events build on initiatives like those of the Australian government which has opened up a wide range of geospatial datasets and resources following the "Declaration of Open Government". This initiative was supported by the creation of the Digital Transformation Agency (DTA) and the data.gov.au portal which has released over 5000 datasets from Geoscience Australia.

Within Australia, Melbourne has taken a leading role in urban Open Data. Well known as one of the most liveable cities in the world according to a variety of indexes, Melbourne has been innovative in taking a research and evidence-based planning approach to maintaining its liveability. The city uses its Open Data portal to communicate its successful planning and design outcomes to cities around the world. The city provides fine-grained spatial data on trees and pedestrian movements to the public. A range of other more sensitive data such as building stock, urban land use and thermal imaging is provided for use by researchers via The Australian Urban Research Infrastructure Network (AURIN), a semi-Open Data platform for researchers.

Similar to AURIN, some Open Data initiatives such as the European Commission's Urban Data Platform, make Open Data easier to navigate and allow interactive data visualisations at a variety of scales. Such initiatives allow analyses but also stimulate the public imagination on future possible applications of the information.

Massive venture capital-backed infotech companies, such as Airbnb and Uber, have collated vast quantities of data that equal or exceed those produced by city or national governments. A necessary debate is emerging around whether these commercial organisations will silo their big data or develop this valuable asset, through techniques such as aggregation and anonymisation, and make it available to cities that could benefit from the rich data.

The publication of Open Data by Uber is an important public engagement and contribution by the company and necessary to understand the impact of such radical technologies on our urban fabric. Uber's data has been used to develop new and more accurate global positioning system (GPS) technologies effective in the difficult urban canyons that exist in inner city areas. Due to requests and engagement by civil society and government, Uber released Open Data for London in 2018 through the Uber Movement platform which allows the public to analyse aggregated journeys and understand patterns of urban movements.

A new set of incentives and guidelines must be imagined, created and legislated to mandate governments and businesses to open up data and to create an Open Data landscape better suited to addressing the world's challenges. The big data scholars, Mayer-Schoneberger and Ramge (2018), argue that the way companies such as Google and Uber are taxed needs to change as many of them pay no tax at all. Instead of trying to close complex tax loopholes and pressing for such companies to contribute the money they owe, it may be more valuable and effective for them to contribute Open Data which could be used by smaller firms, the government and the public (Mayer-Schoneberger & Ramge, 2018, p. 199).

Alongside this more formal push, a culture of Open Data can be created to stimulate a more open and collaborative organisational "DNA" for governments and communities. Of course, in such contexts, individual privacy needs to be preserved and the ethical use of data safeguarded. These are not small endeavours. For example recent research has demonstrated that some rich geospatial datasets have enough information to identify 95% of the sample populations using four spatial and temporal points per person. GPS is a powerful technology and has inadvertently revealed national security threats.

Transnational corporates, start-ups and tech companies alike are situated in today's richest and most innovative cities. However not all of these cities are good places to be with various degrees of social polarisation and quality of life for their residents. Some have unacceptable levels of poverty, shameful homeless problems, violence and environmental degradation and unhealthy conditions. The so-called "Valley of Genius" (Fisher, 2018) is a case in point. Silicon Valley is an entrepreneurial wonderland but a social dystopia that delivers radical technological services to the world but whose own urban backyard decays with social and environmental problems. Within such urban contexts, a new breed of "social" tech entrepreneur is making use of Open Data to address a range of wicked social and environmental problems—those problems that are economic externalities.

Social initiatives such as "Blindsquare", an app that provides audio directions for blind people, would not be feasible without Open Data sources from OpenStreetMap (OSM) and Foursquare. In this case a new social initiative makes use of Open Data from both crowdsourced (OSM) and commercial (Foursquare) sources. Similarly, "Poverty in NYC" uses government Open Data from the American Community Survey and enriches it with data from the Poverty Research Team to build digital maps which help align and implement anti-poverty initiatives to make the city a more equitable place.

The likes of IBM, CISCO and Siemens have been champions of the smart city, but smart does not always translate to the concept of the Just City (Fainstein, 2010). Open Data has the potential to stimulate and engage a lively citizenry in collaboration with such large-scale commercial, government and academic sectors. *Open Data–Open Cities* provides a culmination of different views and critiques of our urban information ecosystem, one which problems are addressed through enhanced information, communication based on this information and actions that result from the collaborative opportunities that arise.

Open Data Strategies for Inclusive Cities, Resilient Cities and Innovative Cities

Cities are the "engines" of growth and "innovative labs" for today's economic cities and have much to gain from the new data economy. Cities are now coming to realise the inherent value of the platforms and patterns which produce the extraordinary concentrations of data that now characterise modern cites. Instead of being passive users of the data and the platform upon which the disruptive economy has extracted value, cities such as London, Amsterdam and Barcelona have created vital and valuable information ecosystems through careful Open Data strategies.

The contributing authors to this *Open Data–Open Cities* volume highlight both the opportunities and the challenges facing us. The research and case studies in this volume provide guidance on how cities and governments can champion the Open Data cause and collectively work towards inclusive urban development in a range of different ways. They highlight the different understandings and variants of openness suggested by Pomerantz and Peek (2016) in their helpful paper "50 shades of Open". Contributions range from strategic high-level insights drawing on the experience of a number of projects, to detailed case studies of specific Open Data projects. Both types of chapter suggest ways to evolve the Open Data landscape through practical and strategic policy approaches. Other chapters are more experimental in nature, dabbling in innovative concepts and ideas associated with Open Data and open cities.

The expanding digital footprint of cities presents new opportunities for better understanding and communicating in our urban world. This book addresses this potential with its specialised focus on Open Data in relation to cities and urban space. It seeks to educate, inspire and inform urban managers, designers, architects, academics and information specialists on this topic.

The book discusses innovative ways to open up data and to use it in new applications to create smarter, more open cities. Various streams of data are currently available to citizens, researchers and communities including crowdsourcing data; data compiled by businesses into open online databases; and data released in an open format by government agencies. This book examines how these multiple streams of Open Data can transform cities into more liveable and productive systems in three thematic sections focusing on (1) social entrepreneurship and urban inclusion, (2) knowledge ecosystems and urban resilience and (3) civic innovation and transparency.

The book is a cross-disciplinary survey of the Open Data landscape as it relates to cities, their citizens and their challenges. It combines the knowledge of a wide range of scholars and practitioners who present their own experiences of what they have accomplished with Open Data and what more needs to be done.

Through the range of case studies, vignettes and critical reflections, we examine what is needed to make sense of the fraught, complex and growing global data landscape in relation to cities. Through such creative investigations, a series of visions and strategies emerge in the ways in which Open Data can create an economic ecosystem which encourages innovation across a wider range of areas and empowers cities and citizens to become equal players in the new data economy.

Urban Inclusion and Social Entrepreneurship

The new information economy and the technological world have now emerged as an everyday condition for much of the world's population. The book's first theme takes on some of the world's largest injustices and through a range of discussions and case studies looks at practical ways inclusive urban development can become an economic reality. Vulnerable people are often outside the reporting and records of existing economic systems and standards. As leading economists such as Stiglitz (2012) have argued, the costs of such exclusion are severe both for the affected individuals and for the society as a whole. The Open Data movement presents a new tool and strategy for developing initiatives for urban inclusion. Social entrepreneurs are motivated by social and economic equity rather than financial gain, and therefore rely on effective access to information and technological tools that can help address the formidable problems that beset the homeless, women, the poor, the ageing and other vulnerable groups of our society.

What is evident with many of the chapters in the first section is that a range of technological and traditional data collection methods are required to develop knowledge on the social and economic problems of urban inclusion.

Pollock (2018), the author of the *Open Revolution*, argues "the result of running the information economy by the old rules of intellectual monopoly rights is spiraling inequality". Chapter 2 discusses the Open Data challenges of one of the most visible faces of urban inequality: homelessness. Fox and Hugh (2019) argue that developing effective strategies for homelessness on a global scale requires information that accurately describe housing exclusion and the reality of homelessness. Using global ISO standards and open city data, the authors highlight that the task for understanding homeless at a global standard is formidable with data and not standardised, consistent or comparable. The authors reviewed openly available homeless data for 14 cities with "good" Open Data resources. These include Calgary, Toronto, New York, Chicago, San Francisco, Miami, London, Paris, Rome, Barcelona, Beijing, Shanghai, Tokyo and Singapore. Cities such as Toronto indicate a model for others to follow; however, what is evident is that even amongst Open Data city leaders, homelessness is a problem that needs much more attention and energy to highlight the global dimensions of the problem and to identify approaches to addressing it.

Chapter 3 by Bakelmun and Shoenfeld (2019) draws on historical archives, and the work of social entrepreneurs, to develop an important case study on the racial segregation that exists within cities of the United States. Many cities within the United States have inherited past legacies of racists planning policies that result in a racially divided urban landscape. Open Data projects can catalyse communities around the legacy of racial covenants and government policies that might address such inequality. Furthermore, the work illustrates the important role that combining traditional data with new types of data, such as crowdsourced information, has in developing Open Data sets for socially progressive causes.

Chapter 4 by Hawken et al. (2019) suggests that "achieving urban safety within different cultural, legal-political and socio-economic contexts is made clearer through the generation of an information ecosystem that is linked to open, crowdsourced data and clear reporting". In particular, it presents a case study of a social entrepreneur which advocates for women's safety using civic technology and crowdsourced and other forms of data. Gender discrimination is the clearest example of how marginalised groups can remain invisible and overlooked within an indifferent population. Because of the violence towards women, they do not experience the right to participate in the city and experience its social, cultural and political life in the same way that men do. The authors (Hawken et al., 2019) argue that this profound everyday imbalance can be addressed through a virtuous cycle of data collection, advocacy and action that they present as a new framework for Open Data integration with policy action.

Drawing on the ideas of participatory urbanism, Chap. 5 by Rahmat (2019) demonstrates the positive role infotech giants can have at a small scale if their Open Data is further developed and used in creative and effective digital tools. In the chapter, Rahmat (2019) presents a model for interactive and responsive planning and collaboration between citizens, experts and public officials using Twitter social media data analytics and network mapping, to better understand the dynamics of open and complex collaboration in the context of urban planning and governance.

The final chapter (Chap. 6) in the inclusive urbanism and social entrepreneurship section, advocates for the creation of Open Data to document and argue for the rights of informal settlements (Ndemo, 2019). A quarter of the world's population live in informal settlements. Such settlements are essential to the provision of shelter in a world where formal urban development cannot keep pace or provide cost-effective housing. In this chapter, Ndemo (2019) argues that the accurate digital mapping of informal situations in Kibera, Nairobi, can create increased transparency and information to generate sustainable upgrading and renewal. For decades informal settlements have been stigmatised and demolished; however, they remain a vital part of many cities and house as many as 50% of the population in cities such as Mumbai. Open Data can inform positive and careful transformation by celebrating their complexity, diversity and dynamism in future plans.

Information Ecosystems for a More Resilient Society

This section of the book sets out to build knowledge around the role of Open Data in empowering communities, governments and business to better understand and adapt to the wide range of challenging changes that now confront society. The consultancy firm Arup has argued that "human wellbeing in cities relies on a complex web of institutions, infrastructures and information". This web can be described as an information ecosystem and it refers to the type of information that is not inert, or siloed, or separate, from the society and communities that need it. Rather it is accessible, open, and part of the working toolkit that societies can draw upon to adapt to change. Without such working and responsive information landscape, people cannot understand or adapt to the challenging events and conditions affecting their lives.

Information ecosystems are the foundation for resilience and form. Within a healthy and resilient society, unexpected shocks and changes are addressed through timely, accurate and rich information that can assist with lateral and critical problem solving. In contrast in an unhealthy or vulnerable society, biased, missing or delayed information can encourage maladaptations or trigger false or ineffective approaches to the shock or challenge no matter how large or small the disturbance is (Lewandowsky, Ecker, & Cook, 2017; Susman-Peña, 2014). Dimensions of Open Data such as accessibility, interactivity, legibility, accuracy and timeliness are key to healthy information ecosystems. Open Data is a key to today's resilient cities helping support information ecosystems with the capacity to create inventive solutions to unknown problems and unexpected challenges.

The importance of Open Data to resilience is evident in the first chapter (Chap. 7) in this section. In this chapter, Hawken, Yenneti, and Bodilis (2019) use resilience theory to consider the impacts of climate and to measure the heat vulnerability of the city of Sydney. Our understanding of climate change has been focused at abstract continental and global scales. This chapter visualises and assesses the concept of heat vulnerability at the scale of urban districts. It argues that a new paradigm of city-level Open Data access is needed to ensure that societies can develop place-based actions to moderate climate change as it is evident that global climate change has drastic consequences for cities as "centres of action". This relationship is explored through a heat vulnerability index interactive platform developed using accessible public data for Sydney. The potential to develop complex but meaningful concepts such as urban resilience relies on healthy urban data and Open Data ecosystems.

Chapter 8 (Athanassiadis, 2019) also explores the challenges of climate change except from the perspective of urban metabolism. Cities are responsible for approximately three-quarters of global greenhouse gas emissions and energy use. Urban metabolism is a research field that studies resource use and pollution emissions of cities which is reliant on an important process of data collection and analysis. It argues that the production of reliable and recent data is one of the main challenges for studying urban metabolism at a global scale.

Chapter 9 by Pettit et al. (2019) presents research on a number of projects around the world that are emblematic of an emerging approach that uses data to link diverse sectors and domains through an analytical information ecosystem. The authors argue that this emerging framework is a new approach that they term "informed urbanization". The idea that Open Data isn't simply responsive but a language and critical method that can direct multiple conflicting agendas is promising.

Whilst Pettit et al. (2019) demonstrate that cities can learn much from each other as living laboratories, Chap. 10 by Zhang et al. (2019) illustrates the profound advances that a synthesis of data in a single city, Sydney, can make to urban liveability. They use three case studies in the area of water utility, smart parking and urban planning to demonstrate the value that sharing data can have. They are, however, sympathetic of the challenges business have in opening up their data and set out some barriers to achieving such a rich open ecosystem.

The final two chapters in the section are industry case studies of different practitioners experimenting with ways to open up data (Harkins & Heard, 2019; Tilley & Pettit, 2019). Chapter 11 demonstrates the power of bringing in some of the cutting-edge technology from the gaming world to help urban managers care for their utilities and assets. Despite the advances in areas such as Building Information Management systems, it makes a limited contribution to cities. Harkins and Heard (2019) show that augmented reality can unlock the value of data throughout the design, development, management and planning cycle. They have a different definition of Open Data—one in which data even within a project is locked within disciplines and cannot be accessed by other trades and professionals.

The final chapter (Chap. 12) in this section by Tilley and Pettit (2019) shows the importance of information ecosystems in the context of disasters. Using real-life case studies of just-in-time data dashboards, this reflection on recent research and projects demonstrates the great interest and importance that citizens place on information access. The tactical dashboards presented were assembled "just in time" for the arrival of Cyclone Debbie which hit the East Coast of Australia in 2017. Such technologies will become ever more important with the increasing frequency and intensity of disasters related to climate change.

Civic Innovation and Transparency

Smart Cities have moved through three generations from the technology-driven corporate vendor visions of Smart Cities 1.0, to the government-focused visions of Smart City 2.0, to the latest generation Smart City 3.0. In the past few years this latest vision of what the contemporary Smart City can be embraces the process of co-creation and citizen engagement (Sepasgozar, Hawken, Sargolzaei, & Foroozanfa, 2018). One of the world's leading smart cities, Barcelona, has demonstrated this model with the inspirational BCN Open Challenge where the city developed six challenges and used an independent platform, Citymart, to engage local and global citizens and business to contribute to ideas and solutions for the development and governance of the city (Carto, 2017).

Vancouver is another city that used such citizen-focused approach to develop its vision for the Vancouver Greenest City 2020 Action Plan. An ambitious collaborative strategy involved working with 30,000 citizens in the co-creation of the plan (Carto, 2017). For such innovative civic approaches to be effective, cities need to open up the information typically hidden within closed bureaucratic systems so that citizens can be true collaborators and develop innovative but feasible, practical plans, initiatives and strategies (Brody, Koester, Markovits, & Phillips, 2016; Mergel, Kleibrink, & Sörvik, 2018; Sieber & Johnson, 2015).

Such approaches and new models of governance and transparency are evermore critical in a time when traditional existing governmental structures and parties, the media landscape, work life and security are all undergoing dramatic transitions and disruption. There has been a loss of faith in traditional institutions including those that we rely on for accurate and timely information and knowledge.

Ojo, Curry, and Zeleti (2015) investigated 18 Open Data initiatives in 5 cities to analyse the emerging convergence of open cities and Open Data. The study found that Open Data initiatives have significant effects on transparency and accountability on areas ranging from economy, education, environment, governance, tourism and transportation. Building an active Open Data ecosystem is an emerging but central issue for governments striving to provide equitable access to public services and stimulate innovation in the economy (Liu et al., 2015; Sieber & Johnson, 2015).

In this section the book presented the five chapters that explore a range of innovative technologies, strategies and governance strategies in Open Data and policy. The first (Chap. 13) of these by Perez, Pettit, Barns, Doig, and Ticzon (2019) examines an Open Data management strategy for a large-scale organisation in an example of action research. The authors write that the "current urban data landscape is characterised by avoidable duplications, large data gaps and spatial or temporal mismatches. Moreover, many data assets remain siloed within and between organisations so their value is far from being fully realized." In response the authors develop a roadmap towards an urban knowledge-sharing platform called the Connected City Data Hub. Using this case study the authors design an information management framework and use it to review various international city data stores and hubs.

One of the measures of success for city Open Data stores is how often the data is applied and used in a range of different projects. Degbelo, Granell, Trilles, Bhattacharya, and Wissing (2019) in Chap. 14 introduce a tool to do just this. The Open city Toolkit is used to measure the actual usage of Open datasets. Such initiatives are important in understanding how well information ecosystems are developing in relation to Open Data initiatives.

Anastasiu, Foth, Schroeter, and Rittenbruch (2019) in Chap. 15 use a participatory approach to identify opportunities for the use of Open Data in community and business applications. The insights are based on the experiences gained from a Code for Australia fellowship working directly with local councils in South-East Queensland and developing co-creation practices and processes in such contexts using Open Data. The result is a novel approach to data custodianship where local councils are not so much stewards of the data but directors and facilitators.

Within cities around the world there is a great hunger for comprehensive and reliable Open Data to develop smart city and urban planning projects (Praharaj et al., 2017, 2018). Chapter 16 by Praharaj and Bandyopadhyay (2019) evaluates the urban Open Data initiatives within the 100 Smart Cities mission in India. The findings reinforce perceptions that it is at the city level where the most sustainable and progressive city data stores are being developed to manage and engage people with Open Data.

Further developing the citizen-centric focus of Smart City 3.0 approaches, the last chapter (Chap. 17) by Burton, Tiernan, Wolski, Drennan, and Morrissey (2019) presents a framework for linking Open Data through user-created narratives for disaster management. Rather than facing such challenges as an abstract problem, they are broken down into personalised, place-based information for building local resilience.

Conclusion: A Critical Moment for Open Cities

This is a critical time for the information economy and cities (Sieber & Johnson, 2015). Cities around the world are demonstrating that they, rather than the nations of which they are part, are the most progressive agents of change in the new data economy and across a range of other global challenges (Katz & Bradley, 2013; Han & Hawken, 2018). The three themes, introduced in this chapter and developed throughout the book, are illustrated by case studies and experiences from progressive cities demonstrating that the information economy is a powerful transformational force. Inequality, resilience and trust are not often considered together in mainstream discussions of the economy. However, it is essential that they are placed front and centre as both idealistic and achievable objectives in the new information-orientated economy. For all its novelty and uncertainty, there is a great collaborative momentum that has been established in the Open Data movement. Cities must be active in positioning, advocating and shaping the Open Data economy. This book suggests that we have made a very good start on the Open Data journey and gestures towards the exciting way ahead.

References

AlphaBeta. (2017). The economic impact of geospatial services: How consumers, businesses & society benefit from location-based information. Retrieved December 12, 2018, from https://www.alphabeta.com/our-research/the-economic-impact-of-geospatial-services-how-consumers-businesses-society-benefit-from-location-based-information/

Anastasiu, I., Foth, M., Schroeter, R., & Rittenbruch, M. (2019). From repositories to switchboards: Local governments as Open Data facilitators. In S. Hawken, S. Han, & C. Pettit (Eds.), *Open Cities Open Data: Collaborative cities in the information era*. Melbourne: Palgrave Macmillan.

Andersson Schwarz, J. (2017). Platform logic: An interdisciplinary approach to the platform-based economy. *Policy & Internet, 9*(4), 374–394.

Athanassiadis, A. (2019). Urban metabolism and Open Data: Opportunities and challenges for urban resource efficiency. In S. Hawken, H. Han, & C. Pettit (Eds.), *Open Cities Open Data: Collaborative cities in the information era*. Melbourne: Palgrave Macmillan.

Bakelmun, A., & Shoenfeld, J. (2019). Open Data and racial segregation: Mapping the historic imprint of racial covenants and redlining on American Cities. In S. Hawken, H. Han, & C. Pettit (Eds.), *Open Cities Open Data: Collaborative cities in the information era*. Melbourne: Palgrave Macmillan.

Benkler, Y. (2011). A free irresponsible press: WikiLeaks and the battle over the soul of the networked fourth estate. *Harvard Civil Rights-Civil Liberties Law Review, 46*, 311.

Bridle, J. (2018). *New Dark Age: Technology, knowledge and the end of the future*. London and New York: Verso Books.

Brody, S., Koester, A., Markovits, Z., & Phillips, J. (2016). Moving the needle: What works cities and the use of data and evidence. *ArXiv:1609.08754 [Cs]*. Retrieved from http://arxiv.org/abs/1609.08754

Burton, P., Tiernan, A., Wolski, M., Drennan, L., & Morrissey, L. (2019). Resilient cities, user-driven planning, and Open Data policy. In S. Hawken, H. Han, & C. Pettit (Eds.), *Open Cities Open Data: Collaborative cities in the information era*. Melbourne: Palgrave Macmillan.

Cadwalladr, C. (2018, March 18). 'I made Steve Bannon's psychological warfare tool': Meet the data war whistleblower. *The Guardian*. Retrieved from https://www.theguardian.com/news/2018/mar/17/data-war-whistleblower-christopher-wylie-faceook-nix-bannon-trump

Carto. (2017). 40 brilliant Open Data projects preparing smart cities for 2018. Retrieved December 12, 2018, from https://carto.com/blog/forty-brilliant-open-data-projects-preparing-smart-cities-2018/

Cassidy, J. (2018, July 20). Why did the European Commission fine Google five billion dollars? Retrieved from https://www.newyorker.com/news/our-columnists/why-did-the-european-commission-fine-google-five-billion-dollars

Castells, M. (1989). *The informational city: Information technology, economic restructuring, and the urban-regional process*. Oxford, UK; Cambridge, MA: B. Blackwell.

Debatin, B., Lovejoy, J. P., Horn, A.-K., & Hughes, B. N. (2009). Facebook and online privacy: Attitudes, behaviors, and unintended consequences. *Journal of Computer-Mediated Communication, 15*(1), 83–108. https://doi.org/10.1111/j.1083-6101.2009.01494.x

Degbelo, A., Granell, C., Trilles, S., Bhattacharya, D., & Wissing, J. (2019). Tell me how my Open Data is re-used: Increasing transparency through the Open City Toolkit. In S. Hawken, H. Han, & C. Pettit (Eds.), *Open Cities Open Data: Collaborative cities in the information era*. Melbourne: Palgrave Macmillan.

Domscheit-Berg, D. (2011). *Inside WikiLeaks: My time with Julian Assange at the world's most dangerous website*. Random House.

Duke, J., & McDuling, J. (2018, December 10). Facebook, Google scramble to contain global fallout from ACCC plan. Retrieved December 11, 2018, from https://www.smh.com.au/business/companies/competition-watchdog-sug-

gests-new-ombudsman-to-handle-google-and-facebook-20181210-p50l80.html
Evans, D. (2011). The internet of things: How the next evolution of the internet is changing everything. *CISCO White Paper, 1*(2011), 1–11.
Fainstein, S. S. (2010). *The Just City*. Ithaca: Cornell University Press.
Finley, K. (2017, July 25). The Google Machine barrels along despite record EU fine. *Wired*. Retrieved from https://www.wired.com/story/the-google-machine-barrels-along-despite-record-eu-fine/
Fisher, A. (2018). *Valley of genius: An uncensored history of Silicon Valley, as told by the hackers, founders, and… freaks who made it boom*. New York: Grand Central Pub.
Fox, M., & Hugh, S. (2019). Homelessness and open city data: Addressing a global challenge. In S. Hawken, H. Han, & C. Pettit (Eds.), *Open Cities Open Data: Collaborative cities in the information era*. Melbourne: Palgrave Macmillan.
Greenfield, A. (2017). *Radical technologies: The design of everyday life*. London; New York: Verso Books.
Greenwald, G. (2015). *No place to hide: Edward Snowden, the NSA, and the U.S. surveillance state*, London, Hamish Hamilton, an imprint of Penguin Books.
Gurin, J. (2014). *Open Data now: The secret to hot startups, smart investing, savvy marketing, and fast innovation*. New York: McGraw-Hill Education.
Han, H., & Hawken, S. (2018). Introduction: Innovation and identity in next-generation smart cities. *City, Culture and Society, 12*, 1–4. https://doi.org/10.1016/j.ccs.2017.12.003
Harkins, J., & Heard, C. (2019). Interfacing the city: Mixed reality as a form of Open Data. In S. Hawken, H. Han, & C. Pettit (Eds.), *Open Cities Open Data: Collaborative cities in the information era*. Melbourne: Palgrave Macmillan.
Hawken, S., Leao, S. Z., Gudes, O., Izadpanahi, P., Viswanath, K., & Pettit, C. (2019). Safer cities for women: Global and local innovations with Open Data and civic technology. In S. Hawken, H. Han, & C. Pettit (Eds.), *Open Cities Open Data: Collaborative cities in the information era*. Melbourne: Palgrave Macmillan.
Hawken, S., Yenneti, K., & Bodilis, C. (2019). Mapping climate vulnerability with Open Data: A dashboard for place-based action. In S. Hawken, H. Han, & C. Pettit (Eds.), *Open Cities Open Data: Collaborative cities in the information era*. Melbourne: Palgrave Macmillan.
Henttu, H., Izaret, J.-M., & Potere, D. (2012). *Geospatial services: A $1.6 trillion growth engine for the US economy*. San Francisco, CA: The Boston Consulting Group.

Jaakola, A., Kekkonen, H., Lahti, T., & Manninen, A. (2015). Open Data, open cities: Experiences from the Helsinki Metropolitan Area. Case Helsinki Region Infoshare www.hri.fi. *Statistical Journal of the IAOS, 31*(1), 117–122. https://doi.org/10.3233/SJI-150873

Katz, B., & Bradley, J. (2013). *The metropolitan revolution how cities and metros are fixing our broken politics and fragile economy*. Washington, DC: Brookings Institution Press. Retrieved from http://public.eblib.com/choice/publicfullrecord.aspx?p=1214202

Lewandowsky, S., Ecker, U. K., & Cook, J. (2017). Beyond misinformation: Understanding and coping with the "post-truth" era. *Journal of Applied Research in Memory and Cognition, 6*(4), 353–369.

Liu, X., Song, Y., Wu, K., Wang, J., Li, D., & Long, Y. (2015). Understanding urban China with Open Data. *Cities, 47*, 53–61.

Manjoo, F. (2016). Tech's 'frightful 5' will dominate digital life for foreseeable future. *The New York Times*.

Manyika, J., Chui, M., Groves, P., Farrell, D., Van Kuiken, S., & Doshi, E. A. (2013). *Open Data: Unlocking innovation and performance with liquid information* (p. 21). McKinsey Global Institute.

Marsal-Llacuna, M.-L., Colomer-Llinàs, J., & Meléndez-Frigola, J. (2015). Lessons in urban monitoring taken from sustainable and livable cities to better address the Smart Cities initiative. *Technological Forecasting and Social Change, 90*, 611–622. https://doi.org/10.1016/j.techfore.2014.01.012

Mayer-Schoneberger, V., & Ramge, T. (2018). *Reinventing capitalism in the age of big data*. New York: Basic Books.

Mergel, I., Kleibrink, A., & Sörvik, J. (2018). Open Data outcomes: U.S. cities between product and process innovation. *Government Information Quarterly, 35*(4), 622–632. https://doi.org/10.1016/j.giq.2018.09.004

Miller, H. J., & Tolle, K. (2016, Autumn). Big data for healthy cities: Using location-aware technologies, Open Data and 3D urban models to design healthier built environments. *Built Environment, 42*(3), 441–445. https://doi.org/10.2148/benv.42.3.441

Ndemo, B. (2019). Slum digitisation, its opponents and allies in developing smart cities: The case of Kibera, Nairobi. In S. Hawken, H. Han, & C. Pettit (Eds.), *Open Cities Open Data: Collaborative cities in the information era*. Melbourne: Palgrave Macmillan.

Ojo, A., Curry, E., & Zeleti, F. A. (2015). A tale of Open Data innovations in five smart cities. In *2015 48th Hawaii International Conference on System Sciences* (pp. 2326–2335). Retrieved from https://doi.org/10.1109/HICSS.2015.280

Perez, P., Pettit, C., Barns, S., Doig, J., & Ticzon, C. (2019). An information management strategy for city data hubs: Open Data strategies for large organisations. In S. Hawken, H. Han, & C. Pettit (Eds.), *Open Cities Open Data: Collaborative cities in the information era*. Melbourne: Palgrave Macmillan.

Pettit, C., Wentz, E., Randolph, B., Sanderson, D., Kelly, F., Beevers, S., & Reades, J. (2019). Tackling the challenge of growing cities: An informed urbanisation approach. In S. Hawken, H. Han, & C. Pettit (Eds.), *Open Cities Open Data: Collaborative cities in the information era*. Melbourne: Palgrave Macmillan.

Pollock, R. (2018). *The open revolution: Rewriting the rules of the information age*. Retrieved from https://openrevolution.net/media/open-revolution.pdf

Pomerantz, J., & Peek, R. (2016). Fifty shades of open. *First Monday, 21*(5). https://doi.org/10.5210/fm.v21i5.6360

Praharaj, S., & Bandyopadhyay, S. (2019). Understanding the Open Data challenge for building smart cities in India. In S. Hawken, H. Han, & C. Pettit (Eds.), *Open Cities Open Data: Collaborative cities in the information era*. Melbourne: Palgrave Macmillan.

Praharaj, S., Han, J. H., & Hawken, S. (2017). Innovative civic engagement and digital urban infrastructure: Lessons from 100 smart cities mission in India. *Procedia Engineering, 180*, 1423–1432.

Praharaj, S., Han, J. H., & Hawken, S. (2018). Towards the right model of smart city governance in India. *International Journal of Sustainable Development and Planning, 13*, 171–186. https://doi.org/10.2495/SDP-V13-N2-171-186

Rahmat, H. (2019). Open online platforms and the collaborative production of micro urban spaces: Towards an architecture of civic engagement. In S. Hawken, H. Han, & C. Pettit (Eds.), *Open Cities Open Data: Collaborative cities in the information era*. Melbourne: Palgrave Macmillan.

Richterich, A. (2018). How data-driven research fuelled the Cambridge Analytica controversy. *Partecipazione e Conflitto, 11*(2), 528–543.

Roberts, A. (2012). WikiLeaks: The illusion of transparency. *International Review of Administrative Sciences, 78*(1), 116–133.

Sepasgozar, S. M. E., Hawken, S., Sargolzaei, S., & Foroozanfa, M. (2018). Implementing citizen centric technology in developing smart cities: A model for predicting the acceptance of urban technologies. *Technological Forecasting and Social Change*. https://doi.org/10.1016/j.techfore.2018.09.012

Sieber, R. E., & Johnson, P. A. (2015). Civic Open Data at a crossroads: Dominant models and current challenges. *Government Information Quarterly, 32*(3), 308–315. https://doi.org/10.1016/j.giq.2015.05.003

Stiglitz, J. E. (2012). *The price of inequality: How today's divided society endangers our future*. New York: WW Norton & Company.

Sulopuisto, O. (2014). How Helsinki became the most successful Open-Data City in the world. Retrieved December 11, 2018, from http://www.theatlanticcities.com/technology/2014/04/how-helsinki-mashed-open-data-regionalism/8994/

Susman-Peña, T. (2014). *Understanding data: Can news media rise to the challenge?* Washington, DC: The Center for International Media Assistance.

The Economist. (2017, May 6). The world's most valuable resource is no longer oil, but data. *The Economist*. Retrieved from https://www.economist.com/news/leaders/21721656-data-economy-demands-new-approach-antitrust-rules-worlds-most-valuable-resource

Tilley, I., & Pettit, C. (2019). A dashboard for the unexpected: Open Data for real time disaster response. In S. Hawken, H. Han, & C. Pettit (Eds.), *Open Cities Open Data: Collaborative cities in the information era*. Melbourne: Palgrave Macmillan.

Townsend, A. (Ed.). (2014). *Data-driven cities*. Tokyo: Japan Architecture + Urbanism.

Townsend, A. M. (2013). *Smart cities: Big data, civic hackers, and the quest for a new utopia*. New York and London: WW Norton & Company.

United Nations, Department of Economic and Social Affairs, Population Division. (2014). *World urbanization prospects: The 2014 revision: highlights*. New York: United Nations.

We're keeping track of all of Facebook's scandals so you don't have to. (2018). Retrieved November 21, 2018, from http://fortune.com/2018/04/06/facebook-scandals-mark-zuckerberg/

World Wide Web Foundation. (2018). 4th Edition | Open Data Barometer. Retrieved December 11, 2018, from https://opendatabarometer.org/4thedition/?_year=2016&indicator=ODB

Yates, D., Keller, J., Wilson, R., & Dodds, L. (2018). The UK's geospatial data infrastructure: Challenges and opportunities. Open Data Institute. Retrieved from https://theodi.org/article/geospatial-data-infrastructure-report/

Zeiss, G. (2015). Geospatial contributes $21 billion to the Canadian GDP. Retrieved December 12, 2018, from https://geospatial.blogs.com/geospatial/2015/05/geospatial-contributed-21-billion-to-the-canadian-gdp.html

Zhang, L., Zhang, B., Guo, T., Chen, F., Runcie, P., Cameron, B., & Rooney, R. (2019). Linking complex urban systems: Insights from cross-domain urban data analysis. In S. Hawken, H. Han, & C. Pettit (Eds.), *Open Cities Open Data: Collaborative cities in the information era*. Melbourne: Palgrave Macmillan.

Part I

Urban Inclusion and Social Entrepreneurship

2

Homelessness and Open City Data: Addressing a Global Challenge

Sonia Hugh and Mark S. Fox

Abbreviations

CIDOM	City Indicator Data Openness Measure
CoCs	Continuum of Care
HUD	Department of Housing and Urban Development
ETHOS	European Typology of Homelessness and Housing Exclusion

Highlights

- Questions whether the validity of city indicator data can be determined without the open publishing of the data used to derive them.
- Suggests the CIDOM metrics as a basis for determining the openness of city data.
- Reviews the availability of shelter/homeless data across 14 cities with significant Open Data portals.
- Discovers that cities publish little shelter/homeless data on the portals.

S. Hugh • M. S. Fox (✉)
Centre for Social Services Engineering, University of Toronto,
Toronto, ON, Canada
e-mail: s.hugh@griffith.edu.au; msf@eil.utoronto.ca

S. Hawken et al. (eds.), *Open Cities | Open Data*,
https://doi.org/10.1007/978-981-13-6605-5_2

Introduction

Homelessness is a social issue that affects both developing and developed nations. Current data is insufficient to monitor and evaluate the trends and extent of homelessness around the globe (Busch-Geertsema, Culhane, & Fitzpatrick, 2016). For the homeless data that does exist, information tends to be segregated into discrete locals, regions or nations that have their own frameworks, vocabularies and methodologies (Busch-Geertsema et al., 2016; Richter & Botha, 2012). This inhibits global collaboration and understanding of the homeless experience by hindering information exchange and sharing of best practices (McCarney, 2015). More cities are moving towards an open and transparent government (open government), making data and information publicly available (open city data), including homeless datasets that could be used to fill in the gaps of global homeless knowledge. Open city data provides a rich resource of information that can be used to promote accountability, engagement and innovation. It has the ability to be improved, standardized and continuously updated creating the potential for a source of robust and reliable homeless data.

Reliable data is a basic need of policymakers to effectively address the problem of homelessness (Springer, 2000). The international standards ISO 37120 provides a standardized set of indicators with definitions and methodologies that allow for global comparison across cities (Deng, Liu, Wallis, Duncan, & McManus, 2017; McCarney, 2015). The indicators of the international standard were developed in order to help cities measure performance of city services; learn from one another by allowing comparison of performance measures and share best practices (ISO 37120, 2014). The ISO 37120 shelter-themed indicators measure the level of a city's homeless and under-sheltered populations, providing a standardized account of homeless numbers. In an attempt to create a reliable and robust homeless dataset, open city data can be used as source inputs for the ISO 37120 shelter-themed indicators. However, the issue of validity of the indicator values arise. By validity, we ask is the indicator true/correct or false/incorrect? Although the ISO 37120 includes certification and third-party verification, the validity of an indicator's value cannot simply be taken at face value. A single indicator value is misleading as the reader has no clue to the provenance of the underlying data. The validity of an indicator is dependent on the publishing of the underlying data used to derive the indicator value (Fox, 2017).

We examine open city data to gain an understanding of the current state of global homeless data and investigate the validity of shelter indicators of the ISO 37120. We review the extent to which homeless data is standardized, consistent and comparable:

Standardized—Datasets that consist of common representation (terminologies, vocabularies, coding schemes) that allows for semantic interoperability. A standard vocabulary that ensures accurate interpretation of data. Data that is described systematically in unambiguous language to make the data machine-readable;

Consistent—How the data is collected and the periodicity of the data collection allowing for longitudinal analysis of homelessness. Data collected conforms to the definition of the indicator in which it is used;

Comparable—Data that is used to derive city indicators (i.e. ISO 37120 shelter indicators) and examined to note similarities and differences (transversal analysis).

We use the ISO 37120 shelter-themed indicators, open city data and City Indicator Data Openness Measure (CIDOM) to compare cities for their data completeness and explore the indicator validity. This allows us to determine the extent to which Open Data city datasets and city indicators can be used in the longitudinal and transversal analysis of global homeless and unsheltered populations.

Background

The impact of having good data on policy is exemplified by a study of income across Toronto's neighbourhoods. Hulchanski (2010) discovered that in 1970 low-income neighbourhoods were located in the centre of the city (Fig. 2.1), but by 2005 low-income neighbourhoods moved into the inner suburbs (Fig. 2.2).

A direct consequence of this analysis was the movement by the United Way Greater Toronto of many of their services from the inner city to the inner suburbs. Hulchanski's analysis relied upon census data. The question is whether relevant data exists for the homeless.

In order to understand global homelessness, we must conceptualize what it means to be homeless. There is no accepted unified global definition of homelessness. Definitions vary locally, nationally and globally, and are usually influenced by political, economic, climatic and social factors (Busch-Geertsema et al., 2016; Springer, 2000; UN-Habitat, 2000). A definition is of particular importance as it determines who will be recognized as homeless and dictates the prioritization of homeless policy (Peressini, McDonald, & Hulchanski, 1996; Springer, 2000).

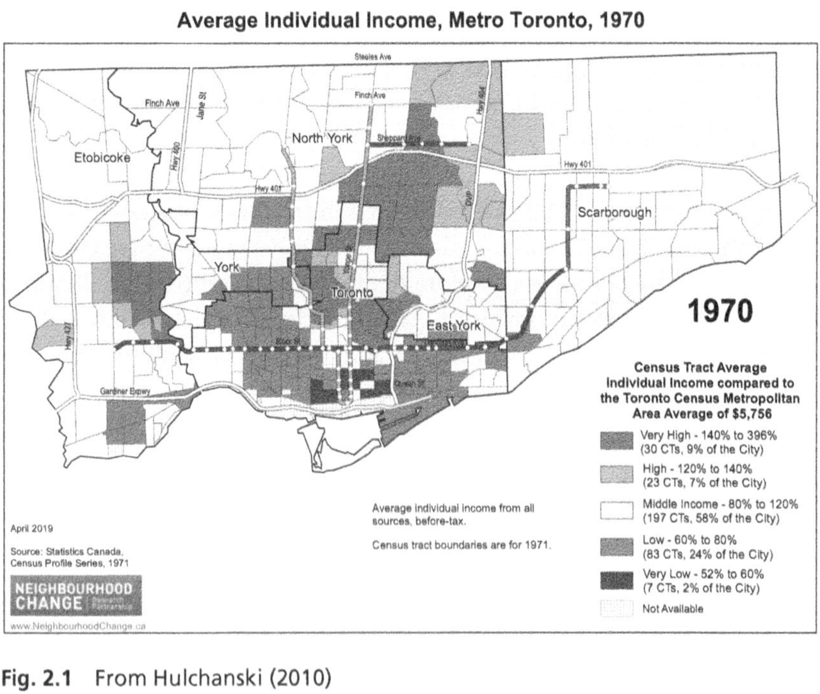

Fig. 2.1 From Hulchanski (2010)

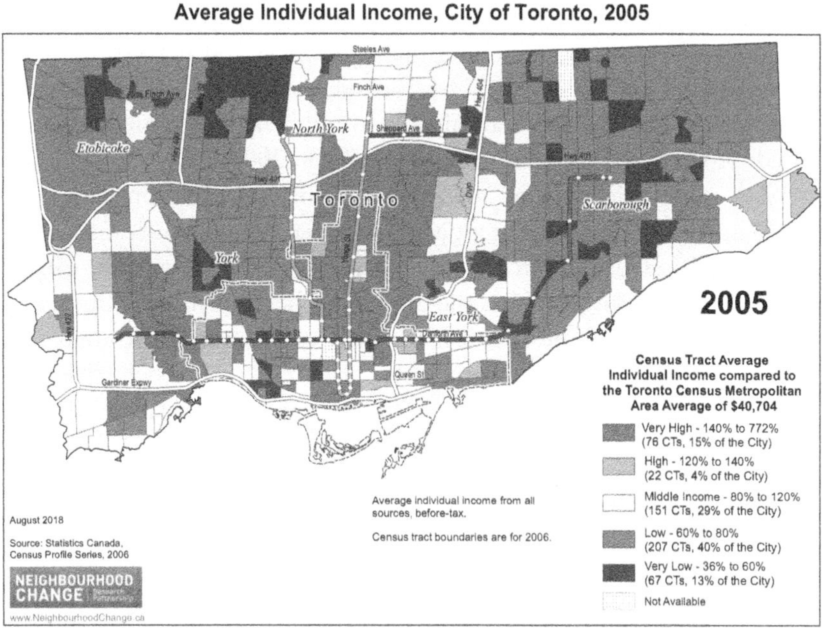

Fig. 2.2 From Hulchanski (2010)

Along with a homeless definition, reliable data about the homeless is needed by policymakers to make informed decisions. There is a general lack of global homeless statistics. Numbers on the homeless are largely drawn from developed nations in North America and Europe. There is a patchwork of national homeless statistics available, but it is not currently possible to calculate a reliable estimate or derivatives of global homelessness (Busch-Geertsema et al., 2016; Gaetz, Gulliver, & Richter, 2013; Springer, 2000).

Most data are obtained through surveys, point-in-time counts and homeless management information systems in shelters and local authorities. Homeless information gathering and dissemination varies across the globe, and there is no standard methodology or periodicity of data collection. Ideally, data collected on a consistent periodic basis is needed to perform longitudinal analysis and generate trend statistics (Busch-Geertsema et al., 2016). Data is fundamental in evaluating the state of homelessness by highlighting the whole system: inputs, services used, outputs and outcomes. The data must reflect the reality of homelessness and housing exclusion, including the experiences and processes leading to becoming homeless (Gaetz, Dej, Richter, & Redman, 2016). This will drive policies that are more sustainable over time and less emergency oriented, proactive rather than reactive (Springer, 2000).

With more cities moving towards policymaking based on data (Fox, 2013), open city data has potential to influence future homeless policy. It makes an ideal data source candidate as it is freely available, can be updated continuously and has the capacity to be standardized. But it also has its drawbacks. Cities publish the data in a variety of formats (i.e. PDFs, spreadsheets, XML), some more accessible than others. The data can also have little utility and are often not comparable as data models are not standardized with no semantic interoperability (Fox, 2013; Fox & Pettit, 2015).

Lack of data and data inconsistency makes it difficult to compare homeless statistics, policies and programmes between cities. ISO 37120 defines an international standard for city indicators that allows cities to measure performance and compare with other cities (McCarney, 2015). ISO 37120 defines three shelter indicators, consisting of one core and two supporting indicators, measuring the homeless and unsheltered populations. The ISO 37120 shelter indicators are defined as:

15.1 Percentage of city population living in slums (Core),
'*The percentage of city population living in slums shall be calculated as the number of people living in slums (numerator) divided by the city population (denominator). The result shall then be multiplied by 100 and expressed as a percentage. The number of people living in slums shall be calculated as the number of slum households multiplied by current average household size.*'
15.2 Number of homeless per 100,000 population,
'*The number of homeless per 100,000 population shall be calculated as the total number of homeless people (numerator) divided by one 100 000th of the city's total population (denominator). The result shall be expressed as the number of homeless per 100,000 population.*'
15.3 Percentage of households that exist without register legal titles.

'*The percentage of households that exist without registered legal titles shall be calculated as the number of households that exist without registered legal titles (numerators) divided by the total number of households (denominator).*' (ISO 37120, 2014)

Along with open city data, ISO 37120 provides the opportunity to create reliable homeless data that is standardized, consistent and comparable. The World Council on City Data (dataforcities.org) offers certification and third-party verification, but the processes are hidden, which raises the issue of indicator validity. Underlying data used to derive indicators are currently not required to be published; homeless indicators are only as good as the openness of the supporting data.

The integrity and validity of a standard city indicator relies on the supporting data being publicly available (Fox & Pettit, 2015). The City Indicator Data Openness Measure (CIDOM) assesses the completeness of open city data in the context of measuring city indicators (Fox & Pettit, 2015). It looks at the extent to which cities openly publish indicator data by quantifying the amount available and assessing the format of the supporting data used to derive the city indicator. CIDOM has three measures:

CIDOM-1: Measure of the completeness of the data published for an indicator. The percentage of nodes in the dependency graph that is openly published by the city.

CIDOM-2: Measure of the 'depth' to which supporting data is completely published. The number of levels of the dependency graph openly published by the city.

CIDOM-3: Determines the dominating format used to openly publish the data. The average of the format type for each node in the dependency graph openly published by the city (Fox & Pettit, 2015). A dependency graph is a diagram of the supporting data that is used in the computation of the indicator. An example of a dependency graph can be seen in Fig. 2.3 for the 15.1 shelter indicator.

Using CIDOM to compare the openness and completeness of city data in the context of the ISO 37120 shelter indicators, we get an

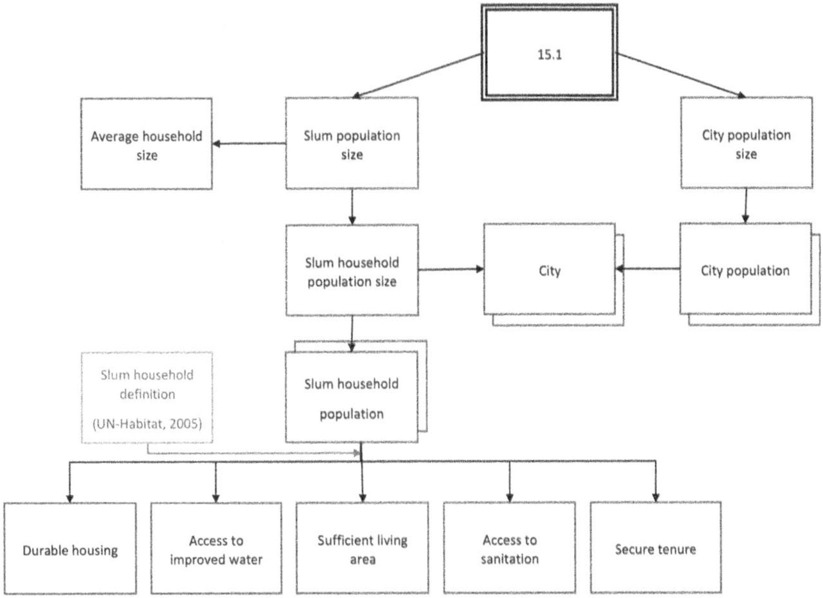

Fig. 2.3 Dependency graph of ISO 37120 shelter-themed indicator 15.1. Percentage of city population living in slums. Blue boxes represent ISO 37120 definitions, while the black boxes represent the actual data. Dependency graph adapted from Wang and Fox (2017). (Colour figure online)

indirect measurement of the validity, quality and reliability of an indicator.

Methodology

We reviewed openly available homeless data for 14 cities chosen for their high degree of openness of their city data: Calgary, Toronto, New York, Chicago, San Francisco, Miami, London, Paris, Rome, Barcelona, Beijing, Shanghai, Tokyo and Singapore (Table 2.1). By openly available, we mean all homeless data publicly available on the city's Open Data website and/or official city website. This includes linked homeless data from the official city website. We acknowledge that cities, like Calgary, have outside organizations that work with the homeless and homeless

Table 2.1 List of city and city Open Data websites investigated

City	City website	Open Data website
Calgary	www.calgary.ca	data.calgary.ca
Toronto	www.toronto.ca	www.toronto.ca/city-government/ data-research-maps/open-data/
New York	www.nyc.gov	opendata.cityofnewyork.us
Chicago	www.cityofchicago.org	data.cityofchicago.org/
San Francisco	sfgov.org	datasf.org/opendata/
Miami	www.miamigov.com	data.miamigov.com/
London	www.cityoflondon.gov.uk	data.london.gov.uk
Paris	www.paris.fr	opendata.paris.fr/
Rome	www.comune.roma.it	dati.comune.roma.it/
Barcelona	www.bcn.cat	www.opendata-ajuntament. barcelona.cat/en/
Beijing	www.egeijing.gov.cn	www.bjdata.gov.cn
Shanghai	www.shangai.cov.cn	www.datashanghai.gov.cn
Tokyo	http://www.metro.tokyo.jp/	
Singapore	www.gov.sg	www.data.gov.sg

Table 2.2 Keyword search for city homeless data

Keywords		
Homeless/ness	Population	Slum
PiT count	Shelter	Clean water
Census	Overcrowding	Housing
Households	Secure tenure	Rough sleeping

data, but were not included in the review as they were outside of our domain of official city websites. This also excludes data from the Department of Housing and Urban Development (HUD), including Continuum of Care (CoCs) data, unless specifically linked or referred to in the official city websites. The official city websites were searched using the keywords found in Table 2.2. The keywords were all related to themes found in the ISO 37120 shelter indicators definitions.

It is acknowledged that the keyword search is not exhaustive and that there is potential to miss homeless data entries. The websites were searched between September and December 2016. Any homeless data found on each city's website was noted and aggregated into categories. For cities with websites in languages other than English, sites were translated, and translated versions of the keywords were used. The review aimed to establish the type of data available on the subject of homelessness for each city and report any standards and consistency of data.

The homeless city data was then compared using the CIDOM. CIDOM assesses the completeness of open city data and is applied in the context of the ISO 37120 shelter-themed indicators. The ISO 37120 shelter-themed indicators are used as a measure of the homeless and unsheltered population of a city. Using the definition of the shelter-themed indicators, three dependency graphs were created (Figs. 2.3, 2.4 and 2.5). The dependency graphs were used as a roadmap to determine the openness of the indicators when calculating the three measures of CIDOM. The indicator's definition and dependency graph are based on the shelter ontology provided by Wang and Fox (2017). CIDOM calculates an indirect mea-

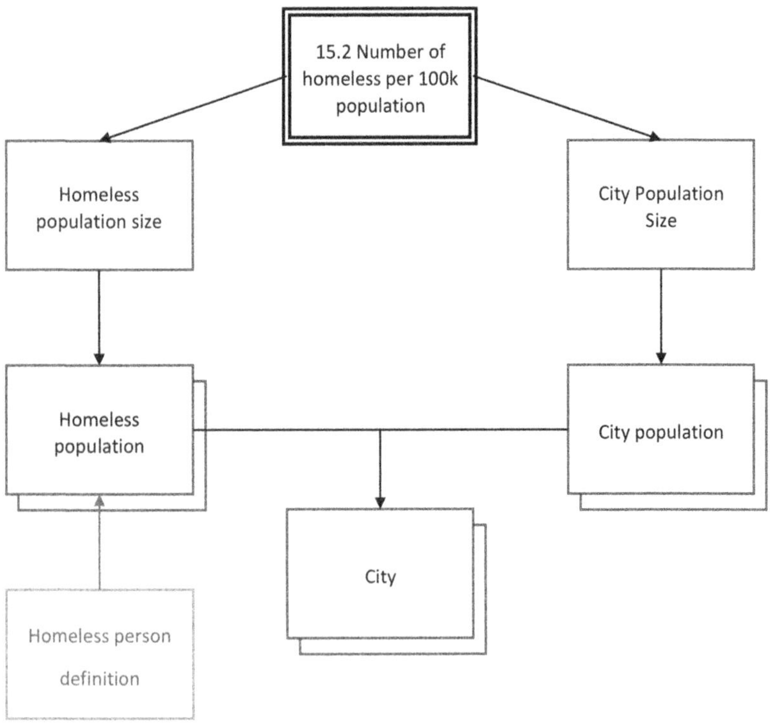

Fig. 2.4 Dependency graph of ISO 37120 shelter-themed indicator 15.2. Number of homeless 100k population

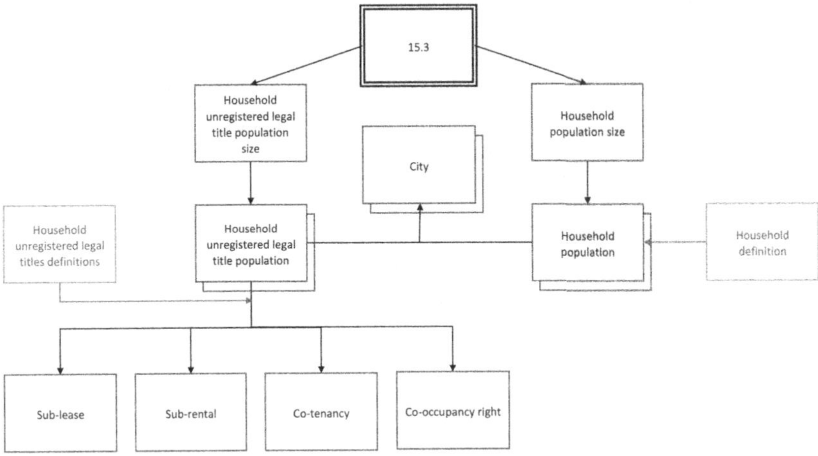

Fig. 2.5 Dependency graph of ISO 37120 shelter-themed indicator 15.3. Percentage of households that exist without registered legal titles

sure of comparability, reliability and validity. The model allows data to be interpreted as standardized and whether the integrity of the indicators can be measured using supporting data.

Results and Discussion

Standardization

Homeless Definition

The definition of 'homeless' is of the utmost importance as it determines how homeless individuals are enumerated and measured. This in turn influences policies that ultimately dictate who receives services and support. There is no globally accepted definition of homelessness, and cities and countries vary in their definition. The North American cities examined tended to go with literal definitions of homelessness, which is usually made up of two groups: unsheltered and sheltered individuals (Busch-Geertsema et al., 2016). The 'relative homeless' (Cooper, 1995) or 'hidden homeless' are not considered (all homeless definitions referred to can be found in Tables 2.5 and 2.6 in the Appendix). All American cities performed homeless surveys according to the US Department of Housing and Urban Development's (HUD) definitions of unsheltered and sheltered homeless, also for chronic homeless for data after 2015. The city of San Francisco extends HUD definitions of homelessness to include inadequate housing (Busch-Geertsema et al., 2016) and those found in correctional, health and treatment facilities (ASR (Applied Survey Research), 2015). Similarly, Toronto's definition includes correctional, health and treatment facilities, but excludes the hidden homeless (Toronto, 2013a). London's homeless definition consists of the statutory homeless and rough sleepers (unsheltered), while the inadequately housed and the hidden homeless are not measured. An individual is considered 'statutory homeless' if they lack a secure place in which they are entitled to live or not reasonably be able to stay in the current accommodation as determined by the local authority, adding an element of subjectivity. Barcelona based their definition on the European Typology of Homelessness and Housing Exclusion (ETHOS), which includes those with insecure and inadequate housing (FEANTSA, 2007; Sales, Uribe, & Marco, 2015).

The data used to inform homeless policies and programmes are based on individuals defined as homeless. The cities examined have similar definitions of homeless; however, there are many subpopulations that are not considered or are missing from city data. The homeless problem is idiosyncratic to each city; thus, it makes sense that a single definition could not be applied uniformly across all cities. Busch-Geertsema et al. (2016) suggest the use of a framework that encompasses a wider range of homeless categories/definitions. The framework would be used as a point of reference for cross-continental discussions and comparisons, providing a means of transparency by clarifying homeless populations included (or not) in the city's definition. There is no current consensus on a global homeless framework. While ISO 37120 offers a starting point for a standard and comparable global homeless dataset, the standard is very coarse. The ISO 37120 shelter indicators use a homeless definition (Table 2.5 in the Appendix), 'those without any physical shelter', that loosely describes the literal homeless, but is much narrower in scope. This definition has the potential to leave many subpopulations out of the derivation of the indicator because it primarily focuses on rough sleepers. A reworking of the definition to be clearer and inclusive of other sheltered subpopulations might lead to more reliable and representative measure of homelessness.

Homeless Data

All of the cities examined had an official city website and open city data website, except for Tokyo. Of the 14 cities, 8 had some form of quantitative homeless data. Most data values were found in tables, text (HTML) or PDFs, and no spatial data was available for any of the cities other than locations of shelters. Over 500 types of homeless data entries were found (but not limited to) on these city websites. The data can be broken down into two broad categories: homeless profiles and homeless services (Table 2.3). Homeless profiles consist of estimates, trends and characteristics of those experiencing homelessness. While the homeless service category focuses on the homeless services offered and utilized by the homeless. Table 2.3 breaks down the homeless data types into broad categories and subcategories. The datasets also vary in data range for each city. For example, San Francisco had approximately 171 quantitative data entries, while Miami had only 27 entries openly available. In the United States, data is readily available on the HUD website but is often not linked to by the city websites.

Eight cities had homeless data (Toronto, New York, Chicago, San Francisco, Miami, London, Barcelona and Tokyo) and a common homeless data category of total homeless. There were no other homeless data categories that matched between the cities. This follows trends of other open city data, where the data has no common representation or semantics (Fox, 2013, 2017; Fox & Pettit, 2015; Nalchigar & Fox, 2014).

Homeless profiles contain major homeless subcategories that further focus on key populations. Most cities begin by estimating the total homeless population followed by counts of the sheltered and unsheltered populations. This is then further broken down into key homeless subpopulations detailing demographics, histories and current situations

Table 2.3 Homeless data types

Category	Subcategories	
Homeless profiles		Total
		Indoor
		Outdoor (unsheltered)
		Rooflessness
		Homelessness
		Insecure housing
		Inadequate housing
		Youth
		Chronic homeless
		Veterans
		Families with children
		Statutory homelessness
Homeless services	Services needs	Housing and homeless services
		Health and treatment services
		Non-housing specific services
		Youth
		Housing assistance
		Wait list for housing
	Services used	Housing and homeless services
		Health and treatment services
		Non-housing specific services
		Government assistance
		Reasons for not receiving government assistance
		Adults
		Youth
		Housing assistance
		Service outcomes
		Homeless prevention enrolments
		Housing placement
		Length of time in temp accommodation
		Housing inventory
		Daily shelter census
		Prevention
		Relief
		City homeless/housing plan

of homeless individuals. Subpopulations are very similar in definition but differ in the scope of who is considered homeless. For example, Barcelona's roofless definition includes those sleeping in the rough and in night shelters but does not include women's shelters, while Toronto's definition of general homeless includes rough sleepers and shelters including women's shelters. This is one of the reasons why there are so many types of homeless data and why it would be difficult to unify the data in the current state.

Homeless services consist of two major categories: service needs and services used. This includes services like housing, health services and governmental assistance. The service needs category details the nature of services that are met and unmet, and the services used describe what and how services are utilized. The services category again varies depending on country and city because each has its own unique response to the homeless problem.

Consistency

Collection of data on a consistent basis is critical in order to generate reliable trend statistics. Trend data highlights shifts and responses to current policy, informing future policy (Busch-Geertsema et al., 2016). In this study, we viewed consistency from a temporal context of data collection, how often data is collected; and from a methodological context, how is the data collected?

The extent and period of the data vary depending on city and category of data. In most cases, data is produced on an annual scale, but is not consistently measured. Some cities, like New York and Toronto, have daily shelter census but perform Point-In-Time (PiT) counts in yearly intervals. London produces quarterly statutory homeless statistics, but annual rough sleeping counts. American cities belonging to the Continuum of Care (CoCs) are required to perform PiT counts on a yearly basis (HUD, 2014). Homeless Open Data produced by cities is still in the initial stages, as the earliest record for Open Data in this study is from 2001 in Tokyo.

Methods to enumerate the sheltered homeless population usually consist of survey or sampling regimes on one given night of the year. Seven of the eight cities with homeless open city data mention point-in-time (PiT counts) as their mode of enumeration; however, PiT methodologies can vary from city to city. PiT counts have been criticized for their lack of consistency and validity. In the United States, varying PiT methodologies have been found to lead to unreliable results (Schneider, Brisson, & Burnes, 2016). Aligning homeless counts from various cities is difficult due to varying details in the enumeration methodologies. Clarity and transparency of count methods will aid in comparability of subsets of the overall homeless population (Smith, 2015). Other methods of homeless estimates include homeless management information systems that collect data at shelters or government facilities. Qualitative data, like interviews of homeless individuals, were found in Barcelona, Calgary and Chicago. Like the definition of homelessness, there is no globally accepted method for measuring homelessness. A systemic approach, adjusted for context, would be required for reliable and comparable homeless estimates (Busch-Geertsema et al., 2016).

Comparability and Validity

We use CIDOM as an indirect measure of validity and comparability of city datasets. The quality and reliability of the indicator data can be measured by looking at the totality of data published for an indicator, the completeness of supporting data and the dominating format used to publish. The calculated CIDOM values can be found in Table 2.4.

Toronto, London, Barcelona and Shanghai have implemented the ISO 37120 standards. However, Toronto is the only city to have shelter indicator values openly available on the city website. It was the most comprehensive of the city datasets reviewed.

CIDOM-1: Totality of the Data Published for an Indicator

The ISO 37120 shelter indicators act as reference points providing standardized values for open city datasets, allowing for comparisons between cities. The supporting data used to calculate the indicators are often not openly available or accessible (Fox & Pettit, 2015), making the value of the shelter indicator an abstract number with no currency. CIDOM-1 is a measure that looks into the totality of published supporting data for indicators. Toronto is the only city that consistently openly published

Table 2.4 CIDOM results

CIDOM	Indicator	Calgary	Toronto	New York	Chicago	San Francisco	Miami	London	Paris	Rome	Barcelona	Beijing	Shanghai	Tokyo	Singapore
	15.1	25	100	25	17	8	17	25	17	0	33	0	0	25	8
1[a]	15.2	40	100	100	100	40	100	80	20	0	100	0	0	100	0
	15.3	0	44	0	0	0	0	0	0	0	11	0	0	0	0
	15.1	0	2	0	0	0	0	0	0	0	0	0	0	0	0
2	15.2	0	2	0	0	0	0	0	0	0	0	0	0	0	0
	15.3	0	2	0	0	0	0	0	0	0	0	0	0	0	0
	15.1	1	1	1	1	1	1	1	1	0	1	0	0	1	1
3	15.2	1	1	1	1	1	1	1	1	0	1	0	0	1	0
	15.3	0	1	0	0	0	0	0	0	0	1	0	0	0	0

[a]CIDOM-1 is measured in %. CIDOM-2 and CIDOM-3 are measured in levels

supporting data for an indicator, publishing data for most of the nodes in the dependency graphs. All CIDOM calculations can be found in Table 2.4. Toronto is one of two cities to have data for indicator 15.3, the other being Barcelona. Calgary, Paris and Singapore show up in the CIDOM analysis because their city websites contained data on household and city populations, fundamental values in the calculation of indicators 15.1 and 15.2. A majority of the cities have totally published datasets for indicator 15.2, but lack data for indicators 15.1 and 15.3. Indicator 15.2 is a direct measure of homelessness and has the potential to provide a standard and reliable homeless estimate for global homeless numbers. Only half of cities reviewed could openly source indicator 15.2.

CIDOM-2: Number of Levels of Supporting Data Openly Published

CIDOM-1 looks at the data from one dimension quantifying how much data is available. CIDOM-2 reviews the integrity of the indicator value and its supporting data. Knowing the extent to which the data is openly published is just as important as knowing how much is openly available. The metadata of supporting data is crucial to the validity of the indicator value. Toronto is the only city to have a level higher than zero (level 2) because the indicator value is openly published. For indicator 15.2, five other cities have all of the openly available data needed to calculate the indicator, but simply have not performed the ISO 37120 calculation or openly published the indicator value.

CIDOM-3: Dominant Format Published

The data was published in a mixture of tables and reports in PDF format, csv (Excel) and html format. The most pervasive format was PDF. In terms of publishing formats, PDF is the least accessible. Homeless values are difficult to extract reducing the opportunity for semantic interoperability.

Toronto is the only city to calculate the ISO 37120 shelter-themed indicators and produce underlying data. It could be argued that indicator

15.2 is the most significant (in the context of homelessness) of the three ISO 37120 indicators as it focuses on the literal homeless, while the 15.1 and 15.3 focus on relative homelessness. Shelter indicator 15.2 could easily be calculated for many cities producing a globally comparable value for homeless numbers, and most of the supporting data was found in half of the cities reviewed. A strength in using the city indicators is also one of its weaknesses, standardized definitions and methodologies. For example, Toronto has a different definition of households living below living standards. Canada's occupancy standard (Toronto, 2013b) states overcrowding at >2 people per bedroom. Indicator 15.1 defines sufficient living spaces as not exceeding three people per room. The slight difference in definition allows Toronto to over report what is considered to be overcrowding by the ISO 37120, skewing the overall value of indicator 15.1. The indicator's definition is based on slum households, defined by the UN-Habitat (2006) in a developing nation context but incorrectly used on developing nations. This showcases the need for openly publishing supporting data; it allows for regional contexts to be transparent and known, aiding in meaningful comparisons of the city data. In addition, publishing underlying data supports the validity of the indicator value.

Conclusion

In this chapter, we reviewed the current state of openly available homeless data by evaluating how standardized, consistent and comparable the data was across cities and examined the validity of ISO 37120 indicators in the context of open homeless data. Improving our knowledge of the complicated dynamics of homelessness requires information that reflects the reality of the homeless. If the ultimate goal is to eradicate homelessness, current homeless data is very limited and informs common mitigation strategies like increased shelter and affordable housing. The drivers of homelessness are multifaceted, and capturing the data required to explain these issues will be a major challenge for future policy. In order to achieve a better understanding of the homeless problem, it is necessary to extend and improve our current local and global database, which includes foundational statistics like homeless counts. We found that there is still a lot of work to be done with open city data to realize a vision of reliable homeless estimates, as data from different nations are not standardized, consistent or comparable.

Fundamental barriers like differences in homeless definitions, methodologies and periodicity of data capture make city data difficult to compare. A global definition/framework for homelessness will aid in the foundation of a reliable global homeless dataset. Global entities like the Institute of Global Homelessness or the United Nation's HABITAT or Human Rights programme are all good candidates to spearhead this initiative as they have already begun dialogue on homeless definitions and measures (Busch-Geertsema, Culhane, & Fitzpatrick, 2015; UN-Habitat, 2000, 2006; UN-Human Rights, 2015). Setting a frame of reference to identify who is homeless, when to perform data captures and permitting room for regional context provides a transparent base to foster meaningful comparisons and information exchange. For example, the US Department of Housing and Urban Development has stringent definitions for homelessness. Methods of enumeration and data capture are standardized allowing for city-to-city comparison. While the HUD system is not flexible, cities like San Francisco still use the predefined standards but also add their own regional interpretation (i.e. expanding on the homeless definitions, measuring other homeless subcategories like youth) in their city reporting.

Using a coarse global standard like the ISO 37120 shelter indicators currently allows for the comparison of city data. The supporting data from which the indicator values are derived must be published. As long as the underlying data is transparent and available, the validity of calculated indicator values can be verified. This gives rise to the issue of how to represent the indicators and the underlying data. Development of a semantic approach will help to unify, link and ground the data. A shelter ontology for homeless city data (Wang & Fox, 2017) helps to achieve computational accessibility allowing for longitudinal and transversal analysis, breaking down the fundamental barriers of homelessness data mentioned above.

This work highlights the need for global standards for homeless data. We were able to show that it is possible to calculate globally standard measures of homelessness (ISO 37120, indicator 15.2) using open city data, ISO 37120 shelter indicators and CIDOM. However, wide data gaps still exist, limiting the utility of the shelter indicators. It also showcases the need for openly published supporting data to prove the robustness and integrity of indicator values. In the global context, open city homeless data is currently not standardized, consistent or comparable. This review contributes to a strategic prioritization for improved Open Data collection, measurement and standardization of homeless data.

Much research remains. To support the standardization of homeless data effort, there are at least two issues researchers need to address:

1. A consensus needs to be developed among homeless researchers in the social sciences as to the various types of data that they need to move their research forward, and
2. Ontologies and standards based on these ontologies need to be developed to precisely, and unambiguously, represent the semantics of the data.

The latter issue is being explored not only by our group at the Centre for Social Services Engineering (csse.utoronto.ca), but more general by standards organizations such as the ISO/IEC (International Standards Organization/International Electrotechnical Commission) Joint Technical Committee Working Group 11 on Smart Cities.

Acknowledgements This research was supported in part by the Faculty of Applied Science and Engineering Dean's Strategic Fund, and the Natural Science and Engineering Research Council of Canada. We also like to acknowledge Prof. C. Pettit's contributions to the development of the CIDOM metrics.

Appendix

Table 2.5 Homeless definitions

Type of homelessness	Definition	Source
Homelessness	Absolute homelessness refers to those without any physical shelter, for example, those living outside, in parks, in doorways, in parked vehicles or parking garages, as well as those in emergency shelters or in transition houses for women fleeing abuse.	ISO 37120
(Toronto)	Any individual sleeping outdoors on the night of the survey, as well as those staying in emergency shelters, in Violence Against Women (VAW) shelters, individuals in health or treatment facilities with no permanent address, as well as those in correctional facilities who are registered in a Toronto court as having no fixed address or a shelter address. Excludes hidden homeless.	(Toronto 2013a)
	Individuals and families who are: Living in a supervised publicly or privately operated shelter designated to provide temporary living arrangement; or with a primary night-time residence that is a public or private place not designed for or ordinarily used as a regular sleeping accommodation for human beings, including a car, park, abandoned building, bus or train station, airport or camping ground.	(HUD, 2014)
(San Francisco)	The definition of homelessness in San Francisco expands HUD's definition to include individuals who are 'doubled-up' in the homes of family or friends, staying in jails, hospitals and rehabilitation facilities, families living in Single Room Occupancy (SRO) units, and in substandard or inadequate living conditions including overcrowded spaces.	(ASR, 2015)
Unsheltered	Individual or family with a primary night-time residence that is a public or private place not designed for or ordinarily used as a regular sleeping accommodation for human beings, including a car, park, abandoned building, bus or train station, airport or camping ground.	(HUD, 2014)
Sheltered	An individual or family living in a supervised publicly or privately operated shelter designated to provide temporary living arrangement (including congregate shelters, transitional housing, and hotels and motels paid for by charitable organizations or by federal, state or local government programmes for low-income individuals).	(HUD, 2014)

(continued)

Table 2.5 (continued)

Type of homelessness	Definition	Source
Chronic	(1) A 'homeless individual with a disability', as defined in section 401(9) of the McKinney-Vento Homeless Assistance Act (42 U.S.C. 11360(9)), who: (i) Lives in a place not meant for human habitation, a safe haven or in an emergency shelter; and (ii) Has been homeless and living as described in paragraph (1)(i) of this definition continuously for at least 12 months or on at least 4 separate occasions in the last 3 years, as long as the combined occasions equal at least 12 months and each break in homelessness separating the occasions included at least 7 consecutive nights of not living as described in paragraph (1)(i). Stays in institutional care facilities for fewer than 90 days will not constitute as a break in homelessness, but rather such stays are included in the 12-month total, as long as the individual was living or residing in a place not meant for human habitation, a safe haven or an emergency shelter immediately before entering the institutional care facility; (2) An individual who has been residing in an institutional care facility, including a jail, substance abuse or mental health treatment facility, hospital or other similar facility, for fewer than 90 days and met all of the criteria in paragraph (1) of this definition, before entering that facility; or (3) A family with an adult head of household (or if there is no adult in the family, a minor head of household) who meets all of the criteria in paragraph (1) or (2) of this definition, including a family whose composition has fluctuated while the head of household has been homeless.	Homeless emergency assistance and rapid transition to housing: Defining 'Chronically Homeless', 80 Fed.Reg. 75791 (4 December 2015). *Federal Register: The daily journal of the United States*. Web. 4 December 2015. url: https://www.gpo.gov/fdsys/pkg/FR-2015-12-04/pdf/2015-30473.pdf
Absolute	People without an acceptable roof over their heads, living on the streets under bridges and deserted buildings.	(Cooper, 1995)
First degree relative	People moving between various forms of temporary or medium term shelter such as refuges, boarding houses hostels or friends.	(Cooper, 1995)
Second degree relative	People constrained to live permanently in single rooms in private boarding houses	(Cooper, 1995)

(*continued*)

Table 2.5 (continued)

Type of homelessness	Definition	Source
Severely inadequate and/or insecure accommodation	3 (a) People sharing with friends and relatives on a temporary basis 3 (b) People living under threat of violence 3 (c) People living in cheap hotels, bed and breakfasts and similar 3 (d) People squatting in conventional housing 3 (e) People living in conventional housing that is unfit for human habitation 3 (f) People living in trailers, caravans and tents 3 (g) People living in extremely overcrowded conditions 3 (h) People living in non-conventional buildings and temporary structures, including those living in slums/informal settlements	(Busch-Geertsema et al., 2016)
Hidden	Individuals living with others in conventional housing but on an emergency basis.	(Busch-Geertsema et al., 2016)
Statutory	Households which meet specific criteria of priority need set out in legislation, and to whom a homelessness duty has been accepted by a local authority. Such households are rarely homeless in the literal sense of being without a roof over their heads, but are more likely to be threatened with the loss of, or are unable to continue with, their current accommodation.	(Department for Communities and Local Government 2013)
Rough sleepers	People sleeping, about to bed down (sitting on/in or standing next to their bedding) or actually bedded down in the open air (such as on the streets, in tents, doorways, parks, bus shelters or encampments) People in buildings or other places not designed for habitation (such as stairwells, barns, sheds, car parks, cars, derelict boats, stations or 'bashes'). The definition does not include people in hostels or shelters, people in campsites or other sites used for recreational purposes or organized protest, squatters or travellers.	(Department for Communities and Local Government 2013)
Literal	People without any accommodation, and those living in temporary or emergency accommodation specifically provided for homeless people.	(Busch-Geertsema et al., 2016)

Table 2.6 ETHOS homeless categories (Amore et al., 2011; Sales et al., 2015)

Homeless	Definition
Roofless	People living rough or in a public space. People sleeping in a night shelter and/or forced to spend the day in a public space.
Houseless	People living in hostels or in accommodation for the homeless. Temporary accommodation. Women's shelter accommodation. People living in temporary accommodation for immigrants or asylum seekers. People who live in housing institutions or penal institutions, prospect of being dismissed in a deadline without shelter housing available. People who live in a continued support accommodation for homeless people.
Insecure housing	People who live in insecure tenancy housing. Without paying rent. People who live under threat of eviction. People who live under threat of family or partner's violence.
Inadequate housing	People who live in temporary or non-conventional structures. People who live in inappropriate housing according to legislation. People who live in overcrowded housing

References

Amore, K., Baker, M., & Howden-Chapman, P. (2011). The ETHOS definition and classification of homelessness: An analysis. *European Journal of Homelessness, 5*(2), 19–37.

ASR (Applied Survey Research). (2015). *San Francisco homeless point-in-time count and survey: Comprehensive report 2015*. San Francisco: Applied Survey Research. Retrieved from http://dhsh.sfgov.org/wp-content/uploads/2016/06/2015-San-Francisco-Homeless-Count-Report_0-1.pdf.

Busch-Geertsema, V., Culhane, D., & Fitzpatrick, S. (2015). A global framework for understanding and measuring homelessness. Retrieved from http://docs.wixstatic.com/ugd/d41ae6_97a693a1aba845058f91e9cf38f7c112.pdf

Busch-Geertsema, V., Culhane, D., & Fitzpatrick, S. (2016). Developing a global framework for conceptualising and measuring homelessness. *Habitat International, 55*, 124–132. Retrieved from http://www.sciencedirect.com/science/article/pii/S0197397515300023

Cooper, B. (1995). *Shadow people: The reality of homelessness in the 90s*. Sydney City Mission. Retrieved from https://www.researchgate.net/publication/228268057_Shadow_People.

Deng, D., Liu, S., Wallis, L., Duncan, E., & McManus, P. (2017). Urban sustainability indicators: How do Australian city decision makers perceive and use global reporting standards? *Australian Geographer, 48*(3), 1–16. https://doi.org/10.1080/00049182.2016.1277074

Department for Communities and Local Government. (2013). *Department for Communities and Local Government annual report and accounts 2012–13 (For the year ended 31 March 2013)*. London, UK. Retrieved July 2019, from https://www.gov.uk/government/organisations/department-for-communities-and-local-government

FEANTSA. (2007). *FEANTSA proposal: A retrospective module on homelessness for household surveys*. Brussels.

Fox, M. S. (2013). *City data: Big, open and linked*. Working Paper, Enterprise Integration Laboratory, University of Toronto. Retrieved from http://www.eil.utoronto.ca/wp-content/uploads/smartcities/papers/City-Data-v5.pdf.

Fox, M. S. (2017). The PolisGnosis Project: Enabling the computation analysis of city performance. In *Proceeding of the 2017 Industrial and Systems Engineering Conference*. Retrieved from http://eil.mie.utoronto.ca/wp-content/uploads/2015/06/IISE-2017-3349.pdf.

Fox, M. S., & Pettit, C. J. (2015). On the completeness of open city data for measuring city indicators. In *2015 IEEE 1st International Smart Cities Conference, ISC2 2015*. Guadalajara.

Gaetz, S., Dej, E., Richter, T., & Redman, M. (2016). *The state of homelessness in Canada 2016*. Toronto. Retrieved from http://homelesshub.ca/SOHC2016.

Gaetz, S., Gulliver, T., & Richter, T. (2013). *The state of homelessness in Canada 2013*. Canadian Homelessness Research Network.

HUD (U.S. Department of Housing and Urban Development). (2014). *Point-in-time count methodology guide*. Retrieved from https://www.hudexchange.info/resources/documents/PIT-Count-Methodology-Guide.pdf.

Hulchanski, J. D. (2010). *The three cities within Toronto: Income polarization among Toronto's neighbourhoods, 1970–2005*. Toronto: Cities Centre, University of Toronto.

ISO 37120. (2014). ISO 37120: Sustainable development of communities—Indicators for city services and quality of life. ISO.org.

McCarney, P. (2015). The evolution of global city indicators and ISO37120: The first international standard on city indicators. *Statistical Journal of the IAOS, 31*(1), 103–110. Retrieved from http://search.ebscohost.com.myaccess.library.utoronto.ca/login.aspx?direct=true&db=crh&AN=101648412&site=ehost-live

Nalchigar, S., & Fox, M. S. (2014). An ontology for open 311 data. In *Workshop on semantic cities: Beyond open data to models, standards and reasoning, AAAI14*. Quebec City: AAAI14.

Peressini, T., McDonald, L., & Hulchanski, D. (1996). *Estimating homelessness: Towards a methodology for counting the homeless in Canada*. Canada Mortgage and Housing Corporation.

Richter, S., & Botha, A. (2012). Happenings / L'Événement—New research initiatives: Addressing global homelessness. *Canadian Journal of Nursing Research Archive, 44*(4). Retrieved from http://cjnr.archive.mcgill.ca/article/view/2379

Sales, A., Uribe, J., & Marco, I. (2015). *2015 diagnosis: The situation of homelessness in Barcelona. Evolution and intervention policies*. Retrieved from http://www.bcn.cat/barcelonainclusiva/ca/2016/2/sense_sostre2015_ang.pdf.

Schneider, M., Brisson, D., & Burnes, D. (2016). Do we really know how many are homeless?: An analysis of the point-in-time homelessness count. *Families in Society: The Journal of Contemporary Social Services, 97*(4), 321–329. Retrieved from https://doi.org/10.1606/1044-3894.2016.97.39

Smith, A. (2015). Can we compare homelessness across the Atlantic? A comparative study of methods for measuring homelessness in North America and Europe. *European Journal of Homelessness, 9*(2), 233–257.

Springer, S. (2000). Homelessness: A proposal for a global definition and classification. *Habitat International, 24*(4), 475–484.

Toronto, C. (2013a). *2013 street assessment results*. Retrieved from https://www.toronto.ca/legdocs/mmis/2013/cd/bgrd/backgroundfile-61365.pdf

Toronto, C. (2013b). Toronto's 2013 results under ISO37120: Indicators of city service deliver and quality of life. Retrieved from https://www.toronto.ca/wp-content/uploads/2017/11/989c-Final-Summary-of-Torontos-WCCD-ISO-37120-Results-6-AODA-.pdf.

UN-Habitat. (2000). *Strategies to combat homelessness*. UN-Habitat.

UN-Habitat. (2006). *The state of world's cities 2006/2007*. UN-Habitat. Retrieved from https://unhabitat.org/books/state-of-the-worlds-cities-20062007/.

UN-Human Rights. (2015). Report of the Special Rapporteur on adequate housing as a component of the right to an adequate standard of living, and on the right to non-discrimination in this context. Retrieved from http://ap.ohchr.org/documents/dpage_e.aspx?si=A/HRC/31/54

Wang, Y., & Fox, M. S. (2017). Households, the homeless and slums towards a standard for representing city shelter Open Data. In *Proceedings of the AAAI Workshop on AI and OR for Social Good*.

3

Open Data and Racial Segregation: Mapping the Historic Imprint of Racial Covenants and Redlining on American Cities

Ashley Bakelmun and Sarah Jane Shoenfeld

Abbreviations

HOLC	Home Owners' Loan Corporation
GIS	Geographic Information System
OCR	Optical Character Recognition

Highlights

- Open Data is not just relevant to current demographic trends but can improve our understanding of historic inequity and racial segregation.
- Historic mapping projects catalyse a community dialogue around the legacy of racial covenants and government policies that enforced racial

A. Bakelmun (✉)
Urban Equity Lab, New York City, NY, USA
e-mail: ashley@urbanequitylab.com

S. J. Shoenfeld
Prologue DC, LLC, Washington, DC, USA
e-mail: sarah@prologuedc.com

S. Hawken et al. (eds.), *Open Cities | Open Data*,
https://doi.org/10.1007/978-981-13-6605-5_3

segregation—helping to guide current efforts towards equalising access to housing.
- Geolocating previously archived racial covenant data onto openly available maps improves our understanding of its relationship to discriminatory lending (redlining) and land use policies, paving the way for more informed, equitable development.
- Citizen participation, or crowdsourcing, can be used to assist in the time-intensive efforts of capturing data from archived records and to deepen participants' understanding of how racial segregation was historically maintained in the United States.

Introduction: Using Open Data to Further Urban Equity

We live in a time where complex sets of Open Data are creating new opportunities for innovation in cities (Barns, 2016; McArdle & Kitchin, 2016) while also providing new streams of information for evidence-based policy. Municipal efforts in cities such as Rio, New York City, Dublin, and London have brought together diverse sets of Open Data to spot trends and inform interdisciplinary conclusions (Kitchin, 2014). In the United States, for example, New York's *Roadmap for the Digital City* (City of New York, 2011) reported 350 sets of government data as of 2011, Chicago's *The City of Chicago Technology Plan* reported 400 datasets as of 2013 (City of Chicago, 2013), and Washington DC was providing some 475 datasets by the end of 2017 (Open Data DC, 2017). These numbers have continued to grow with the addition of new datasets to their Open Data portals (see https://opendata.cityofnewyork.us/data; https://data.cityofchicago.org; http://opendata.dc.gov). Innovative applications of open datasets have influenced the real-time operation of cities via analyses of traffic patterns, weather, amenity locations, and other factors affecting urban quality of life (Kitchin, 2014).

However, while "smart cities" and Open Data provide exciting technological possibilities for resolving complex urban development challenges, if technology is "coarsely applied to complex problems," it often falls short (Townsend, 2013, p. 17). Data that is not equally accessible to all can serve to increase existing power imbalances (Townsend, 2013, p. 13); for example, data related to increased surveillance is more likely to result in over-policing of racial/ethnic minorities, and information provided to real estate developers but not to individual property owners, is likely to perpetuate housing injustices. While some city governments have effectively used big data to improve services for underserved residents (Goldsmith, 2017), citizen-led movements are far more likely to use technology as a tool for increasing equity (Townsend, 2013). For example, the grassroots group Code for DC engaged over 100 volunteers in merging and visualising data to help preserve affordable housing in Washington DC (Hernandez & O'Maley, 2017).

Whether initiated by city governments or by activist-minded citizens, investigating structural inequity through Open Data can assist the broader strategic goals of cities. For example, a primary goal of Washington DC's 2017 *Economic Strategy* is to reduce employment disparities; 13.5% of black residents are currently unemployed as compared to just 2.6% of white residents (DC Office of the Mayor, 2017b). Addressing the racial employment gap, which correlates with residential settlement patterns and the location of jobs (Carter & Reardon, 2014), requires a broader understanding of the interdisciplinary factors behind segregation today and a more accurate historical account of how we got to this point. Open Data can be used to expose the historic injustices that underlay racial inequities in Washington DC and in cities across the United States.

This chapter looks at several grassroots projects using historical data for drawing attention to the inequities embedded in urban landscapes. These efforts—all of which involve mapping data that reveals historical patterns of racial segregation—may contribute to equitable development by pushing city planners and policymakers to ensure future development does not maintain or preserve these patterns. When such data is publicly exhibited, the role of legally sanctioned efforts to enforce racial segregation is far less likely to be overlooked.

Using Historical Data to Address Equity

Table 3.1 lists five online projects that have been undertaken throughout the United States to document the explicit historic use of race as a category for creating and enforcing urban settlement patterns that persist today. With a primary focus on the project *Mapping Segregation in Washington DC*, this chapter shows how datasets can be combined to expose the legacy of racialised land use policies. In addition, by correlating existing data, specifically U.S. Census data and Geographic Information System (GIS) layers, with historical land use data and with new open datasets documenting present-day rates of poverty and home ownership, for example, these projects serve as valuable resources for a multidisciplinary approach to advancing equitable urban policy today.

Background

In the United States, racial segregation cannot fully be understood—or addressed—without acknowledging its roots in policies designed to maintain a racial order institutionalised by 250 years of chattel slavery (Bernasconi, 2017; National Advisory Commission on Civil Disorders, 1968; Pettigrew, 1979). Racially restrictive deed covenants came into common use in cities across the United States during the early decades of the twentieth century. These agreements—either inserted by real estate developers into deeds of sale for land or buildings or added to deeds by white property owners—barred African Americans and often other racial/ethnic minorities from future ownership or occupancy (Gotham, 2000a, 2000b). Upheld by American courts as private contracts between neighbours, racial covenants served to legally maintain exclusively white neighbourhoods throughout the United States. As millions of black migrants relocated from the U.S. South to cities across the North during this period, they were confined to the oldest, most poorly located housing available (Lowe, 1948). By associating property values with race and enforcing residential segregation (Gotham, 2000b; Howell, 2006), racial covenants laid the foundation for longstanding patterns of racialised land use. The federal government institutionalised racial covenants, beginning

Table 3.1 Links to the five case studies reviewed in this chapter and their associated cities' Open Data portal

Mapping project name	City	Mapping platform used	Link to project	Municipal Open Data portal
Mapping Segregation in Washington DC	Washington DC	ArcGIS/Esri StoryMap	http://mappingsegregationdc.org	http://opendata.dc.gov
Mapping Prejudice	Minneapolis	ArcGIS/Carto	https://www.mappingprejudice.org	City: http://opendata.minneapolismn.govCounty (housing deed data): https://www.hennepin.us/residents/property/property-information-search
Redlining Louisville	Louisville	ArcGIS/Esri StoryMap	http://www.arcgis.com/apps/MapSeries/index.html?appid=a73ce5ba85ce4c3f80d365ab1ff89010	http://portal.louisvilleky.gov/service/data/6/
Segregated Seattle	Seattle	N/A	http://depts.washington.edu/civilr/covenants.htm	https://data.seattle.gov
Mapping Inequality	Multicity	Panorama/custom mapping app (using React, Carto, d3, Leaflet)	https://dsl.richmond.edu/panorama/redlining	N/A

in the 1930s, by upholding them as a criterion for determining where to provide federally guaranteed mortgage loans for new housing (Jones-Correa, 2000). Although the U.S. Supreme Court finally ruled, in the 1948 case *Shelley v. Kraemer*, that racial covenants were legally unenforceable (Jackson, 1985; Ming Jr., 1949), their use continued at least into the 1960s (Brooks & Rose, 2013; Rose, 2016).

In addition, between 1935 and 1940, the U.S. government's Home Owners' Loan Corporation (HOLC) produced colour-coded "Residential Security" maps of 239 American cities that classified sections of each city according to their perceived levels of financial risk for insuring mortgages (Aaronson, Hartley, & Mazumder, 2017; Crossney & Bartelt, 2005; Greer, 2012). The maps were based on surveys conducted by local real estate officials that used race along with a range of other factors to designate risk levels (Greer, 2012). Areas occupied by African Americans, or close to black populations, were coloured red for "hazardous" (Fig. 3.1), making these areas off-limits for federally insured home loans and, in turn, ensuring disinvestment and decline (Gotham, 2000b; Grove, Cadenasso, Pickett, Burch, & Machlis, 2015; Grove et al., 2018; Oliveri, 2015). As a legacy of these maps, the term "redlining" continues to refer to racially discriminatory practices by banks and insurers that have served to maintain racial segregation and economic inequality in American cities (Jackson, 1985; Nier III, 1999; Woods, 2012). Seventy-four per cent of formerly redlined areas remain low-to-moderate income today, with median family incomes at 80% or less than the average area income; 64% of formerly redlined areas are majority-minority (National Community Reinvestment Coalition, 2017).

The 1965 establishment of the U.S. Department of Housing and Urban Development, followed by the passage of the Fair Housing Act (1968), the Home Mortgage Disclosure Act (1975), and the Community Reinvestment Act (1977), comprises notable federal government efforts to counteract persistent residential segregation (Nier III, 1999). These and other efforts to "affirmatively further fair housing," as the 1968 Fair Housing Act requires (U.S. Department of Housing and Urban Development, 2014), have been numerous and varied, but have not fulfilled their intended impact of decreasing racial and economic

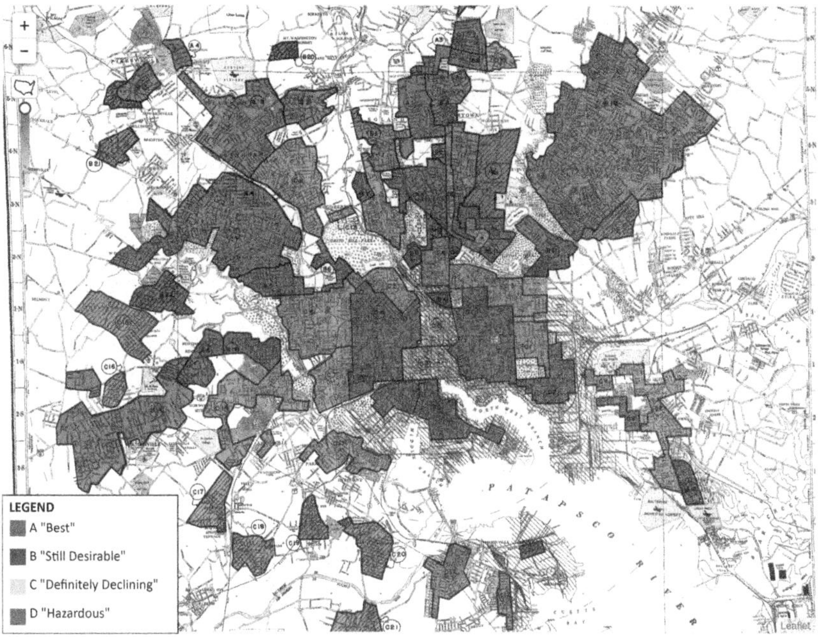

Fig. 3.1 A HOLC Residential Security map (Baltimore). Red areas were designated "hazardous" and denied home loans. Source: Nelson et al., 2017. (Colour figure online)

segregation. The act has not been adequately enforced (Massey, 2015), while tax policies, banking practices, and local governments' reliance on real estate development for funding city budgets have continued to reinforce racial disparities in access to capital and housing (Howell, 2006; Nier III, 1999; Pettigrew, 1979; Rothstein, 2017; Woods, 2012). In fact, the black/white racial gap in wealth and property ownership—in the United States, this is the primary tool for long-term economic advancement—has increased since the 1960s. It is hoped that the projects described here will inform a wide audience about the history of racial segregation and will simultaneously equip advocates and policymakers with the data needed to justify reparative efforts to equalise access to housing. This chapter assembles five case studies of work being done across the United States in this regard.

Combining Existing and Archived Datasets

The projects outlined in this chapter are digitising archival data and combining it with existing Open Data towards two ends: (1) to exhibit previously underrecognised information on the historical racial segregation of American cities, and (2) to expose relationships between racially segregationist policies, resistance to these policies, and the historical and current demographics of these cities.

Using open GIS layers to geolocate historical data that was largely inaccessible until now also allows urban designers, planners, policymakers, and others to utilise the data for addressing historical inequities in access to urban space. Guided by a shared vision of creating a more informed dialogue around the structural racism that has shaped their cities, the leaders of these projects see their work as fundamental for informing public policies around equitable development and inclusion.

Mapping and Visualising Existing Open Data

In accordance with its current *Data Policy*, Washington DC requires each "dataset that directly supports the mission of one or more public bodies" to be open by default, unless reasons for restriction are provided (DC Office of the Mayor, 2017a). The city's Open Data portal, listed in Table 3.1, includes multiple formats of datasets for things like business licences, building permits, capital projects, and current demographics. *Mapping Segregation in Washington DC* has utilised the portal's open GIS data, which defines property boundaries, to map properties once subject to racially restrictive covenants. Most of the other projects profiled in this chapter are also using open GIS data, along with open GIS base maps—available via ArcGIS and other online mapping platforms—to incorporate historical data into maps showing current geographical boundaries and names. By doing so, their audiences are able to more easily make connections between the cities they live in today and the historical factors that have shaped them.

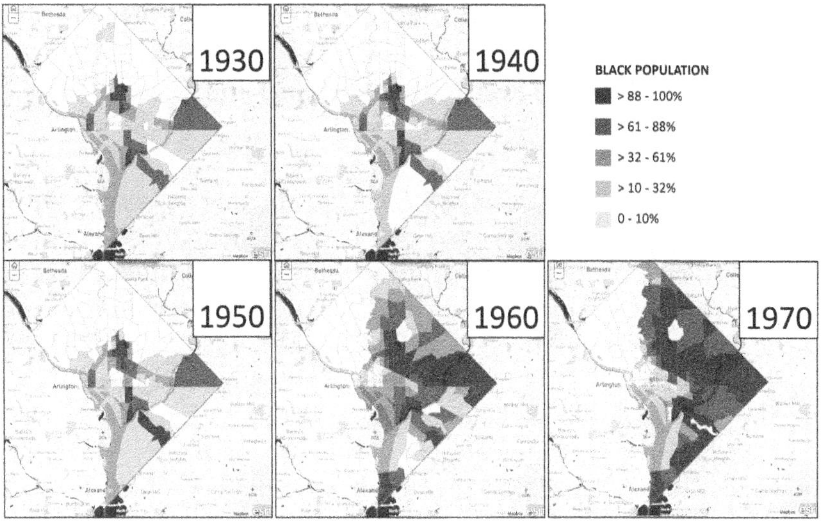

Fig. 3.2 Expansion of black population in DC, 1930–1970. Source: Prologue DC, 2018a

Mapping Segregation in Washington DC is also using open U.S. Census data made available as source tables (summary statistics) and GIS boundary files (shapefiles) by the University of Minnesota's National Historic Geographic Information System (see NHGIS.org). Mapping this data shows the dramatic racial transformation of Washington DC between 1930, when just over 70% of residents were white, and 1970, when the same percentage of residents were black (Fig. 3.2). The maps show how rapidly African Americans claimed space that had formerly been off-limits after racial covenants became legally unenforceable in 1948, and more importantly, as the federal government subsidised white out-migration to racially exclusive suburbs during this period; black long-time residents of rural enclaves were simultaneously pushed out of emerging suburbs and into the city. The pace of the so-called white flight quickened after the Supreme Court desegregated public schools in 1954 (Prologue DC, 2018b).

Compiling, Opening, and Visualising Historical Data

Documenting Racially Restrictive Covenants and Their Demographic Impact

Mapping Segregation in Washington DC relies not only on existing Open Data but on archival datasets that were inaccessible to the wider public before the project's staff and volunteers compiled and digitised them. Federal Housing Administration data documenting racial makeup and other demographic information for individual city blocks in 1934 was available only in the form of a map with handwritten numbers indicating the percentage of households per block for each category of information collected (% non-white, % owner-occupied, % "unfit for occupancy," etc.) (Fig. 3.3). Project staff entered a portion of this data into spreadsheets and then geocoded it using DC open property data to assign current geographical identifiers to each block (DC uses a Square, Suffix, and Lot number to identify individual properties, with Squares

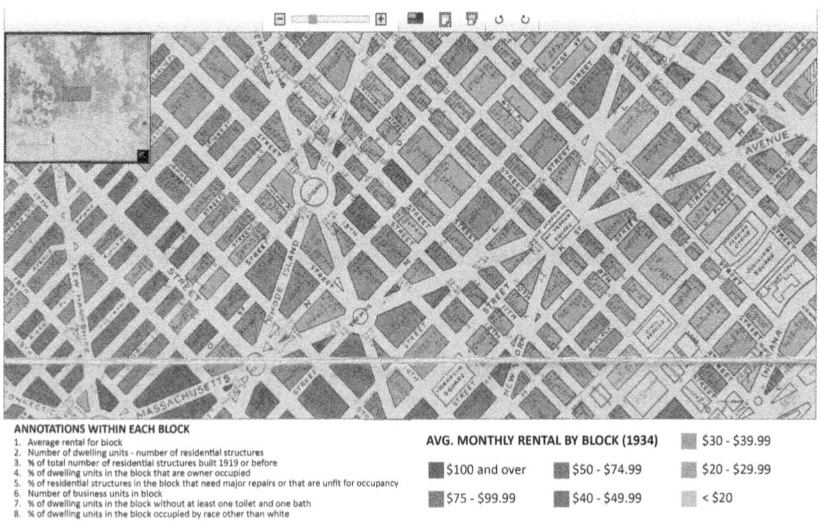

Fig. 3.3 Map of DC indicating demographic compositions of each block in 1934. Source: Library of Congress, 2019, original produced by US federal government, ca. 1937

being equivalent to a square block), so the data could be incorporated into a present-day open base map via Esri's ArcGIS platform. Project staff have also compiled decennial U.S. Census data for city blocks, covering 1940, 1950, 1960, and 1970, and have incorporated this data for one large section of the city thus far (Prologue DC, 2018b). Much of this block-level data could not be accessed, and none was visualised, prior to this project's ongoing work to digitise and "open" these valuable datasets.

Most importantly, *Mapping Segregation in Washington DC* is documenting and digitising all of the historic property deeds with racially restrictive covenants. Thus far, the team has documented around 10,000 properties that were racially restricted, of around 150,000 total properties. A similar effort is underway in Minneapolis, where the project *Mapping Prejudice* has located and mapped over 14,000 property deeds with racial covenants; the team expects to add between 5000 and 10,000 more (Delegard, Ehrman-Solberg, & Petersen, 2017). Moving beyond housing, the DC team has also compiled smaller datasets of racially restricted subdivisions, racially segregated public schools and recreation areas, and properties subject to lawsuits challenging racial covenants—and plans to map segregated public housing and properties financed by the Federal Housing Administration, which used racial covenants as a criterion for underwriting loans.

Housing deeds from the early to mid-1900s are typically found in city archives, often as printed documents that may be handwritten, or are housed as public records by municipal agencies responsible for recording transactions related to property deeds. For *Mapping Segregation in Washington DC*, the team began by paging through massive printed volumes of deed transactions (about eighty-five 500-page volumes per year) from around 1909 to 1912, but due to time constraints and the limitations of the facility in which these records are housed, they eventually shifted their focus to deeds available online, from 1921 and later. The team is currently using a publicly accessible database to review deeds one by one for racial covenants, focusing on individual housing developers and small sections of the city at a time. With assistance from the *Mapping Prejudice* team in Minneapolis, they hope to implement a more efficient and crowdsourced approach to data-gathering moving forward.

The Minneapolis team incorporated citizen participation into their data collection from the outset, via an open software programme designed for crowdsourcing. They began by receiving the county's approval for using Optical Character Recognition (OCR) to "scrape" historic property deeds—all of which were already digitised—for keywords related to racial covenants ("Negro," "African," etc.). Based on these keywords, around 240,000 deeds were flagged for further review and uploaded to Zooniverse, a platform built for citizen participation in gathering data (Zooniverse, 2017). To ensure accuracy, four different volunteers must review each deed, and each must answer a set of nine queries, requiring them to enter information identifying the precise location of the property in question and to transcribe the racially restrictive language each deed contains (see https://www.zooniverse.org/projects/kevinesolberg/mapping-prejudice). The team then exports this data for review, cleans, and appends to a master dataset used for mapping. A volunteer who has been with the project since its inception, Penny Petersen, has reviewed more than 13,000 deeds herself, but the project counts over 1100 additional registered volunteers (as of early 2018), who typically review 10–20 deeds each (Delegard et al., 2017).

A third project documenting racial covenants online is *Segregated Seattle*, which began in 2005 and features a publicly accessible database of just 416 racially restrictive deeds—each covering a subdivision, single property, or multiple properties—compiled from county property files. Thirteen university students scrolled through microfilmed deeds for Seattle and surrounding suburbs, focusing on the years 1927–1948 but only getting through half of these years (Gregory, 2017). They made digital copies of any racial deed restrictions found. Because some of the scanned deeds are difficult to read, and due to other priorities of this project—which goes well beyond documenting racial covenants (Gregory, 2006)—*Segregated Seattle*'s racial covenant database remains incomplete.

Visualising Racially Restrictive Covenants and Their Legacy

Mapping Segregation in Washington DC has used Esri Story Map templates for mapping, displaying, and contextualising the data that the

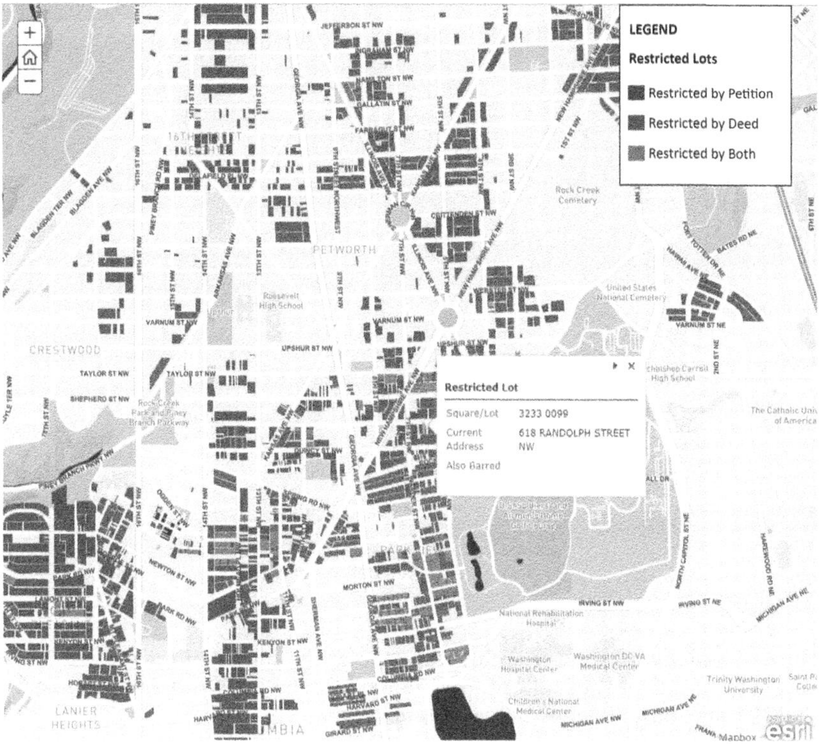

Fig. 3.4 *Mapping Segregation in Washington DC*: location of racially restricted properties (via deeds of sale or neighbour petitions). Source: Prologue DC, 2018a

project's staff has collected and digitised (Shoenfeld & Cherkasky, 2015). Narrative text as well links to historic maps and photographs, and oral history audio clips accompany maps showing locations of racially restrictive deed covenants (Fig. 3.4) combined with other data described earlier. This reveals, for example, that housing, schools, and parks restricted to whites only displaced communities where African Americans lived and owned land (Prologue DC, 2018b). In addition, plotting the location of court challenges to racial covenants—waged by both white property owners and African Americans seeking to buy houses on restricted blocks—reveals that these lawsuits arose primarily along geographical racial borders. As shown via the project's maps, the courts were integral to maintaining segregation at the neighbourhood level by upholding

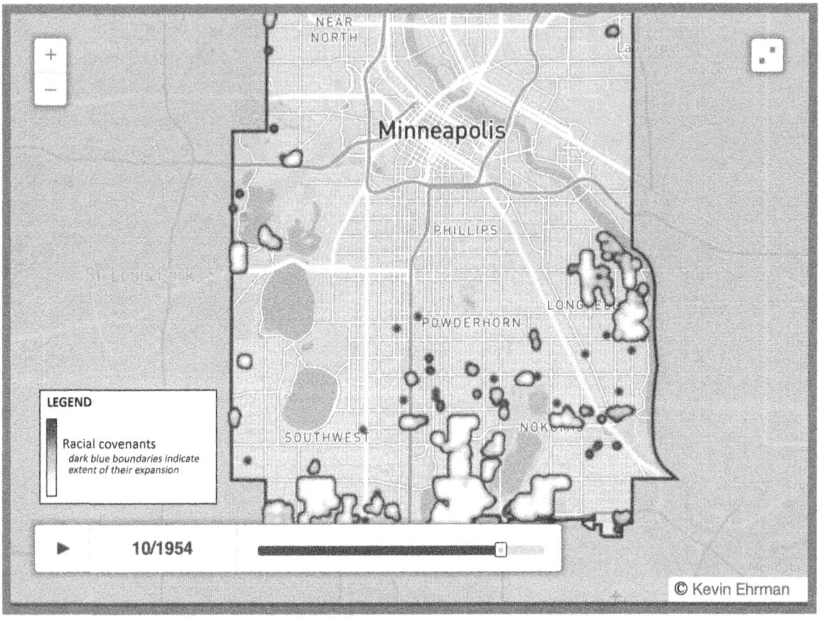

Fig. 3.5 *Mapping Prejudice*: interactive map of the spread of racially restrictive covenants in Minneapolis (this shows 1954). Created by: Kevin Ehrman-Solberg [2017], Data Source: *Mapping Prejudice* Project

racial covenants (Prologue DC, 2018a). Most significantly, the story maps demonstrate the extent to which restrictive covenants were used during the decades when Washington DC expanded well beyond its original boundaries—controlling where people lived during a period of major growth. Racial covenants enforced patterns of housing segregation that would have a long-term impact on the city.

Whereas the DC team has moved forward with building online exhibits while data collection is ongoing, Minneapolis' *Mapping Prejudice* has thus far primarily focused on compiling a dataset of restrictive covenants, which will be made public once complete. However, a time lapse map on the project's home page shows the dramatic spread of racial covenants across Minneapolis between 1911 and 1962 (Fig. 3.5) and is accompanied by a map showing the extent and location of African American settlement in 1910, when racial covenants first came into use there. The team's growing dataset also allows for static maps to be produced (Fig. 3.6)

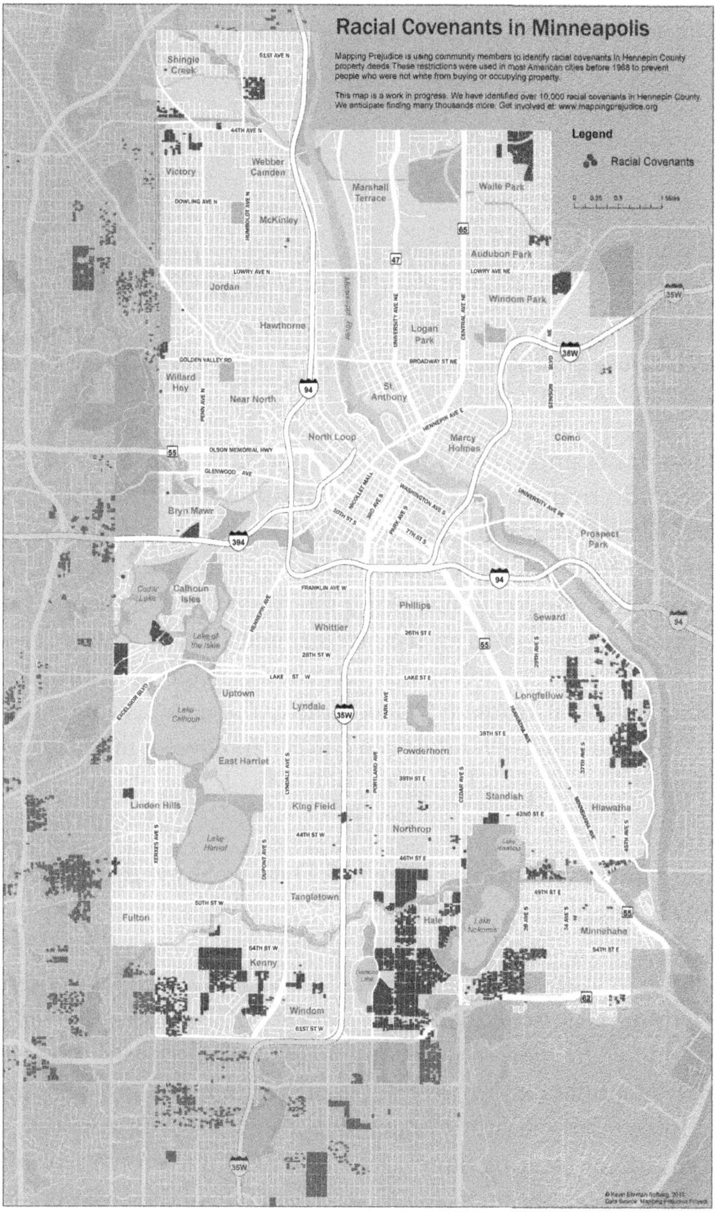

Fig. 3.6 An in-progress map of racial covenants in Hennepin County property deeds. Created by: Kevin Ehrman-Solberg [2018], Data Source: *Mapping Prejudice* Project

Database of 416 racial restrictive covenants from King County Recorder's Office. Copyright © Seattle Civil Rights and Labor History Project 2006-																	
neighborhood	plat name	restriction	Covenant pdf link	seller	date	tracts	Vol #	identifying page or number	Case #	researcher	notes	city	full tract covenant	Gregory checked	dataset	researcher	notes
Seattle																	
Alki	Williams Alki Addition	The lot, nor any part thereof, shall not be sold to any person either of whole or part blood, of the Mongolian, Malay, or Ethiopian races, nor shall the same nor any part thereof be rented to persons of such races.		Williams, Franklin	3/18/1929	lot 3	1414	6052525973		Carolyn Lind		city			fall2006	Carolyn Lind	
Ballard/Sunset Hill	Westhaven	No part of said property hereby conveyed shall ever be used or occupied by any person of the Ethiopian, Malay, or any Asiatic race, and		Westlands Investment co.	1928	lot 4, block 5	1393			James Gregory		city			sp2005	James Gregory	

Fig. 3.7 *Segregated Seattle*: screenshot of database of 416 racially restricted covenants. Source: Gregory, 2006

that show, for example, how much of the housing around city parks were restricted, effectively barring access for African Americans to public green spaces (Delegard & Ehrman-Solberg, 2017).

Segregated Seattle's website contains numerous historic and more recent maps exhibiting the racial makeup of the city's neighbourhoods, but the project's racial covenant data is not visualised (Fig. 3.7). On the other hand, this is the only project that has already made its entire database accessible online, so while it is incomplete, the data can potentially be transferred and used by other researchers interested in pursuing the topic further.

Opening and Visualising Historical Data to Show the Impact of HOLC "Redlining" Maps

In addition to institutionalising the use of racial covenants as a criteria for underwriting real estate loans—a practice that lasted from 1934 until 1950—the U.S. government produced colour-coded "Residential Security" maps (also known as HOLC or "redlining" maps) beginning in 1935. These maps—nearly 250 in number—classified perceived levels of financial risk for insuring mortgages and served to reinforce existing patterns of racial segregation in American cities. Two of the projects listed in Table 3.1 focus on digitising these maps and correlating them with present-day urban conditions.

The *Mapping Inequality* project began as a collaboration between the University of Richmond's Robert Nelson and the University of North Carolina-Chapel Hill's Richard Marciano, now at the University of Maryland, then expanded to include Virginia Tech's LaDale Winling and Johns Hopkins University's Nathan Connolly. At the heart of this multicity project are more than 150 digitised Residential Security maps, formerly available only in print at the National Archives. Each colour-coded area on the maps is clickable to allow for reading survey data compiled by local real estate officials hired by the federal government. These "area descriptions" were georeferenced—again, using existing open GIS data—and are currently being transcribed by university students (Nelson, 2017). The clickable maps also display correlations between the financial "grades" assigned to land, relative distance from the urban core, and approximate density (Fig. 3.8). This is done by overlaying Burgess concentric circles (based on a highly influential and now discredited theory that naturalised the racial/ethnic segregation that segregationist policies created) onto many of the maps, helping to enrich users' understanding of how these maps might have been used and their long-term impact on American cities.

The project *Redlining Louisville* focuses on a single city, using Louisville, Kentucky's HOLC map as the basis for making connections between the

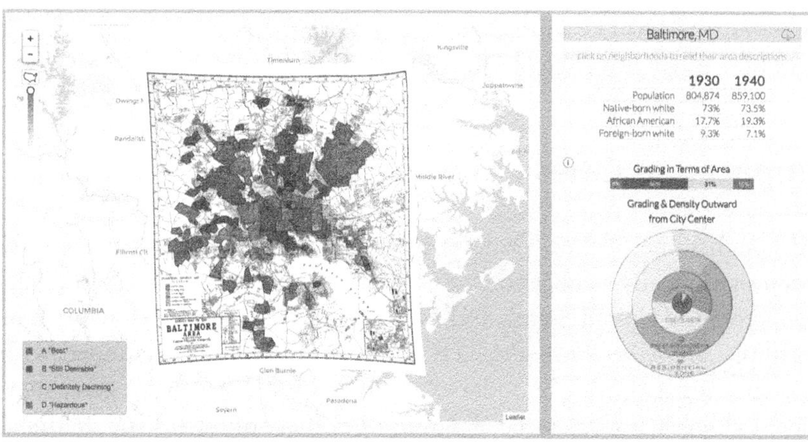

Fig. 3.8 *Mapping Inequality*: the Residential Security map for Baltimore, next to demographic and density analysis. Source: Nelson et al., 2017

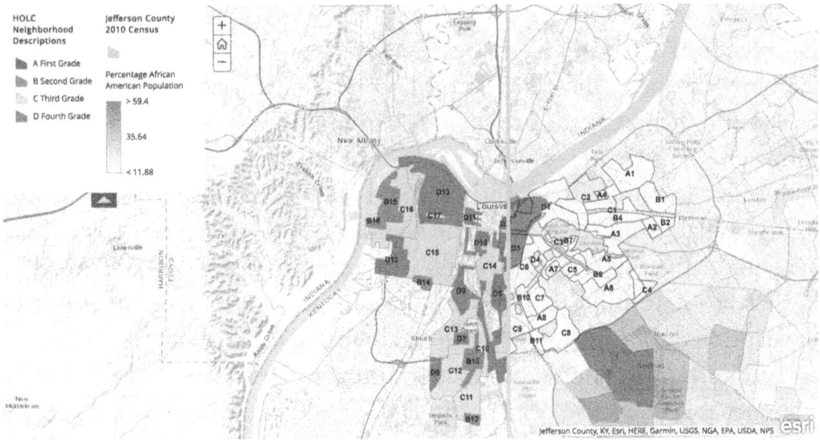

Fig. 3.9 *Redlining Louisville*: Residential Security map neighbourhood classifications (left) compared to racial distribution in 2010. Source: Poe, 2015

past and present. The project uses an Esri Story Map template for displaying side-by-side images comparing the HOLC map with open municipal and U.S. Census data from 2010 to 2014 (Fig. 3.9); data showing racial makeup, property values, building permits, and mortgage denials are among the categories mapped. Each dataset is separately paired with a map showing 1937 HOLC data, allowing for users to see the relationship between the HOLC appraisers' determinations regarding financial risk and how those same neighbourhoods are faring today.

Impacts on Cities

Although some of these projects are still in progress, they have already been significantly publicised and used by journalists, scholars, and teachers across the United States, and their impacts are cross-disciplinary. For example, *Mapping Inequality's* HOLC maps have been combined with Congressional district boundaries to show how districts are gerrymandered along historical racial dividing lines that still exist. The National Community Reinvestment Coalition has also used the maps, in combination with open U.S. Census data, to look at the relationship between historic redlining and gentrification (Abello, 2017).

More substantially, economists with the Federal Reserve Bank of Chicago recently merged *Mapping Inequality's* geocoded data with open U.S. Census data covering 1910–2010 and city block-level credit risk scores from 1999 to 2016 to show the long-term impact of discriminatory lending—an outcome of the HOLC maps—on the racial wealth gap and neighbourhood development. The report shows "a long-run decline in home ownership, house values, and credit scores along the lower graded side of HOLC borders that persists today" (Aaronson et al., 2017, p. 1). The federal Department of Justice also used a digitised HOLC map in its 2016 report on policing in Baltimore, Maryland, as evidence of the federal government's historic role in the persistence of extreme racial and economic segregation (U.S. Department of Justice, 2016, pp. 72–73). (Notably, because Baltimore City provides open GIS data on police stops, the report's authors were also able to map the profoundly disproportionate impact of policing on black neighbourhoods.)

Once the HOLC's "area descriptions" (the surveys used by property appraisers to assign colour-coded grades to each section of each city) have been fully transcribed and opened via *Mapping Inequality's* website, the project will become even more valuable for understanding the profound legacy of this federal government programme. In addition, the team has recently launched a new project, *Renewing Inequality*, which visualises historical data on the massive urban renewal projects that transformed American cities in the late 1950s and 1960s. These maps will exhibit the profound demographic impact of razing and redeveloping entire neighbourhoods that were formerly occupied primarily by African Americans.

These projects have also ignited new conversations among city planners and policy makers. The *Mapping Prejudice* team in Minneapolis was recently invited to meet with the directors of several city government agencies, including the police, to educate them on the historical factors behind the city's racial divide. Minneapolis has the lowest rate of black homeownership in the United States (Buchta, 2017). The Minneapolis team also notes a clear correlation between the segregated settlement patterns created by racial covenants and data showing, for example, point source pollution, the presence of trees, and health indicators such as birth weight (Delegard et al., 2017). Areas in which African Americans were forced to live, and where many remain, are far more likely to be afflicted by unhealthy environmental conditions. "We want to change the grand narrative of the past," says the project's director, Augsburg University's Kirsten Delegard. "We're helping build the political will that is going to drive policy" (Delegard et al., 2017). As another example of this, the Minneapolis team successfully worked with policymakers on legislation that now allows homeowners to permanently remove any type of restrictive covenant from their property titles (State of Minnesota, 2018).

The *Mapping Prejudice* team has also been approached by the Minneapolis Community Planning and Economic Development Office as it prepares to update the city's Comprehensive Plan to strategically guide Minneapolis' development through 2040. The Comprehensive Plan update will likely result in many revisions to the city's zoning code (Worthington, 2018). The zoning code, which has not been updated since the 1950s, reinforced patterns of racial segregation that were frozen into place by HOLC's Residential Security maps and subsequent redlining by banks and insurers, says Kevin Ehrman-Solberg, but the patterns were originally created by real estate developers' use of racial covenants (Delegard et al., 2017). Historic reports estimate that 37% of Minneapolis' real estate businesses did not offer services to racial/ethnic minorities and—based on a survey of 48 real estate firms covering new development only—40% of new subdivisions in Minneapolis contained racial restrictions in their deeds (Real Estate and Housing Committee, 1948). The team expects to find "that racial covenants worked with other mechanisms of discrimination to prevent people who were not white from living in huge swathes of the metropolitan area" (Delegard et al., 2017).

Segregated Seattle has also engaged policymakers and public agency officials, with 3.5 million page views recorded by the *Seattle Civil Rights & Labor History Project* since 2008 (Gregory, 2017). The project helped drive the passage of a 2006 law to help remove old racial covenants from property deeds (State of Washington, 2006). It has also been used in a range of classroom settings (Gregory and Griffey, 2007)—many local teachers have used the database with their students, and the University of Washington School of Law offered a course based on the data, for example. In DC, the *Mapping Segregation in Washington DC* team has presented its project at the federal Department of Housing and Urban Development.

All of these projects have had a significant impact on raising public awareness. As evidenced, for example, by articles in local newspapers that pop up across the United States when residents express shock and dismay at the discovery of old racial covenants in their property deeds (Furst, 2017; Garrison, 2008; Thomas, 2016), very few Americans know about these historical legal mechanisms for shaping exclusive neighbourhoods, and even fewer understand how widespread the practice was. The Minneapolis teams sees their research process, which engages citizen volunteers to review and enter data on racial covenants, as central to catalysing a community dialogue around this history. "The power of our methodology is you can read these resources yourself," explains Delegard. "We're amplifying the message of the project in a different way," she continues, noting that volunteers have had "transformative experiences" from reading the racially restricted deeds themselves (Delegard et al., 2017). Joshua Poe has commented that his project on Louisville's racial history has especially catalysed younger audiences, who he believes feel "a sense of obligation to do something with this information" (Poe, 2017). The Louisville Office of Redevelopment Strategies has also indicated plans for using the project to help generate community dialogue around redlining and its impacts by appointing "volunteer ambassadors" to help disseminate project information (City of Louisville, 2017). In DC, around 1000 local citizens have attended presentations of *Mapping Segregation in Washington DC*, mostly at public libraries, and the team is regularly asked by teachers to bring the project to their classrooms.

Numerous other projects are opening, merging, and visualising historical and contemporary data that will increase awareness around the historical roots of social inequities and guide us towards a more equitable future. Although not otherwise profiled in this chapter, San Francisco, California's *Anti-Eviction Mapping Project* (antievictionmap.com) is currently among the best American examples of a project using Open Data towards the goal of achieving social justice. The wide-ranging project includes oral histories, essays, educational curriculum, and resources for resisting residential displacement. Maps detailing, for example, police stops for non-violent offences combined with racial demographics; maps showing the location of absentee-owned property; and maps exhibiting the relationship between historic redlining and property foreclosures today, meet the project's goal of visualising "displacement and resistance" in the San Francisco Bay area (Anti-Eviction Mapping Project, 2017).

Another project, Princeton University's Eviction Lab (evictionlab.org), has compiled and mapped data on evictions throughout the entire United States for 2000–2016, and encourages local activists and policymakers to download the raw data for their own use (Eviction Lab, 2018). With such data becoming more widely available via open city data portals across the United States (City of Denver, 2015; City of LA, 2017), the future is promising for exhibiting the linkages between our past and present in this regard.

Conclusion

Historic data should be an essential component for informing housing and planning policy, especially to mitigate the reification of racial segregation that results from rapid, developer-driven urbanisation. Each of the projects described in this chapter has already been successful in "opening" historic data in accessible formats; they are also supplemented by interpretive text, and in some cases, photographs, documents, and audio provide additional historical context and multiple modes of engagement. In so doing, the projects have increased the transparency of historic and contemporary land use policies; property ownership and real estate; banking and insurance practices; and other factors that have served to racially and economically segregate American cities.

The next step is to more fully disseminate this information by, for example, making the data and shapefiles available for public download to encourage their innovative use by others. Historic data, provided openly to the public, can then further inform contemporary discussions around equitable urban development and influence concrete policy decisions. As a result, all of these projects will help combat the systemic racial segregation that pervades American society.

Acknowledgements The authors would like to thank several people for their contribution to the case studies presented in this chapter: Kirsten Delegard, Kevin Ehrman-Solberg, and Penny Petersen (*Mapping Prejudice*); Joshua Poe (*Redlining Louisville*), James Gregory (*Segregated Seattle*), and Robert Nelson (*Mapping Inequality*).

References

Aaronson, D., Hartley, D. & Mazumder, B. (2017). The effects of the 1930s HOLC "Redlining" maps. *Federal Reserve Bank of Chicago: Working Paper*, No. 2017-12.

Abello, O. P. (2017). These 5 neighborhood maps show roots of gentrification. Retrieved January 22, 2018, from https://nextcity.org/daily/entry/new-maps-show-redlining-gentrification-neighborhoods.

Anti-Eviction Mapping Project (2017). Anti-Eviction Mapping Project: About. Retrieved from https://www.antievictionmap.com/about.

Barns, S. (2016). Mine your data: Open data, digital strategies and entrepreneurial governance by code. *Urban Geography, 37*(4), 554–571.

Bernasconi, R. (2017). Race as a historico-spatial construct: The hermeneutical challenge to institutional racism. In B. B. Janz (Ed.), *Place, space and hermeneutics, Contributions to Hermeneutics* (Vol. 5, pp. 493–504). Cham, Switzerland: Springer.

Brooks, R. W., & Rose, C. M. (2013). *Saving the neighborhood: Racially restrictive covenants, law, and social norms*. Cambridge: Harvard University Press.

Buchta, J. (2017). Already-low homeownership of rates of Twin Cities minorities fall further. Retrieved January 22, 2018, from http://www.startribune.com/already-low-homeownership-rates-of-twin-cities-minorities-fall-further-down/441087863.

Carter, P. L. & Reardon, S. F. (2014). Inequality matters. *A William T. Grant Foundation Inequality Paper*. Retrieved January 22, 2017, from https://ed.stanford.edu/sites/default/files/inequalitymatters.pdf.

City of Chicago. (2013). The City of Chicago Technology Plan. Retrieved October 17, 2017, from http://techplan.cityofchicago.org/wp-content/uploads/2013/09/cityofchicago-techplan.pdf.

City of Denver. (2015). Open data catalog: Foreclosures. Retrieved from https://www.denvergov.org/opendata/dataset/city-and-county-of-denver-foreclosures.

City of LA. (2017). 2017 Registered foreclosure properties. Retrieved January 22, 2018, from https://data.lacity.org/A-Well-Run-City/2017-Registered-Foreclosure-Properties/p88b-3h5k.

City of Louisville. (2017). Redlining community dialogue. Retrieved May 15, 2018, from https://louisvilleky.gov/government/redevelopment-strategies/redlining-community-dialogue.

City of New York. (2011). Road map for the digital city: Achieving New York city's digital future. Retrieved August 6, 2017, from http://www.nyc.gov/html/media/media/PDF/90dayreport.pdf.

Crossney, K. B., & Bartelt, D. W. (2005). The legacy of the Home Owners' Loan Corporation. *Housing Policy Debate, 16*(3/4), 547–574.

DC Office of the Mayor. (2017a). District of Columbia Data Policy, Mayor's Order 2017-115. Office of the Chief Technology Officer. Retrieved November 15, 2017, from https://octo.dc.gov/page/district-columbia-data-policy.

DC Office of the Mayor. (2017b). DC's economic strategy: Strategy report, March 2017. Retrieved November 15, 2017, from http://dceconomicstrategy.com/wp-content/uploads/2016/10/DC-Economic-Strategy-Strategy-Report-FULL-May-1-2017-1.pdf.

Delegard, K. & Ehrman-Solberg, K. (2017). Playground of the people'? Mapping racial covenants in twentieth-century Minneapolis. *Open Rivers: Rethinking The Mississippi, no 6*. Retrieved from http://editions.lib.umn.edu/openrivers/article/mapping-racial-covenants-in-twentieth-century-minneapolis.

Delegard, K., Ehrman-Solberg, K & Petersen, P. (2017). Interview with Ashley Bakelmun and Sarah Shoenfeld, 8 November.

Eviction Lab. (2018). Eviction Lab. Retrieved May 29, 2018, from https://evictionlab.org.

Furst, R. (2017). Massive project works to uncover racist restrictions in Minneapolis housing deeds. *Star Tribune*. Retrieved February 16, 2017, from http://www.startribune.com/massive-project-underway-to-uncover-racist-restrictions-in-minneapolis-housing-deeds/441821703.

Garrison, J. (2008). Living with a reminder of segregation. *Los Angeles Times*. Retrieved February 16, 2018, from http://articles.latimes.com/2008/jul/27/local/me-covenant27.

Goldsmith, S. (2017). How civic tech can address urban inequality. *Governing the States and Localities*. Retrieved October 13, 2017, from http://www.governing.com/blogs/bfc/col-civic-technology-urban-inequality.html?AMP.

Gotham, K. F. (2000a). *Race, real estate, and uneven development*. Albany: State University of New York Press.

Gotham, K. F. (2000b). Racialization and the state: The Housing Act of 1934 and the creation of the Federal Housing Administration. *Sociological Perspectives, 43*(2), 291–317.

Greer, J. (2012). The Home Owners' Loan Corporation and the development of the Residential Security maps. *Journal of Urban History, 39*(2), 457–473.

Gregory, J. (2006). Seattle Civil Rights & Labor History Project: Restrictive covenants database. Retrieved October 13, 2017, from http://depts.washington.edu/civilr/covenants.htm.

Gregory, J. (2017). Email sent to Ashley Bakelmun, 13th October.

Gregory, J. N. & Griffey, T. (2007, April). Teaching a city about its civil rights history: A public history success story. *Perspectives on History*. Retrieved October 13, 2017, from https://www.historians.org/publications-and-directories/perspectives-on-history/april-2007/teaching-a-city-about-its-civil-rights-history-a-public-history-success-story.

Grove, J. M., Cadenasso, M. L., Pickett, S. T. A., Burch, W. R., & Machlis, G. E. (2015). *The Baltimore school of urban ecology: Space, scale, and time for the study of cities*. New Haven, CT: Yale University Press.

Grove, M., Ogden, L., Pickett, S., Boone, C., Buckley, G., Locke, D. H., … Hall, B. (2018). The legacy effect: Understanding how segregation and environmental injustice unfold over time in Baltimore. *Annals of the American Association of Geographers, 108*(2), 524–537.

Hernandez, T., & O'Maley, D. (2017). New "housing insights" digital mapping tool provides insight into DC's affordable housing context. *Greater Greater Washington*. Retrieved October 13, 2017, from https://ggwash.org/view/amp/65097.

Howell, B. (2006). Concentrated subprime lending as housing discrimination. *California Law Review, 94*(1), 101–147.

Jackson, K. T. (1985). *Crabgrass frontier: The suburbanization of the United States*. New York: Oxford University Press.

Jones-Correa, M. (2000). The origins and diffusion of racial restrictive covenants. *Political Science Quarterly, 115*(4), 541–568.

Kitchin, R. (2014). The real-time city? Big data and smart urbanism. *GeoJournal, 79*, 1–14.

Library of Congress. (2019). Map of the District of Columbia. Retrieved January 15, 2019, from https://www.loc.gov/item/87694106.

Lowe, R. G. (1948). Racial restrictive covenants. *Alabama Law Review, 15*, 15–39.

Mapping Prejudice. (2017). Mapping Prejudice. Retrieved January 15, 2018, from https://www.mappingprejudice.org.

Massey, D. S. (2015). The legacy of the 1968 Fair Housing Act. *Sociological Forum, 30*(S1), 571–588.

McArdle, G., & Kitchin, R. (2016). Improving the veracity of open and real-time urban data. *Built Environment, 42*(3), 457–473.

Ming, W. R., Jr. (1949). Racial restrictions and the Fourteenth Amendment: The restrictive covenant cases. *The University of Chicago Law Review, 16*(2), 203–238.

National Advisory Commission on Civil Disorders. (1968). *Report of the National Advisory Commission on Civil Disorders*. New York: E. P. Dutton.

National Community Reinvestment Coalition. (2017). HOLC maps. Retrieved November 8, 2017, from http://maps.ncrc.org/holc.

Nelson, R. (2017). Email sent to Ashley Bakelmun, 26th October.

Nelson, R. K., Winling, L., Marciano, R., & Connolly, N. (2017). Mapping Inequality: Redlining in New Deal America. In R. K. Nelson & E. L. Ayers (Eds.), *American Panorama*. Retrieved October 13, 2017, from https://dsl.richmond.edu/panorama/redlining/#loc=9/31.9603/-86.8002&opacity=1&city=montgomery-al.

Nier, C. L., III. (1999). Perpetuation of segregation: Toward a new historical and legal interpretation of redlining under the Fair Housing Act. *The John Marshall Law Review, 32*, 617–665.

Oliveri, R. C. (2015). Setting the stage for Ferguson: Housing discrimination and segregation in St Louis. *Missouri Law Review, 80*, 1053–1075.

Open Data DC. (2017). Open Data DC: Connecting you with government data. Retrieved October 13, 2017, from http://opendata.dc.gov/datasets.

Pettigrew, T. F. (1979). Racial change and social policy. *Annals of the American Academy of Political and Social Science, 441*, 114–131.

Poe, J. (2015). Redlining Louisville: The history of race, class, and real estate. Retrieved October 13, 2017, from http://www.arcgis.com/apps/MapSeries/index.html?appid=a73ce5ba85ce4c3f80d365ab1ff89010.

Poe, J. (2017). Interview with Ashley Bakelmun, 24 October.

Prologue DC. (2018a). Mapping segregation in Washington DC: Legal challenges to racially restrictive covenants. Retrieved May 28, 2018, from http://mappingsegregationdc.org.

Prologue DC. (2018b). Mapping segregation in Washington DC: How racially restricted housing shaped ward 4. Retrieved May 28, 2018, from http://mappingsegregationdc.org.

Real Estate and Housing Committee. (1948). Report and Recommendations of the Real Estate and Housing Committee of the Minneapolis Community Self-Survey of Human Relations. Minneapolis City Hall municipal archives.

Rose, C. M. (2016). Racially restrictive covenants—Were they dignity takings? *Law & Social Inquiry, 41*(4), 939–955.

Rothstein, R. (2017). *The color of law: A forgotten history of how our government segregated America*. New York: Liveright Publishing Corporation.

Shoenfeld, S., & Cherkasky, M. (2015). Mapping Segregation in Washington DC. Retrieved October 18, 2017, from http://forum.savingplaces.org/blogs/special-contributor/2015/06/22/mapping-segregation-in-washington-dc.

State of Minnesota. (2018). Minnesota State House Bill HF 2767, Amendment to Minnesota Statutes 2016, section 507.18, subdivision 2. Retrieved February 20, 2018, from https://www.revisor.mn.gov/bills/text.php?number=HF2767&version=latest&session=ls90&session_year=2018&session_number=0.

State of Washington. (2006). Engrossed Senate Bill 6169. Retrieved January 22, 2018, from http://depts.washington.edu/civilr/images/segregated/covenants/6169.PL.pdf.

Thomas, J. L. (2016). 'Curse of covenant' persists—Restrictive rules, while unenforceable, have lingering legacy. *The Kansas City Star*. Retrieved February 16, 2018, from http://www.kansascity.com/news/local/article92156112.html.

Townsend, A. M. (2013). *Smart cities: Big data, civic hackers and the quest for a new utopia*. London: W.W. Norton & Company.

U.S. Department of Housing and Urban Development. (2014). Affirmatively furthering fair housing proposed rule. Washington, DC: U.S. Department of Housing and Urban Development. Retrieved May 16, 2018, from http://www.huduser.org/portal/affht_pt.html.

U.S. Department of Justice. (2016). Investigation of the Baltimore City Police Department. US Department of Justice Civil Rights Division. Retrieved January 22, 2018, from https://www.justice.gov/crt/file/883296/download.

Woods, L. L. (2012). The Federal Home Loan Bank Board, redlining, and the national proliferation of racial lending discrimination, 1921–1950. *Journal of Urban History, 38*(6), 1036–1059.

Worthington, H. (2018). Email sent to Kirsten Delegard, 9 February.

Zooniverse. (2017). Mapping racial covenants in Minneapolis. Retrieved November 10, 2017, from https://www.zooniverse.org/projects/kevinesolberg/mapping-prejudice.

4

Safer Cities for Women: Global and Local Innovations with Open Data and Civic Technology

Scott Hawken, Simone Z. Leao, Ori Gudes, Parisa Izadpanahi, Kalpana Viswanath, and Christopher Pettit

Abbreviation

SDGs	Sustainable Development Goals

Highlights

- Critical overview of urban safety in relation to vulnerable people
- Scalable, universal set of safety metrics generated through Open Data
- Distinctive methodology for data collection for government decision-making

---

S. Hawken (✉)
Urban Development and Design, Faculty of the Built Environment, University of New South Wales, Sydney, NSW, Australia
e-mail: s.hawken@unsw.edu.au

S. Z. Leao
Urban Modelling and Simulation, University of New South Wales, Sydney, NSW, Australia
e-mail: s.zarpelonleao@unsw.edu.au

S. Hawken et al. (eds.), *Open Cities | Open Data*,
https://doi.org/10.1007/978-981-13-6605-5_4

- Insights on the localisation of global approaches to urban safety
- Urban safety dashboard generated based on Open Data collected through the social tech startup 'SafetiPin'

Introduction

Urban Safety: A Universal Goal

Safe public spaces that are universally accessible for enjoyment by cycling and walking have become important goals for cities around the world. Safety is a fundamental requirement for cities to become sustainable and inclusive and a key target for global initiatives such as the Sustainable Development Goals (SDGs), specifically the SDG 11: Sustainable Cities and Communities. Safety is a difficult performance criterion to measure at the scale of the local neighbourhood and street. Coarse statistics are available for cities and urban districts, but the safety and accessibility of cities can differ dramatically from one street to another. Such fine-grained information on safety is rarely available from conventional sources. Furthermore, much of the existing data is gender biased and

O. Gudes
University of New South Wales, Sydney, NSW, Australia
e-mail: o.gudes@unsw.edu.au

P. Izadpanahi
Design and Built Environment, Curtin University, Perth, WA, Australia
e-mail: Parisa.Izadpanahi@curtin.edu.au

K. Viswanath
SafetiPin, Gurgaon, India
e-mail: kalpana.viswanath@safetipin.com

C. Pettit
Urban Science, Faculty of the Built Environment, University of New South Wales, Sydney, NSW, Australia
e-mail: c.pettit@unsw.edu.au

underestimates the scale of street-based crimes such as sexual harassment (Belknap, 2007; Palermo, Bleck, & Peterman, 2014). Safety information can have life-and-death consequences for vulnerable people such as women, children or urban newcomers who are unfamiliar with local dangers and risks. How to collect, understand and improve urban safety is therefore a fundamental research challenge for urban planners, sociologists, economists and policymakers. Technological advances within today's emerging smart cities are creating a new information landscape involving data production, collection and analysis at an immense scale. This 'big data', as it is known, can be used to describe urban patterns in fine spatial and temporal scales. Commercial big data applications such as Uber and Airbnb are well known; however, civic applications initiated by civic and social startups have also evolved to address urban challenges and empower citizens.

Globally there is a great need for civic applications to address safety. The SDGs have been set up as a collaborative and consensus-building enterprise to achieve such aims. The success of such initiatives rests upon linking the new information and data to projects and organisations which can assess the current situation and swiftly act upon it. This chapter reports on practical efforts to develop safety-related metrics based on Open Data. Further the chapter discusses the civic entrepreneurship necessary to use such metrics to inform the management and design of safer urban spaces. This discussion is facilitated through a case study of innovative social and civic tech startup 'SafetiPin' and the presentation and conceptualisation of their safety data model. SafetiPin works to improve safety in 30 cities around the world. One of these cities, Bogota, is discussed to highlight the challenges in generating and using data to achieve the urban transformation necessary to make urban public spaces safer for vulnerable urban people. Two additional innovations are contributed by the authors using Bogota's existing Open Data on safety. A statistical analysis using cumulative odds ordinal logistic regression with proportional odds was run to determine the effect of street lighting, openness, visibility, walk path and public transport on the safety feeling in Bogota. Finally an open and interactive map visualisation has also been created by the authors using the SafetiPin data collected in Bogota and Tableau analytic software. These innovations further improve the openness of Bogota's safety data and increase its value through integrating it with a range of useful urban metrics.

Monitoring Urban Safety

The Promise of Big Data for Urban Safety

With the advent of the internet of things (IoT) and social media, there are now more data sources and a greater volume of data than ever before. This data landscape is generated within a variety of social contexts including governments, business, the public and autonomous machines. There are also a range of approaches to capture and interpret data such as citizen scientists, crowdsourcing and public-private partnerships between business, government and social activists. Together this network of data producers and collectors forms a network of astounding complexity within our cities. Studies have indicated that big data from various social network services such as Twitter and Facebook can enhance our understanding of sustainable development progress in cities by enabling real-time analysis and monitoring of urban activities and change (Ong et al., 2016; Vallicelli, 2017). Despite such promise, the current civic value of such social networks is uncertain. Even though increasingly large amounts of urban data are being collected, there is a lack of clarity around how such data can be analysed and interpreted to address the many wicked problems involved in urban sustainability. Zheng, Capra, Wolfson, and Yang (2014) argue that modern urban computing must evolve to connect urban sensing, data management and data analytics into a linked, cyclic process for the continuous improvement of people's lives and urban systems. This innovation is not inevitable and will require innovative civic and social initiatives to drive positive change.

Location-specific decision-making support tools related to urban planning and management are essential to gain understanding at different urban scales: from the scale of the street, to the metropolitan area. As adoption of big data increases and understanding of the technology grows in the wider community, new business, government and personal value propositions will continue to emerge (Ong et al., 2016). These values will not always be in alignment. Although some of this data is being made available through Open Data portals, many big datasets remain closed to civic uses. There is a risk that big data will be used for narrow commercial agendas and that the civic needs of the kind associated with the SDGs and the urban agenda will be left unaddressed. For this reason, a concentrated focus on Open Data and civic applications is needed (Han & Hawken, 2018). The next section considers how big data can be sourced from citizens, business and government to address demanding urban problems such as safety and inclusion in a collaborative way.

Linking Crowdsourced Data with Urban Governments

Existing national metrics and statistics on safety often do not have the required resolution to describe safety in our cities. The available data shows that current standards fall short of the targets set by the SDGs. This lack of rigour and focus in the reporting reflects legal and political shortcomings. According to UN Women (2011), for example, many countries in the world do not have legislation against sexual harassment, or do not even record data on the topic. This applies to 76% of the countries in the Middle East and North Africa; 60% of the countries in East Asia and the Pacific; 45% in Asia and Sub-Sahara Africa; 33% in Latin America and the Caribbean; and 25% in Central and Eastern Europe and Central Asia. The UN-Habitat monitoring framework (UN-Habitat, 2016) suggests the following metrics to monitor safety targets: the percentage of girls and women aged 15+ who are subjected to physical or sexual harassment in the last 12 months. Two sources of data are identified in the framework for this metric: the Demographic and Health Survey (DHS) developed by United States Agency for International Development (USAID) every five years; and the Multiple Indicator Cluster Survey (MICS) developed by countries with assistance of UNICEF. While the first is more related to domestic violence and incidence of selected health conditions, the second is focused on children and youth. Such metrics are difficult to collect data on due to the stigma and shame associated with being a victim of sexual assault. Sexual harassment, and especially sexual assault, is well known to be underreported by women (Fitzgerald, 2017; Hlavka, 2014).

Personal technologies have shown potential to address this difficult problem. A growing proportion of the world is intimate with personal technologies on a daily and continuous basis. Personal technologies can empower individuals in the process of monitoring and understanding their environment. Khan, Xiang, Aalsalem, and Arshad (2013) present an extensive review of current mobile sensing applications, covering areas such as health and wellbeing, transportation, environment, social interaction and commerce. In the domain of urban safety, some examples include HARASSmap (Grove, 2015), HearMe (Akash, Md, Adhikary, Md, & Sharmin, 2016) and SafetiPin (Viswanath & Basu, 2015), which is the focus of this chapter. Unlike other applications, SafetiPin is focused on achieving positive environmental change rather than on reinforcing negative spatial associations. It does this by linking crowdsourced information with urban governments and communities to inform and advocate for environmental and social change in a cyclical approach (see Fig. 4.1). In contrast to other safety apps, SafetiPin is a solution-focused, rather than a symptom-focused, tool.

Crowdsourcing, participatory sensing, community sensing and citizen science are all interrelated concepts under the umbrella of public participation (Leao & Izadpanahi, 2016). These concepts involve citizens and community groups documenting their lives in varying environments. Environments and contexts covered include workplaces, homes, streets, parks and other public and private spaces. Information recorded ranges from intimate observations to collaborative efforts involving thousands of participants. Together, such environmental sensing reveals extraordinarily detailed socio-spatial patterns across entire cities (Goldman et al., 2009).

However, crowdsourcing is not a perfect solution and relies on individuals being motivated enough to accurately record both positive and negative observations. When it comes to urban trauma, recording details maybe the last thing on someone's mind. Crowdsourcing is still in its initial stages and needs to develop beyond the current model of individuals acting together in isolation. For crowdsourcing to be effective, it needs to be better socialised and form part of the resilience strategies of cities. Such initiatives therefore need to build on existing physical support networks rather than replace them. Until this occurs crowdsourced data

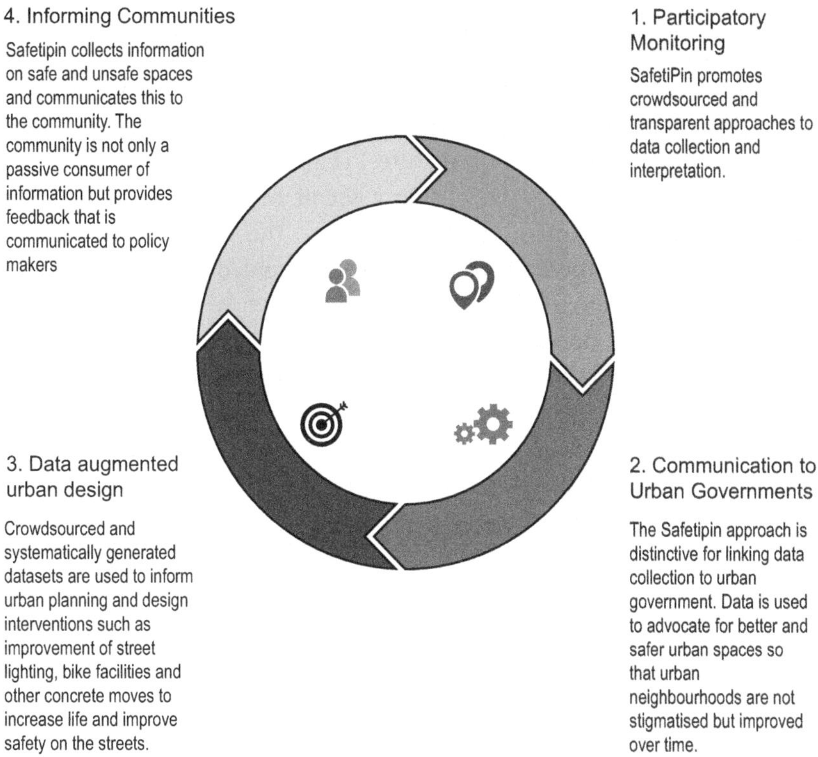

Fig. 4.1 SafetiPin uses a cyclical approach linking crowdsourced information with urban governments and communities to inform and advocate for environmental and social change

must be complemented with commercial enterprises or through systematic civic collection approaches such as via IoT networks. This has been the approach of startups such as SafetiPin which supplement crowdsourced urban data with systematically collected commercial urban data. Both types of data can be analysed and processed to form accessible open datasets which are helpful for informing decisions that can change urban environments for the better.

SafetiPin started its social entrepreneurial business in 2013 in India in response to the exclusive and unsafe character of neighbourhoods for vulnerable demographics such as women and children. The two premises of SafetiPin are that 'large-scale data collection can lead to change, and that safety will ensue when more people become engaged with the issue' (Viswanath & Basu, 2015, p. 46). This setting and support from UN led SafetiPin to achieve a global reach, being applied in 30 cities across 10 countries in the world, collecting a large dataset (more than 150,000 audit pins and almost 1 million pictures) (Fig. 4.2).

SafetiPin rates public spaces using a set of key parameters of safety using a rubric-based method of assessment. The rubric, described in Table 4.1, includes the following parameters: light during night; openness of public space; visibility of the space through windows and doors; presence of people; presence of security professionals; quality of walking path; proximity to public transport; balance of gender usage of public space; and feeling/perception of safety.

Two methods are used for data collection in SafetiPin. The first is based on voluntary participation of citizens using the SafetiPin mobile phone app (My SafetiPin app). The second uses local trained professionals to assess night time photographs of public spaces (SafetiPin Nite app). Both methods use the same safety audit rubric presented in Table 4.1.

The principle of SafetiPin's safety audit is that the user of a space is the expert and has the best understanding of why the space is unsafe or inaccessible. The My SafetiPin app allows any user of the city to rate its infrastructure and social usage through the rubric, thereby generating bottom-up knowledge. Mobile phone resources allow SafetiPin users to

Country	City
Bangladesh	Dhaka
Brazil	Rio de Janeiro
Colombia	Bogota
India	Bengaluru
	Bhopal
	Chandigarth
	Chennai
	Delhi
	Faridabad
	Ghaziabad
	Gurugram
	Guwahati
	Hazaribagh
	Hyderabad
	Karnal
	Kochi
	Kolkata
	Kota
	Mumbai
	Noida
	Pune
	Ranchi
Indonesia	Jakarta
Iran	Tehran
Kenya	Nairobi
Malaysia	Kualalumpur
Mexico	Guadalajara
	Mexico City
	Monterrey
Philipines	Manila

Fig. 4.2 Countries which have SafetiPin data collected and analysed

Table 4.1 The SafetiPin safety audit rubric. A safety score is calculated as a result of the combination of the nine safety audit parameters

	0	1	2	3
Light (night)	*None*. No street or other lights	*Little*. Can see lights, but there is low visibility in the area	*Enough*. Lighting is enough for clear visibility	*Bright*. Whole area brightly lit
Openness	*Not open*. Many blind corners and non-clear sightline	*Partially open*. Able to see a little ahead and around	*Mostly open*. Able to see in most directions	*Completely open*. Can see clearly in all directions
Visibility	*No eyes*. No windows or entrance of shops or residences overlook this point	*Few eyes*. Less than 5 windows or entrances overlook the point	*More eyes*. Less than 10 windows or entrances overlook the point	*High visibility*. More than 10 windows or entrances overlook this point
People	*Deserted*. No one in sight	*Few people*. Less than 10 people in sight	*Some crowd*. More than 10 people visible	*Crowded*. Many people within touching distance
Security	*None*. No guards or police visible in surrounding area	*Minimal*. Some private security visible in surrounding area but nor nearby	*Moderate*. Private security within hailing distance	*High*. Police/ reliable security within hailing distance
Walk path	*None*. No walking path available.	*Poor*. Path exists but in very bad conditions	*Fair*. Can walk but not run	*Good*. Easy to walk fast or run
Public transport	*Unavailable*. No metro or bus stop, auto/ rickshaw within 10 min walk	*Distant*. Metro or bus stop, auto/rickshaw between 5 and 10 min walk	*Nearby*. Metro or bus stop, auto/ rickshaw between 2 and 5 min walk	*Very close*. Metro or bus stop, auto/ rickshaw available within 2 min walk

(*continued*)

Table 4.1 (continued)

	0	1	2	3
Gender usage	*Nor diverse*. No one in sight, or only men	*Somewhat diverse*. Mostly men, very few women or children	*Fairly diverse*. Some women and children	*Diverse*. Balance of all genders or more women and children
Feeling	*Frightening*. Will never venture here without sufficient escort	*Uncomfortable*. Will avoid whenever possible	*Acceptable*. Will take other available and better routes when possible	*Comfortable*. Can take this route even at night.

Source: Viswanath and Basu (2015)

provide automatically a multi-attribute record: the GPS allows users to geotag their selected locations; date and time are recorded by the phone calendar; the inbuilt camera provides means for photographs to be associated to audit pins; and finally, the entry of text can extend the record to more qualitative and descriptive information. People, and particularly women, are therefore given an opportunity to generate knowledge from their experience. This is a process of democratising data both in its production and in its consumption. All data that is gathered is instantly available for others to see and use both at an individual level to make safer decisions, as well as at a collective level for advocacy and change.

In addition to this data from citizen users, SafetiPin adds a layer of data through capturing and assessing night time pictures of the city which supplements the user-generated data and makes it much more extensive and large scale. This has resulted in large datasets being made available for analysis, and also a more even coverage of urban areas, filling gaps of data in areas not pin-pointed by voluntary citizens. As an example, in Delhi just over 10,000 audits were collected through crowdsourcing, whereas the analysis of photographs resulted in over 60,000 points in the city having a safety score. This methodology has been used in many cities in India as well as globally. In Delhi a large number of photos were taken during the night in targeted areas by Uber drivers, and then assessed by members of SafetiPin team. In Bogota, night time photos were taken by members of the local government and the public while riding bicycles across the city.

SafetiPin goes beyond safe navigation to work with local stakeholders to change the dangerous spaces themselves. The impact of SafetiPin's initiatives has already been felt in cities such as Delhi where urban spaces have been transformed to be safer and more social. SafetiPin has identified over 8000 poorly lit public spaces in the city, which have been subsequently been improved by the city government. SafetiPin has engaged in partnership with local government and non-government organisations to make the data robust, consistent, open and accessible. The datasets have been used for advocacy, better planning and decision-making towards safer and more inclusive urban living to all citizens. SafetiPin's approach points a way forward for a global approach scaling up from the 30 cities it is already operating in. The significance of this global project is addressed in the next section which contextualises women's safety within cities. Urban safety is then deconstructed to provide insights into how this complex phenomenon can be measured and interpreted at the urban scale

The Global Significance of Women's Urban Safety

Safety is one of the core components of SDG 11, the 'urban goal'. Target 11.7 specifically states that by 2030, cities and nations should 'provide universal access to safe, inclusive and accessible, green and public spaces, in particular for women and children, older persons and persons with disability' (UN-Habitat, 2016). Moreover, SDG 5 emphasises the need for 'gender equality', and target 5.1 states the need to 'end all forms of discrimination against all women and girls everywhere' (UN-Habitat, 2016). Despite the global recognition of safety's importance, the scale and specific dynamics of the problem, as they relate to place, are not well understood.

The immense scale of the urban safety challenge is not well represented by existing data sources. While violence and fear may effect the whole population, marginalised groups that remain invisible and overlooked can be especially effected. The clearest example of this is the gender discrimination within cities. Gender remains a central axis of discrimination that magnifies poverty, disability and other vulnerabilities. For women, sexual violence is a constant threat and limits their ability to move around, their ability to work and their general wellbeing. Because of the feeling of insecurity, women do not experience the right to either use the city or participate in its social, cultural and political life. This denial of the right to the city is experienced by women in their everyday lives (Fenster, 2005). It is important to identify factors which lead to women's exclusion and their lack of access to the city. While an incident of violence or harassment may not be an everyday phenomenon, the feeling of insecurity or lack of safety is. Girls and women in many cities around the world experience fear while navigating the city, often after dark. This fear often shapes women's ability to use public spaces, their comfort and eventually their sense of freedom and inclusion in the city. Thus, the ramifications of this fear go well beyond the actual experience of violence. Therefore, addressing women's safety is in fact addressing women's rights as citizens of the city.

In the past decade there have been several studies that have tried to explore the nature and manifestations of the problem of violence against women in cities. A defining characteristic is its ordinary and continuous nature. While there are occurrences of horrific and violent crimes, it is the everyday nature of violence and its normalisation that characterise it. A study by Hollaback and Cornell University in 2014 that interviewed over 16,000 women reported that over 50% in Europe and 75% in the US reported facing their first incident of harassment before the age of 17. Over 70% reported anger and anxiety on facing street sexual harassment. Over 81% of women interviewed had experienced some form of sexual harassment. The multi-country study, 'Gender Inclusive Cities Project', found that over 70% of women interviewed in Delhi (India), Dar es Salaam (Tanzania) and Rosario (Argentina) had experienced some form of sexual violence in the city (WICI, 2012). It is important to note that:

> *Although feeling unsafe is not confined to women, the fear that women feel in urban areas is quite particular. It is to do with physical and psychological honour. … Although not all women have been raped or attacked, all have felt at*

some point that indescribable feeling of unease which ranges from merely feeling uncomfortable to paralysis. (Smaoun, 2000, p. 29)

The causes of violence against women are both structural and systematic, making urban spaces discriminatory and exclusionary. The safety audit, adapted by SafetiPin, is a tool which has been used to unpack the causes of the lack of safety that women experience in public spaces in cities. It was originally developed in the late 1980s (Whitzman, Andrew, & Viswanath, 2014) and has been extensively used by many different groups and organisations and is suitable for advancing the SDGs.

SDG 11 aims at making cities inclusive, safe, resilient and sustainable. Target 11.7 focuses on the 'provision of universal access to safe, inclusive and accessible, green and public spaces, in particular for women and children, older persons and persons with disabilities'. The presence of 'People' on streets and the balance of gender of this presence ('Gender Usage') captured by SafetiPin are suggested as metrics for SDG target 11.7. SDG target 11.2 aims at 'providing access to safe, affordable, accessible and sustainable transport system for all, improving road safety, notable expanding public transport, with special attention to the needs of those in vulnerable situations, women, children, persons with disabilities and older persons'. Proximity to 'Public Transport' as assessed by SafetiPin is therefore recommended as the metrics for SDG Target 11.2. SDG Target 11.3 states the need to 'enhance inclusive and sustainable urbanisation and capacity for participatory integrated and sustainable urban planning and management'. In terms of enhancing inclusive urbanisation, metrics associated to urban structure such as 'Openness', 'Visibility', 'Walk Path' and 'Light' captured by SafetiPin are therefore recommended as metrics for SDG target 11.3. In future to enhance participatory planning, metrics enabling the 'citizens' voice' in urban design and planning could be also used.

Local Urban Transformation and Open Data

Although every city generally aspires to be safe, this universal aspiration is far from straightforward. Safety is made up of a range of interrelated factors, or metrics, that play out in different ways at the street level and within different cultural contexts. To make a difference on the street level, the interplay of the metrics in space can be assessed through a range of analytic and digital mapping approaches that can be accessed and understood by government and citizen alike.

Colombia suffers from a lack of urban safety due to violence associated with social and economic inequalities. Within Bogota, possible specific causes of urban violence are connected to cultural and political events, poverty, state absenteeism and lack of civilian participation (Montenegro 1994). Camacho (2002) and Merchán and Cárdenas (2007) describe various urban policies for preventing crime with projects that work with citizen's culture, public space, participation, knowledge, prevention, mediation, justice, nationwide dialogue, decentralisation and police vigilance. However, Bogota still suffers from serious urban safety issues. The '2017 Safe Cities Index' has assessed 60 cities across the globe in terms of urban safety. Bogota's performance was listed as very low for all categories of urban security. For personal security, which is partially associated with harassment, assaults and crimes in public spaces, Bogota ranked in the bottom five countries (The Economist, 2017).

In the city of Bogota, Colombia, urban safety data has been collected through crowdsourcing in a partnership between SafetiPin and the Local Government from January 2016. The urban safety data collected is made available for analysis and visualisation through the government Open Data portal (https://www.ideca.gov.co/, Fig. 4.3), and it is integrated into their city's geographic information system (GIS) platform.

Bogota's 'security index' developed with the information, gathered by SafetiPin and available as Open Data, has allowed the government and other local stakeholders to integrate different sources of information, in a social cartography. The city's cadastral authority has included a layer based on the index in the official cartography of the city that can be contrasted and layered in relation to all the cadastral and cartographic information of the city. Moreover, the Secretary for Women of the City of Bogota has used the 'index of night security' to inform authorities in charge of urban investments. This has been an important source of information for the prioritisation of local and municipal investments in infrastructure. The Open Data generated through SafetiPin's approach has been used to improve urban safety by identifying the need for lighting in

Fig. 4.3 Bogota's existing Open Data portal showing night time security/safety mapping based on SafetiPin data

streets, squares and parks; and strategic locations for CCTV cameras; the improvement of footpaths for better and more inclusive accessibility; and the promotion of mixed land use development to improve the visibility and use of public spaces.

Urban interventions by committed leadership in the city of Bogota have radically altered the cityscape with efficient mass transit mobility, the renewal of public spaces, mixed-use design and a focus on people and pedestrians. The city's aim was to create a model of urban development which reduced violence, increased social access, and a greater sense of civic safety and pride through events and social programmes. For example, an annual 'Women's Night Out' has been initiated that raises awareness and has begun to turn the tide on gender violence (CityLab, 2014).

A New Safety City Dashboard for Bogota

SafetiPin provides rich data, to explore which factors affect people's feeling of safety in public urban areas, and to what extent. In this study we have further developed the analysis and visualisation of Bogota's SafetiPin data. A statistical analysis using cumulative odds ordinal logistic regression with proportional odds was run to determine the effect of street lighting, openness, visibility, walk path and public transport on the safety feeling in Bogota. These were selected as they are useful for informing improved urban planning and urban design interventions in the city. The results of this analysis indicated that all named factors, with exception of walk path, have a statistically significant effect on whether people feel safe in public spaces of Bogota. Visibility had the most significant influence on safety. Based on our analysis of visibility, one is 500 times less likely to feel safe if there are 'no eyes' on the street than if it is highly visible. This improves to 83 times if there are 'few eyes'; and then to 9 times if there are 'more eyes'. The presence of public transport was the second most influential factor to safety in Bogota. Based on the analysis of public transport, one is 31 times less likely to feel safe if the public transport is distant than if it is very close, improving to 5 times if it is nearby. In relation to the presence of street lighting, one is 29 times less likely to feel safe in a poor lit place than where is brightly lit, improving to 4 times if it has 'enough' lighting. And finally, one is 15 times less likely to feel safe if the place is narrow than if it is completely open; and this improves to 5 times if the place is mostly open. These metrics demonstrate the relevance of using crowdsourcing data to better understand public perceptions of public spaces that influence the way they use those spaces, and ultimately to provide evidence for urban planners and designers of the potential consequences of their interventions in the city.

An open and interactive map visualisation has been created by the authors using the SafetiPin data collected in Bogota and Tableau analytic software. This visualisation allows users to query different dimensions of the data and get an insight on the spatial patterns in the city and the interrelations among varied factors affecting urban safety on the streets. For example, one can search for the most or the least safe areas in the city, locate them on the map and understand the characteristics of these areas in terms of range of variables including street lighting, walk path, openness, visibility, public transport, public presence, gender balance and so on. Specific characteristics, such as identifying areas with the best gender balance in Bogota, can be reviewed in relation to a range of other factors. Fig. 4.4 shows that all factors, apart from walk path, are strongly associ-

Fig. 4.4 Interactive online visualisation of Bogota's SafetiPin data

ated with more gender-balanced areas when compared to the overall data for Bogota. These provide insights on what composition of factors may lead to support better gender-balanced public spaces. Importantly this open online visualisation does not require any expert skills in cartography or geographical information systems to interact with, making it useful for a wide audience interested in urban safety in the city.

Conclusion

Urban Safety: A Multidimensional Challenge

The complexity of achieving urban safety within different cultural, legal-political and socio-economic contexts is made clearer through the generation of an information ecosystem that is linked to open, crowdsourced data and clear reporting. Further, initiatives such as those by SafetiPin, distribute responsibility by using Open Data to communicate to both government and citizen users. Through Open Data everyone becomes a decision-maker of sorts, able to make choices that impact on the quality of their lives. Public pressure and momentum exert pressure on decision-makers within urban agencies to contribute resources to improve public urban space.

The multi-dimensional metric-based approach of social tech startups such as SafetiPin links global initiatives with local realities in a practical way. Understanding global and local progress on the SDGs is challenging due to a lack of suitable, focused data. The SDGs emphasise the need to monitor goals through objective targets and indicators based on common denominators; therefore, the ability to collect and maintain relevant standardised data is essential. This is particularly important considering the network-based nature of the SDGs. Each SDG is interrelated and shares synergies and trade-offs with other SDGs. When acting within this complex web of targets and goals, integrated strategies that use Open Data show considerable promise.

For local initiatives such as those of Bogota, the challenge is to use global initiatives, such as the SDGs, to generate local momentum, civic aspirations and inclusive standards. Bogotanos have a big challenge ahead to overcome exclusive urban space and to give women, children and vulnerable people equal access to safe urban spaces. How we replicate these types of initiatives at a micro scale to focus on creating inclusive neighbourhoods where all citizens—including women—feel safe and improve their quality of life is an ongoing experiment that requires continued attention and scrutiny. Open, crowdsourced data and clear reporting play a critical role in this process.

Open, crowdsourced data and associated technologies such as city dashboards (Pettit & Leao, 2017; Pettit, Lieske, & Jamal, 2017) can counter the uneven representation of violence and safety. Dashboards that are easy to use, usable, intuitive, trustworthy and accessible to decision-makers and urban citizens are tools for both experiencing cities more carefully and affecting change within cities. Today's media landscape not only makes the 'safety landscape' within cities more difficult to understand but also makes them more unsafe through the promotion of negative anti-social values and landscapes of fear. Open Data, big data and mobile web 2.0 technologies are powerful and unconventional tools in addressing this bias. Such tools are about initiating fundamental changes to culture rather than crime-focused policing.

In this study, we have attempted to suggest practical metrics and measures that underpin SDG 11 goals and enable the conceptual modelling of safety within our cities. This initial research is distinct and unfortunately unique in an area that has global significance for more than half the world's population of women and children. As the data collection model is based on crowdsourced data and openly available open source tools, it is replicable and scalable to cities around globe.

Acknowledgements The authors acknowledge the Australia Korea Foundation and the Department of Foreign Affairs and Trade Australia for funding related to this project.

References

Akash, S. A., Md, A.-Z., Adhikary, T., Md, A. R., & Sharmin, A. (2016). HearMe: A smart mobile application for mitigating women harassment. In *2016 IEEE International WIE Conference on Electrical and Computer Engineering (WIECON-ECE)* (pp. 87–90). IEEE.

Belknap, J. (2007). *The invisible woman: Gender, crime, and justice*. Belmont, CA: Thomson Wadsworth.

Camacho, A. (2002). *Laseguridad ciudadana enelmunicipio colombiano: Undiagnóstico sobre eldiseño institucional y unanálisis delcomportamiento*. Bogotá: Centrode Estudios Socioculturales e Internacionales CESO, Facultad de Ciencias Sociales, Universidad de los Andes.

CityLab. (2014). How urban design could help reduce rape in India. Retrieved May 8, 2018, from https://www.citylab.com/equity/2014/06/how-urban-design-could-help-reduce-rape-in-india/372612/.

Fenster, T. (2005). The right to the gendered city: Different formations of belonging in everyday life. *Journal of Gender Studies, 14*(3), 217–231.

Fitzgerald, L. F. (2017). Still the last great open secret: Sexual harassment as systemic trauma. *Journal of Trauma & Dissociation, 18*(4), 483–489. https://doi.org/10.1080/15299732.2017.1309143

Goldman, J., Shilton, K., Burke, J., Estrin, D., Hansen, M., Ramanathan, N., … Srivastiva, M. (2009). *Participatory sensing: A citizen-powered approach to illuminating the patterns that shape our world.* White Paper, Foresight & Governance Project.

Grove, N. S. (2015). The cartographic ambiguities of Harassmap: Crowdsourcing security and sexual violence in Egypt. *Security Dialogue* (Special issue on Questioning security devices: Performativity, resistance, politics), *46*(4), 345–364.

Han, H., & Hawken, S. (2018). Introduction: Innovation and identity in next-generation smart cities. *City, Culture and Society, 12*, 1–4. https://doi.org/10.1016/j.ccs.2017.12.003

Hlavka, H. R. (2014). Normalizing sexual violence: Young women account for harassment and abuse. *Gender & Society, 28*(3), 337–358. https://doi.org/10.1177/0891243214526468

Khan, W. Z., Xiang, Y., Aalsalem, M., & Arshad, Q. (2013). Mobile phone sensing systems: A Survey. *IEEE Communications, Surveys and Tutorials, 15*(1), 402–427. https://doi.org/10.1109/SURV.2012.031412.00077

Leao, S. Z., & Izadpanahi, P. (2016). Factors motivating citizen engagement in mobile sensing: Insights from a survey of non-participants. *Journal of Urban Technology, 23*(4), 85–103.

Merchán, M., & Cárdenas, M. (2007). *El concepto de seguridad ciudadana; un enfoque práctico a partir de la experiencia en Bogotá (Colombia).* Bogotá: Secretaria de Gobierno de la Alcaldia Mayor de Bogotá y Friedrich Ebert Stiftung.

Montenegro, A. (1994). Justicia y desarrollo económico. *Revista Planeación y Desarrollo, XXV*, 31–45. Colombia: DNP. Retrieved from https://colaboracion.dnp.gov.co/CDT/RevistaPD/1994/pd_vXXV_1994_art.1.pdf.

Ong, K.-L., De Silva, D., Boo, Y. L., Lim, E. H., Bodi, F., Alahakoon, D., & Leao, S. Z. (2016). Big data applications in engineering and science. In S. Yu & S. Guo (Eds.), *Big data concepts, theories, and applications.* Cham, Switzerland: Springer.

Palermo, T., Bleck, J., & Peterman, A. (2014). Tip of the Iceberg: Reporting and gender-based violence in developing countries. *American Journal of Epidemiology, 179*(5), 602–612. https://doi.org/10.1093/aje/kwt295

Pettit, C. J., & Leao, S. Z. (2017). Dashboard. In L. A. Schintler & C. L. McNeely (Eds.), *Encyclopedia of big data* (pp. 1–6). Cham: Springer International Publishing.

Pettit, C. J., Lieske, S. N., & Jamal, M. (2017). CityDash: Visualising a changing city using Open Data. In *Planning support science for smarter urban futures* (pp. 337–353). Springer International Publishing.
Smaoun, S. (2000). *Violence against women in urban areas: An analysis of the problem from a gender perspective*. United Nations Urban Management Programme, Working Paper 17.
The Economist Intelligence Unit Limited. (2017). *Safe Cities Index 2017: Security in a rapidly urbanising world*. Report, London. Retrieved November 27, 2017, from https://dkf1ato8y5dsg.cloudfront.net/uploads/5/82/1-safe-cities-index-2017-eng-web.pdf.
UN Women (United Nations Entity for Gender Equality and the Empowerment of Women). (2011). Matrix of violence against women prevalence surveys, by country. Retrieved from http://www.endvawnow.org/uploads/browser/files/vaw_prevalence_matrix_15april_2011.pdf.
UN-Habitat, United Nations. (2016). A guide to assist national and local governments to monitor and report on SDG Goal 11 indicators: Monitoring framework, definitions, metadata. UN-Habitat Technical Report.
Vallicelli, M. (2017). Smart cities and digital workplace culture in the global European context: Amsterdam, London and Paris. *City, Culture and Society, 12*(1), 25–34.
Viswanath, K., & Basu, A. (2015). SafetiPin: An innovative mobile app to collect data on women's safety in Indian cities. *Gender & Development, 23*(1), 45–60.
Whitzman, C., Andrew, C., & Viswanath, K. (2014). Partnerships for women's safety in the city: Four legs for a good table. *Environment and Urbanisation, 26*, 443–456.
WICI. (2012). Tackling gender exclusion: Experiences from the Gender Inclusive Cities Program. Retrieved from http://safecities.economist.com/safe-cities-index-2017.
Zheng, Y., Capra, L., Wolfson, O., & Yang, H. (2014). Urban computing: Concepts, methodologies, and applications. *ACM Transactions on Intelligent Systems and Technology (TIST), 5*(3), 38.

5

Open Online Platforms and the Collaborative Production of Micro Urban Spaces: Towards an Architecture of Civic Engagement

Homa Rahmat

Introduction

Social media has come to be known as a valuable source of data generated by people. While representing varying degrees of openness (depending on user's privacy choice), social media platforms channel the flow of information and facilitate multidirectional dialogues that in turn open up space for inclusive participation and the collaboration of a diverse set of individuals. The more accessible and participatory these platforms are, the more opportunities they will offer for decentralising the process of decision-making and problem solving and generating new forms of organisation and control. In the context of urban planning and governance, how to harness this potential to improve citizen-expert dialogue and collaboration has opened new directions in research and practice. City officials have started using these platforms to engage the public in different stages of a planning process. However, little is known about the

H. Rahmat (✉)
UNSW Built Environment, Sydney, NSW, Australia
e-mail: h.rahmat@unsw.edu.au

S. Hawken et al. (eds.), *Open Cities | Open Data*,
https://doi.org/10.1007/978-981-13-6605-5_5

prospects of social media data and analytics to develop and continually refine open and collaborative urban systems, which constitute the main focus of this chapter. A new architecture of civic engagement is discussed that is informed by data to leverage non-linear models of participation of many and diverse participants through open online platforms.

To illustrate opportunities to amplify collaboration and mobilise citizen initiatives through social media data analytics, this chapter presents a case study of parklets' online discussion and diffusion on Twitter. Parklets, a temporary transformation of urban spaces often driven by citizens, provide an example of small-scale urban improvement efforts known as tactical/DIY urbanism. As an urban innovation that has been adopted in many cities around the world and experienced a transition to an official urban strategy in several cities, parklets represent an interesting example of experimental models of citizen-expert collaboration. It is first demonstrated how Twitter facilitated collaborative production of parklets and second how analytics help to understand the underlying dynamics of open and non-linear collaboration. This chapter concludes with findings that can assist the planning profession engage with new information flows and to more effectively engage in civic conversations and community activity.

Research Context: Revisiting Participatory Urbanism

Participatory approaches to urban planning took root in the 1960s when critiques of urban planning policy arose most famously by Jane Jacobs against the mentality of the profession of planning in which "every significant detail must be controlled by the planners from the start and then stuck to" (Jacobs, 1961, p. 28). The alternative view of planning as an iterative process constantly modified and improved by not only experts but citizens is in contrast with the essence of twentieth-century planning that according to Sandercock (2004, p. 136) was "regulatory, rule bound, procedure driven, obsessed with order and certainty: in a word, inflexible." Jacobs (1961) argues for a model of city planning and public policy that allows for a great range of unofficial plans, ideas, and opportunities by incredible numbers of different people and different private organisations to flourish.

Participatory planning that refers to the power of citizens to affect the outcome of a planning process (Arnstein, 1969) is often defined and used as a strategy from the top, to engage the public in different stages of a planning process. In a more recent study by Innes and Booher (2004, p. 419) to reframe public participation in the twenty-first century, participation is defined "as a *multi-way* set of interactions among citizens and other players who together produce outcomes." Developing a collaborative model of planning, they extend the conventional forms of participation as the formal and at most two-way interaction of citizens and government, to "a multi-way interaction in which citizens and other players work and talk in formal and informal ways" (Innes & Booher, 2004, pp. 428–429). Collaborative planning is in contrast to the traditional approach to urban change as a relatively linear process that as Ermacora and Bullivant (2016, p. 64) state involves "commissioners, then designers, followed by public consultation, before entering legal, financial and formal execution." In the context of placemaking, they argue that participatory approaches are more cyclical in which participants can join the cycle in different steps.

To embrace and take advantage of the multiplicity and diversity of participants and operate complex and non-linear dynamics of collaboration, social media platforms that are open and participatory in nature offer new opportunities. Building on the context of collaborative planning, the potential of social media data to enable more interactive and responsive planning that adopts open forms of collaboration between citizens, experts, and public officials will be argued in the following section.

Social Media Data and Analytics for Sensing and Supporting Citizen Initiatives

Using digital tools and services and interacting online, people generate data trails in great quantity and variety. The recorded evidence of activities undertaken through online information systems including social media platforms is known as digital trace data (Howison, Wiggins, &

Crowston, 2011). Although not controlled by a public administration, social media data represents varying degrees of availability depending on the property rights of public or private actors (Severo, Feredj, & Romele, 2016). Highlighting the bottom-up source of data available on social media, Severo et al. (2016, p. 359) argue that "they can be treated as traces of public opinion and they can facilitate the involvement of citizens in the definition of the public action." As compared with traditional methods such as survey and interviews, social media data collection is much simpler, cheaper, and faster and allows us to map the constantly shifting landscape of citizens' needs, interests, and demands.

Besides reflecting everyday life as manifested online, social media data is a valuable source to observe and measure collective human behaviour (Kleinberg, 2008). As an informal and horizontal means of communication and peer-to-peer interactions, social media is widely used by citizens to organise and promote grassroots activities including citizen-led initiatives. Therefore, it offers opportunities for the planning profession and public officials to become aware of and involved in the existing civic conversations and actions of the grassroots and the creative class. As social media has helped citizens raise their voices and claim their right to the city (Tayebi, 2013), it is needed to consider new ways of listening to citizens. Yet, extracting useful information from conversations in the noisy environment of social media requires data analytics and computational techniques. Integrated with traditional methods of urban analysis, social computing techniques enable us to identify creative efforts and influential individuals, when exploring a large database of citizens' comments collected from various social media and user-generated content platforms. By sensing and identifying the agglomeration of grassroots efforts and emerging trends, the informed planning systems are then able to seek ways to appropriate them to their own and communities' benefit. This paves the way for an open and non-linear process of collaborative planning in which experts get engaged in existing civic conversations and actions initiated and/or developed by individual citizens and local communities.

As Foth, Tomitsch, Satchell, and Haeusler (2015) suggest, new technologies of communication can be used to improve the connection, exchange, and dialogue between creative efforts of communities, civic activism, and official urban policy. Understanding the social dynamics of collaboration as appear on social media has implications for policy and practice to develop effective creativity strategies that according to Peck (2005) attempt not only to connect public officials to the creative class but also to develop ties between individual units of the creative class. He argues for the challenge of civic leaders in knowing where the creative sparks will ignite and suggests that their duty is to make sure that a *spatial* network connects neighbourhoods and districts that creative individuals inhabit. However, more recently, online networks of creative citizens offer greater opportunities for a global scale of connectivity.

Technologies of social networking and content creation and sharing can be used as a means of spreading information about an urban innovation as well as a channel of influence to persuade other citizens to adopt the idea. An effective system of communication can play a key role in the success and durability of pop-up and experimental tactics and facilitate their transition to a formal urban design strategy (as noted in the case of parklets below). As an inclusive conversation space, social media facilitates a multidirectional exchange of information between experts and non-experts, individuals, and organisations. This accelerates the diffusion of urban innovations. Diffusion helps to make the benefits of these temporary activities visible, and as Mould (2014) and Lydon, Bartman, Garcia, Preston, and Woudstra (2012) suggest, once they prove to be beneficial, they become part of the city's strategies and are used by urban policy institutions. The online manifestation of temporary activities to improve urban spaces often lasts for a longer time than the physical change. Further, it can reach a wider audience through social networking platforms. Therefore, it contributes to demonstrating the success and value of local actions and temporary urban improvements.

There is a growing interest in the use of open online platforms to support spontaneous participation and self-organisation of citizens in urban planning; for example, Sawhney, de Klerk, and Malhotra (2015, p. 343) suggest that new forms of globally accessible technologies can be used to "support localized, community organizing that builds on the work of 'active' citizens rather than simply deferring to governments through problem reporting platforms." Yet, there has been little empirical evidence and data-driven research that investigates how social media can be employed to amplify and mobilise citizen initiatives.

Case Study Context: Citizen Initiatives and Micro Urban Space

Over the last decade, there have been a growing number of spontaneous efforts and small-scale interventions aiming to improve urban spaces. These micro actions, experimental tactics, and incremental changes that are mostly driven by individual citizens and local communities represent a new urban paradigm in which the right to appropriate urban spaces, rather than a request or demand, is declared and verified in practice (Iveson, 2013). This has been recently widely discussed as DIY (Do It Yourself) and tactical urbanism (Douglas, 2014; Lydon et al., 2012; Sawhney et al., 2015). These new trends of urbanism emphasise the role of citizens in improving urban spaces and highlight community-oriented initiatives on a smaller scale, often conducted outside the official capacity of the city by people who have a desire to change and reconfigure their city (Mould, 2014). Although small in scale, these urban innovations have been observed to proliferate in the sense that the idea is adopted by communities in cities around the world (Lydon et al., 2012). Transmittable and adaptable, these interventions are simple enough to accomplish without the aid of experts and reflect the new DIY culture in which the boundaries between producers and consumers are blurred and a social practice of learning and creativity takes place.

Diffusion of Parklets on Twitter[1]

Among the wide range of small-scale citizen-driven efforts, including pop-up cafes, pop-up shops, yarn bombing, guerrilla gardening, and food trucks, parklets are chosen as a case study. Parklets, which refer to an

[1] The data analysis provided here was first presented at 15th International Conference on Computers in Urban Planning and Urban Management 2017.

innovative way of transforming on-street parking spaces into green and public spaces (Fig. 5.1), emerged in San Francisco in 2005 and spread widely as the annual event, Park(ing) Day, and later as semipermanent installations. Representing varying degrees of formality and informality from pilot temporary projects to established parklet programmes, the parklet is known as "the most famous example of tactical urbanism turned city policy" (Davidson, 2013, p. 27), and provides an interesting example in which "the line between community activism and urban strategy has been blurred" (Littke, 2016, p. 165). Studying the diffusion of parklets on Twitter enables us to illustrate how transmittable urban innovations have the potential to diffuse in social networks, inspire other communities, get implemented in different places, and integrate into formal planning processes.

The initial analysis of the rate of Twitter messages per day containing the little-known word of "parklet" over a three-month period (February–April 2015) revealed an average of 79 Twitter links per day in relation to "parklet." To model the structure of social relationships as well as the

Fig. 5.1 Parklet example in Sydney 2014. Source: author

flows of information, social network analysis was used. Twitter data collected over a six-month period (February–August 2015) was visualised as a network in which nodes represent Twitter users who tweeted, or were replied to, or were mentioned in one of the tweets that contain parklet. As schematically shown in Fig. 5.2, there is a link between two nodes if they have mentioned or replied to each other in a tweet. To measure the influence of a node, the graph metric, node in-degree, that is, the number of connections that point inward at a node was used. This is regarded as the simplest actor-level measure of prestige (Wasserman & Faust, 1994). In a network of Twitter conversation, nodes with high in-degree are users who have been highly mentioned by other users: a measure of the attention a user received and therefore they can be regarded as a major source of information (Isa & Himelboim, 2018).

Filtering techniques were used to reduce the complexity of the graph as shown in Fig. 5.3. Layout algorithms were then applied on the largest connected component to place nodes on the canvas based on attractive and repulsive forces that non-linked and linked nodes put to each other (Jacomy, Venturini, Heymann, & Bastian, 2014). Next, the community detection algorithm, known as modularity in Gephi (Blondel, Guillaume, Lambiotte, & Lefebvre, 2008), was used to identify clusters of highly interconnected nodes. A cluster was used as a unit of analysis to investigate network properties as well as the content of messages. Of the 39 clusters, the 5 largest ones were analysed to illustrate how subconversations made up the larger parklet discussion. As shown in Table 5.1 and visualised in Fig. 5.3, clusters are ranked by the number of nodes or users assigned to each. Based on the number of nodes, the five largest clusters constitute 40.81% of the whole network.

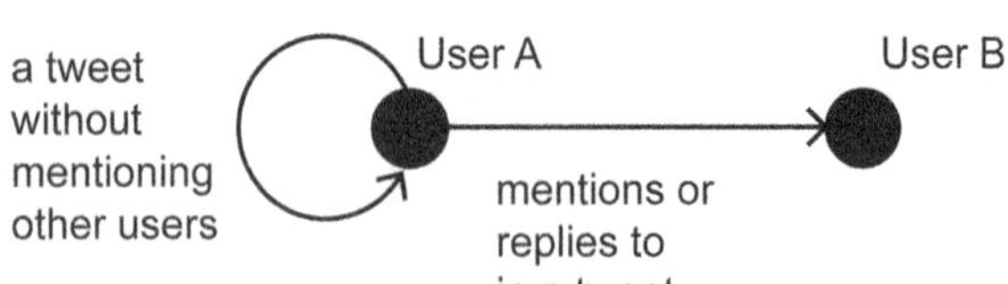

Fig. 5.2 Twitter data visualised as a network

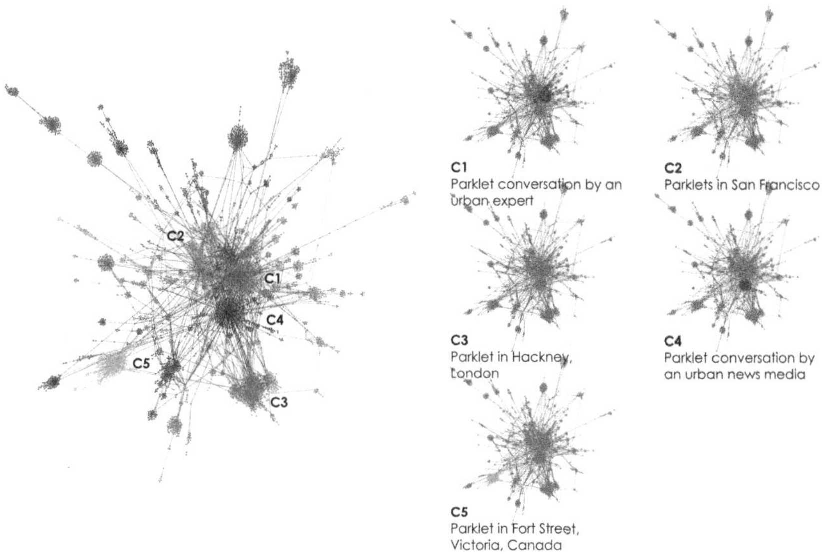

Fig. 5.3 The largest connected component in the parklet network and five largest clusters

Table 5.1 Properties of the five largest clusters in the parklet Twitter network

Cluster	Title	Nodes (percentage in the whole network)	Edges (percentage in the whole network)
1	Parklet conversation by an urban expert	525 (*11.8%*)	844 (*9.21%*)
2	Parklets in San Francisco	393 (*8.83%*)	848 (*9.25%*)
3	Parklet in Hackney, London	335 (*7.53%*)	949 (*10.35%*)
4	Parklet conversation by an urban news media	327 (*7.35%*)	409 (*4.46%*)
5	Parklet in Victoria, Canada	236 (*5.3%*)	605 (*6.6%*)
Sum	–	1816 (*40.81%*)	3655 (*39.87%*)

As highlighted in Fig. 5.3, the five largest clusters in the whole network were chosen for further investigation of the structure and content of each. Clusters are labelled with the topic of most messages in that cluster, identified through a content analysis of tweets. A brief description of the content of each cluster is given below.

Cluster 1, the largest cluster in the whole network, is a central one in the parklet network and includes a hub user, *@BrentToderian*, a city planner who shared general information about parklets and played a curator role by aggregating parklet ideas together and contributed to steering the conversation. The second cluster which is also positioned at the centre of the network is about parklet conversations mainly in San Francisco. Messages in this community of Twitter users included not only information about new parklets but also the parklet programme in San Francisco as well as ideas and articles on parklets in general. In Cluster 3, the largest peripheral cluster in the network, the dominant topic of conversation was about a new parklet in Hackney, London. This community of Twitter users shared information about the first parklet in London before, during, and after the construction. The conversation in Cluster 4 centred on a tweet by *@CityLab*, the online magazine about city planning and design. This tweet linking to an article, entitled "Why Some Parklets Work Better Than Others," was shared by many users because of the popularity of *@CityLab*. Finally, Clusters 5 was concerned with a specific parklet in Victoria, Canada. Users in this cluster included local businesses and community groups that contributed to the parklet making.

This initial cluster analysis revealed that the main content of tweets in each cluster was related to either a specific parklet (Clusters 3 and 5) or a general interest in parklets (Clusters 1 and 4), with an exception for Cluster 2 in which the conversation addressed both. In what follows, a closer look at the pattern of connection in each cluster through a network structure analysis is provided. To analyse the structure, each cluster is filtered out of the overall network. Then, layout algorithms are applied to visualise the shape of conversations in each cluster. Drawing on network typology presented by Baran (1964) and Siegel (2009), the structure of each cluster is analysed. As shown in Fig. 5.4, Cluster 1 represents a centralised network in which individuals were connected to a focal user and rarely interacted with each other. It also partially resembles a hierarchical network (Siegel, 2009) in which star-shaped networks can be seen at various levels of expansion, that is, while this cluster represents a single-agent structure, there are multiple local agents at a lower level. These focal users have contributed to a wider diffusion of information shared by the central user, *@BrentToderian*, the city planner mentioned earlier.

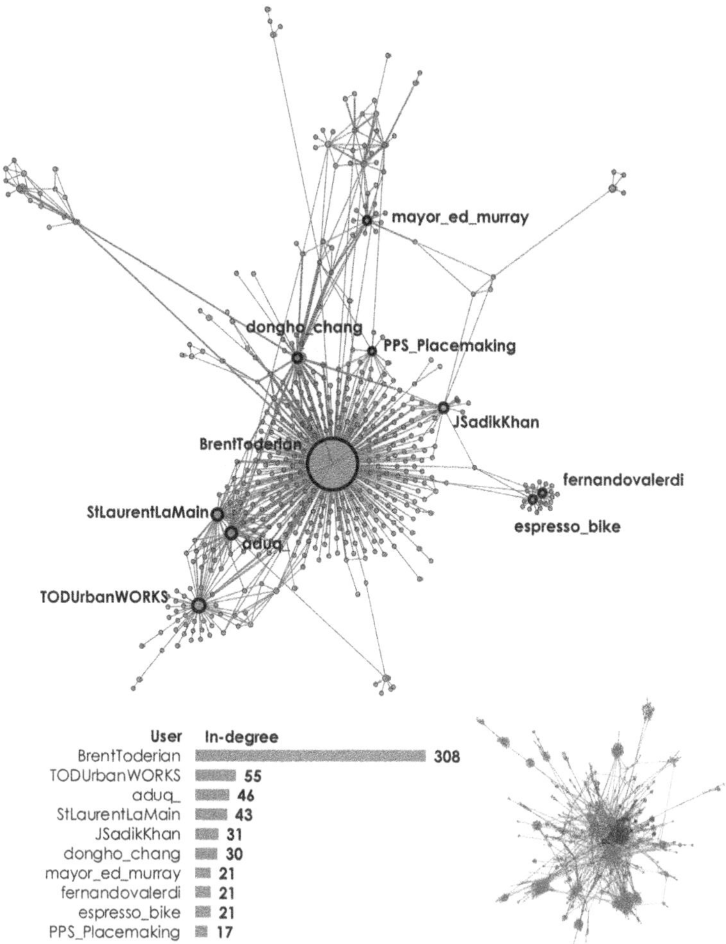

Fig. 5.4 Cluster 1: parklet conversation by an urban expert; size of nodes reflects the in-degree of connectivity as also shown in the bar chart. The position of this cluster in the overall network is highlighted in the diagram at the bottom

The second cluster that contains predominately messages about parklets in San Francisco exhibits a decentralised network. As shown in Fig. 5.5, rather than a central user, this cluster formed around several individuals that were relatively equally important in the spread of information about parklets. These focal individuals contributed to the diffusion

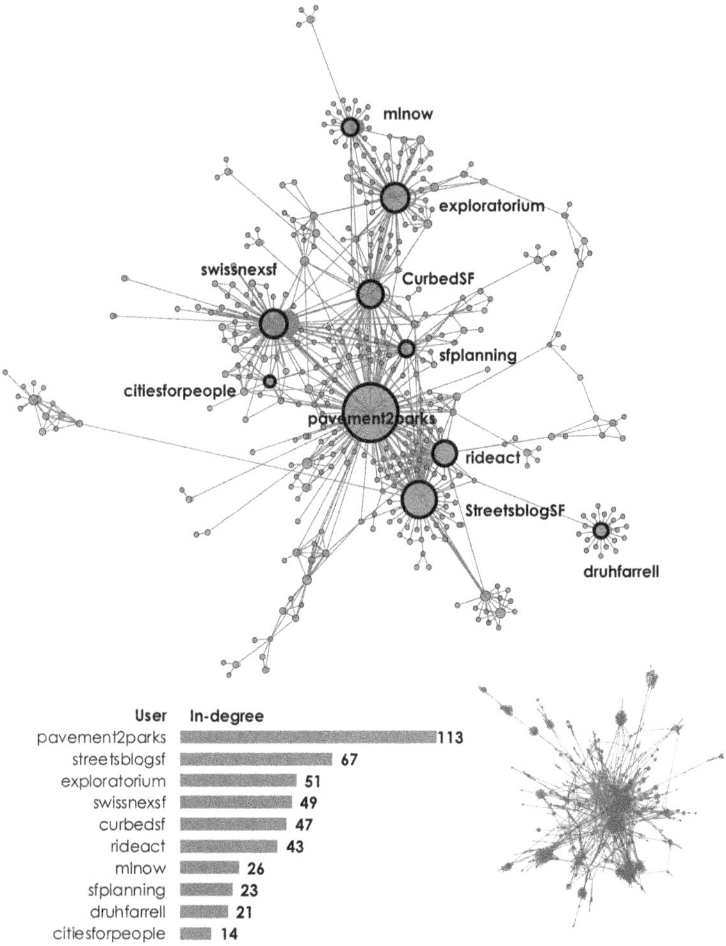

Fig. 5.5 Cluster 2: parklets in San Francisco

of parklets in a lower level of sharing information to a local audience. The most frequently mentioned user was affiliated with the government: @ *pavement2parks*, the official Twitter account of the Pavement to Parks Program that is part of the City of San Francisco. There were two other users that were related to the government including *@sfplanning*, the San Francisco Planning Department, and *@rideact*, the Alameda-Contra Costa Transit in Oakland. The significant role of governmental users in shaping the conversation in this cluster indicates the formal quality of parklets in San Francisco. The other type of users that mediated the conversation was local news media including *@streetsblogsf, @exploratorium, @curbedsf,* and *@mlnow*.

Cluster 3 (Fig. 5.6) is a dense structure that contains an interconnected set of nodes. Like Cluster 2, it resembles a decentralised network. However, most individuals in this cluster interacted with more than one local central node that resulted in a dense and interlocking network. The

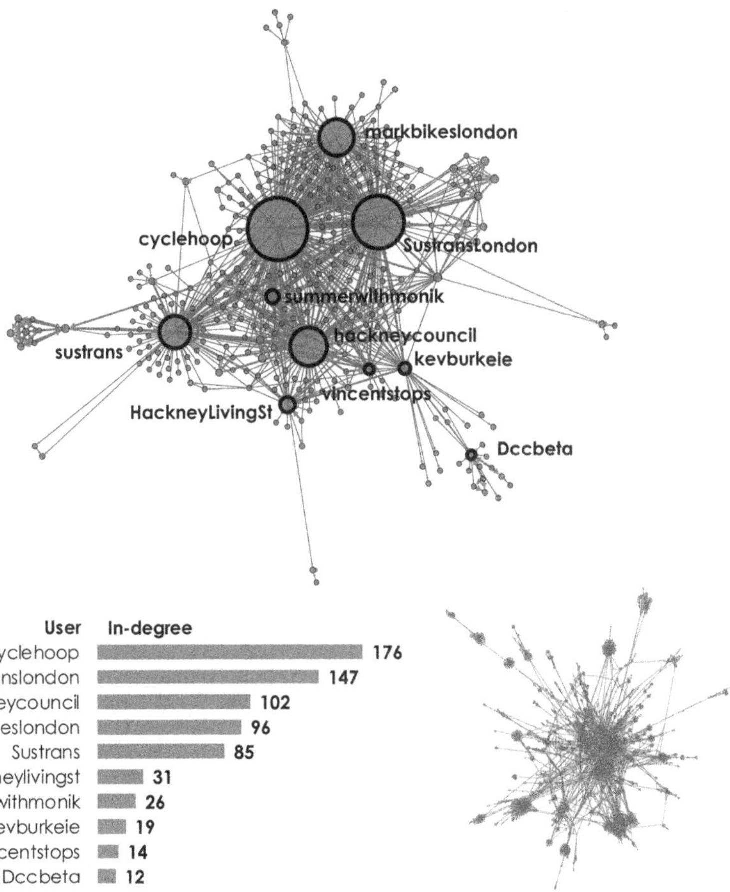

Fig. 5.6 Cluster 3: parklet in Hackney, London

in-degrees of top five nodes show insignificant differences compared to those of Cluster 1, reiterating the presence of more than one central node. As content analysis shows, focal nodes highlighted in Fig. 5.6 were users involved in the implementation of a parklet in Hackney, London: *@cycle-hoop*, which was a design firm that designed and built the parklet while it was funded by *@SustransLondon* (a charity organisation) and *@hackney-council* (Hackney Council).

Cluster 4, similar to Cluster 1, resembles a centralised structure in which *@CityLab* is the focal user (Fig. 5.7). This urban news media was connected to many users that rarely interacted with each other. This structure emerged as *@CityLab* shared an article about parklets that was retweeted by many others. Moreover, conversation was expanded by replying to and sharing the tweet of *@CityLab*. As such, participants, each connected to its own community, sparked multiple subconversations.

Cluster 5 represents a dense structure that is semicentralised (Fig. 5.8). The central user in this cluster was *@FabulousFort* which was a local community in charge of the Victoria parklet that actively generated content about that particular parklet over the six-month period of data collection. The other main participant involved in the construction of this parklet was *@cityspaces*, an urban design and architecture firm that, although located in a central position in the cluster, neither generated much content nor had a long-lasting presence in the conversation as compared to *@FabulousFort*. The right side of the graph represents a denser and interconnected structure with a few focal users among which there are local communities and businesses in Victoria such as *@FortProp*, *@ DutchBakery1956*, and *@cfax1070*. This part of cluster is also formed around *@lisahelps*, the mayor of Victoria, and *@CityOfVictoria* that contributed to the conversation.

The cluster analysis revealed how centralised and decentralised structures were coupled with each other which channelled the flow of information and therefore the spread of parklets that are being adopted in other cities. On the one hand, the centralised structure of Clusters 1 and 4 in the core of the whole network was in fact the best way for the extensive exchange of information according to Rogers (1995, p. 308) who argues "radial personal networks are less dense and more open, and thus allow the focal individual to exchange information with a wider

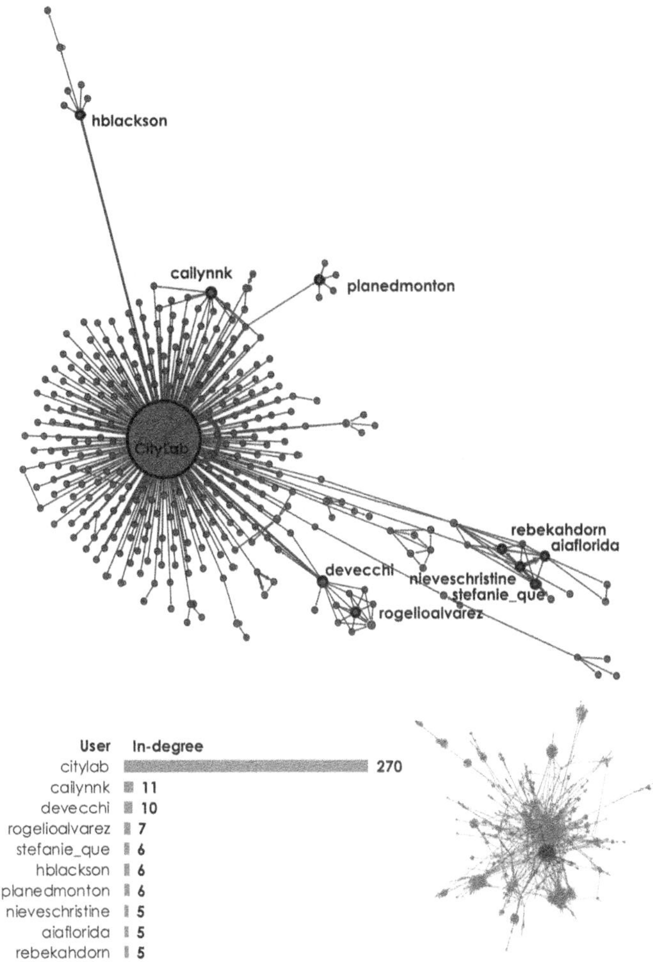

Fig. 5.7 Cluster 4: parklet conversation by an urban news media

environment." On the other hand, the relatively unstructured and non-linear distribution and exchange of information in peripheral Clusters 3 and 5 that included a larger number of grassroots participants played a role as a channel of influence persuading others to adopt the idea.

As illustrated in Fig. 5.9, the overall network consisted of a core of users that connected a much larger periphery. This indicates a global

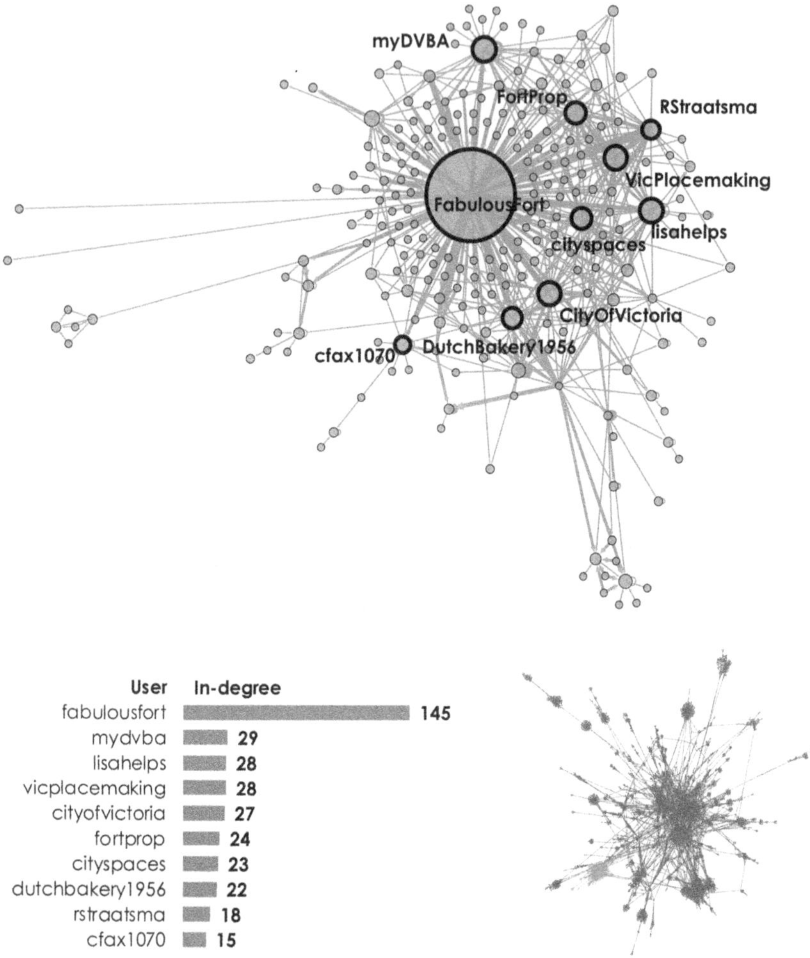

Fig. 5.8 Cluster 5: parklet in Victoria, Canada

coordination of the conversation beyond local-level interactions by users such *@BrentToderian* and *@pavement2parks*. The significant breadth of central clusters observed in the parklet network suggests the extent to which the conversation about this urban intervention is self-organised: In fact, micro interactions turned into a complex network quite dependent on a group of central clusters. However, there is no centre in the dense

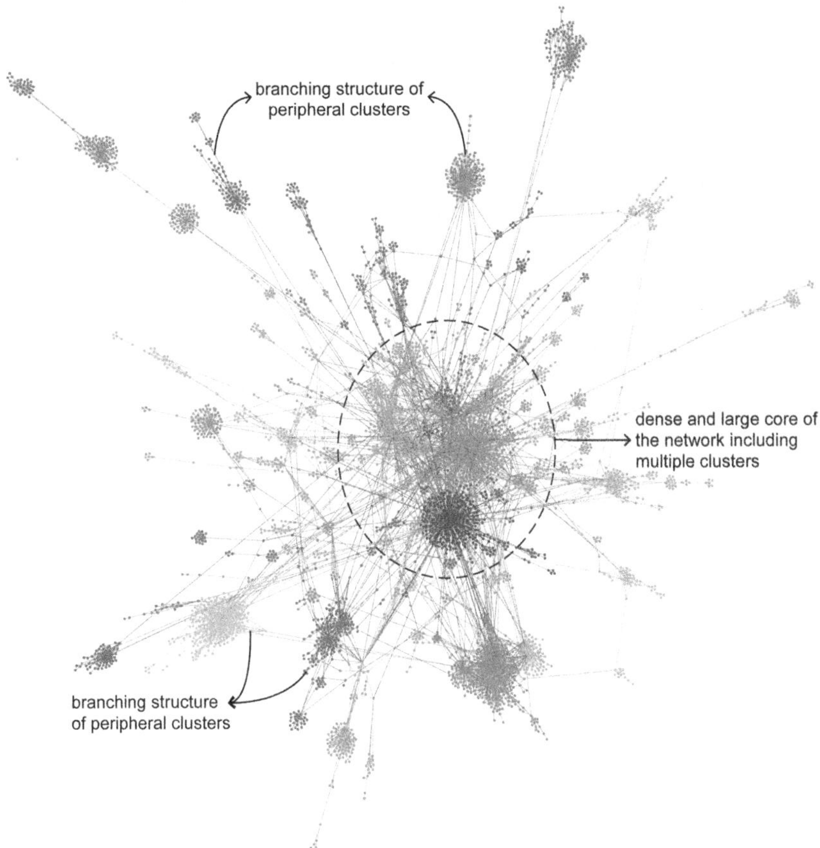

Fig. 5.9 Structural properties of the largest connected component

and large core of the network, but a collection of three main clusters as subsystems, that suggests the decentralised nature of the system. A moderate self-organised quality for the overall parklet network indicates how controlled and autonomous processes simultaneously exist and intricately relate to each other. Moreover, a self-similar pattern was identified for the peripheral clusters that were largely concerned with a specific parklet project. This is consistent with Rheingold's (1993) remark about the structure of grassroots virtual communities that is a self-similar branching structure in which central participants grow a branching set of connections, and then many more subconversations grow off those. This insight into the dynamics of peer-to-peer networks explains the growth of the conversation about an urban intervention in a reciprocal exchange of information between local and global participants contributing to the conversation even though the majority of them were not actively involved in any parklet projects.

Open and Complex Collaboration: Insights from the Case Study

The case of parklet conversation on Twitter revealed a complex hybrid dynamic in which grassroots initiatives in different cities were integrated with official sources and efforts: the central clusters that formed around government agencies and influential experts were incorporated into peer-to-peer networks of individual citizens. It should be noted that Twitter as an open and accessible platform mediated such an integration by bringing formal and informal players together in an inclusive space of conversation. This is consistent with Ermacora and Bullivant's (2016, p. 60) remarks about the welcome effect of "hybrid forums" that reduces the distance and provides interdependent relationship between professional specialists and lay people. The diversity of participants is also consistent with Innes and Booher's (2004) remarks about effective participation that is supported and built on the interactions of all involved entities in increasingly complex contemporary society, including public sector agencies, non-profits, business organisations, and advocacy groups. They note that diversity of participants is one of the principal characteristics of a collaborative network. Social media platforms that are widely used by laypeople as well as experts and public officials provide new opportunities to bring together the wide range of involved individuals and organisations.

Findings of this research also have implications for the use of social media platforms by city councils that are currently mainly to inform the public by disseminating information and promoting council events and activities. In addition to these, as Howard (2012, p. 38) outlines, social media has the benefit of allowing "council staff to learn about the issues that are important to the community without the need to conduct official market research. By listening to the social media conversations, and participating in, as appropriate, council staff can tap into the topics that people are discussing." Openly accessible social media data is critical to ensure seeking out existing conversations that do not necessarily take place in the designated social media pages, for example, council's Facebook or Twitter pages. These civic conversations often happen in ad hoc ways, outside the formal social media engagement processes. Ongoing monitoring enables agility in communication strategies to benefit from these dynamic populations. Informed by data, councils' social media activities would be able to engage the targeted audience and ensure that the voice of citizens is heard and those who feel heard would be encouraged to engage further and actively participate. The findings of this study also demonstrate how Twitter facilitated the diffusion of urban tactics and channelled the exchange of information that allowed for referring to similar practices and opened spaces for local experimentation and implementation of urban innovations. These insights into non-linear models of citizen participation can be applied to develop new forms of urban governance that are mediated by open digital platforms and leverage the power of peer-to-peer self-organised networks. However, to fulfil an open model of collaboration and fully exploit the opportunities of inclusive citizen-expert dialogue that leverages Open Data, a shift in policy is required to accommodate and further mobilise experimental and innovative efforts in urban development processes.

Furthermore, this case study demonstrated the utility of social media data analytics for advancing our understanding of citizen-oriented planning processes. Illustrating the structure of social connections as the hidden force behind bottom-up urban processes offered new insights into the effective patterns of connectivity that foster citizen participation. These insights can be employed to develop an effective communication mechanism in urban governance. While the focus of this research was on parklets as a small-scale community effort, the techniques used here scale well and are applicable to larger problem instances. Future research on other urban planning cases and broader participatory processes would help to determine how observations of a network graph can be used in practice to develop a more democratic, decentralised planning process.

The work presented in this chapter is the first step towards building a model of inclusive citizen-expert dialogue that leverages Open Data. It was a pilot study on a problem that will become more significant with time, that is, how to interpret social media data using network analysis in the context of urban studies. Yet, it provides a basis for future research using larger data sets to scale up and further refine the techniques presented here. In future urban research, the use of social network analysis along with survey-based methods would represent new way of listening to citizens to better understand a range of civic issues and public opinion by extracting information from conversations in the noisy environment of social media platforms.

Conclusion

This chapter was an attempt to rethink civic engagement in an increasingly digitally mediated society. Key features of the social dynamics observed in the case study, such as inclusive space of conversation, a hybrid system of formal and informal efforts, and the multidirectional exchange of information through social media, can be utilised by public officials as an enabling strategy. This chapter discussed and provided evidence on the potential of social media data and analytics to mediate complex and open forms of collaboration in which rather than providing solution, planners inform agents and communities, open the process, create and formulate social relations, and decentralise control.

References

Arnstein, S. R. (1969). A ladder of citizen participation. *Journal of the American Institute of planners, 35*(4), 216–224.

Baran, P. (1964). On distributed communications networks. *IEEE Transactions of the Professional Technical Group on Communications Systems, 12*(1), 1–9.

Blondel, V. D., Guillaume, J. L., Lambiotte, R., & Lefebvre, E. (2008). Fast unfolding of communities in large networks. *Journal of statistical mechanics: Theory and experiment, 2008*(10), P10008.

Davidson, M. M. (2013). *Tactical urbanism, public policy reform, and 'innovation spotting' by government: From Park(ing) Day to San Francisco's parklet program.* Doctoral dissertation, Massachusetts Institute of Technology.

Douglas, G. C. C. (2014). Do-it-yourself urban design: The social practice of informal "improvement" through unauthorized alteration. *City & Community, 13*(1), 5–25.

Ermacora, T., & Bullivant, L. (2016). *Recoded city: Co-creating urban futures.* Routledge.

Foth, M., Tomitsch, M., Satchell, C., & Haeusler, M. H. (2015, December). From users to citizens: Some thoughts on designing for polity and civics. In *Proceedings of the Annual Meeting of the Australian Special Interest Group for Computer Human Interaction.* ACM, pp. 623–633.

Howard, A. E. (2012). *Connecting with communities: How local government is using social media to engage with citizens.* ANZSOG Institute for Governance at the University of Canberra and Australian Centre of Excellence for Local Government.

Howison, J., Wiggins, A., & Crowston, K. (2011). Validity issues in the use of social network analysis with digital trace data. *Journal of the Association for Information Systems, 12*(12), 767.

Innes, J. E., & Booher, D. E. (2004). Reframing public participation: Strategies for the 21st century. *Planning Theory & Practice, 5*(4), 419–436.

Isa, D., & Himelboim, I. (2018). A social networks approach to online social movement: Social mediators and mediated content in# FreeAJStaff Twitter Network. *Social Media + Society, 4*(1), 1–14.

Iveson, K. (2013). Cities within the city: Do-it-yourself urbanism and the right to the city. *International Journal of Urban and Regional Research, 37*(3), 941–956.

Jacobs, J. (1961). *The death and life of great American cities.* Vintage.

Jacomy, M., Venturini, T., Heymann, S., & Bastian, M. (2014). ForceAtlas2, a continuous graph layout algorithm for handy network visualization designed for the Gephi software. *PloS one, 9*(6), e98679.

Kleinberg, J. (2008). The convergence of social and technological networks. *Communications, ACM, 51*(11), 66–72.

Littke, H. (2016). Revisiting the San Francisco parklets problematizing publicness, parks, and transferability. *Urban Forestry & Urban Greening, 15*, 165–173.

Lydon, M., Bartman, D., Garcia, T., Preston, R., & Woudstra, R. (2012). *Tactical urbanism: Short term action, long term change* (Vol. 2). Miami and New York: Street Plans Collaborative.

Mould, O. (2014). Tactical urbanism: The new vernacular of the Creative City. *Geography Compass, 8*(8), 529–539.
Peck, J. (2005). Struggling with the creative class. *International Journal of Urban and Regional Research, 29*(4), 740–770.
Rheingold, H. (1993). *The virtual community: Homesteading on the electronic frontier*. MIT Press.
Rogers, E. M. (1995). *Diffusion of innovations*. Simon and Schuster.
Sandercock, L. (2004). Towards a planning imagination for the 21st century. *Journal of the American Planning Association, 70*(2), 133–141.
Sawhney, N., de Klerk, C., & Malhotra, S. (2015). Civic engagement through DIY urbanism and collective networked action. *Planning Practice & Research, 30*(3), 337–354.
Severo, M., Feredj, A., & Romele, A. (2016). Soft data and public policy: Can social media offer alternatives to official statistics in urban policymaking? *Policy & Internet, 8*(3), 354–372.
Siegel, D. A. (2009). Social networks and collective action. *American Journal of Political Science, 53*(1), 122–138.
Tayebi, A. (2013). Planning activism: Using social media to claim marginalized citizens' right to the city. *Cities, 32*, 88–93.
Wasserman, S., & Faust, K. (1994). *Social network analysis: Methods and applications*. Cambridge University Press.

6

Slum Digitisation, Its Opponents and Allies in Developing Smart Cities: The Case of Kibera, Nairobi

Bitange Ndemo

Abbreviations

GIS	Geographic Information Systems
GSMA	Global Systems for Mobile Communications Association
ICTs	Information Communications Technologies
NGO	Non-Governmental Organizations
UN	United Nations

Introduction

In the future, most people across the world will live in urban areas. Virtually every report on demographic shifts predicts an urban population explosion with exponentially growing inequality. The African Development Bank (2012) report estimated that by 2050, the urban population in Africa will exceed 60%, up from 36% in 2010.

B. Ndemo (✉)
Department of Business Administration, School of Business, University of Nairobi, Nairobi, Kenya
e-mail: bndemo@bitangendemo.me; bndemo@uonbi.ac.ke

S. Hawken et al. (eds.), *Open Cities | Open Data*,
https://doi.org/10.1007/978-981-13-6605-5_6

Additionally, large cities in developing countries will account for more than 90% of the future share of the world's urban population. This urban population explosion will likely change the continent's demographic landscape and pressurise urban infrastructure, especially if socioeconomic inequalities persist and policymakers fail to leverage data for decision-making.

This demographic shift already poses numerous challenges related to the proliferation of slums (UN-Habitat, 2003a). Studies (UN-Habitat, 2016; Amer et al., 2016) suggest that congestion hinders the development of adequate infrastructures, economic development, social inclusion, security, sustainable housing and transport. It also contributes to rising inequality. To tackle these challenges citizens cant utilise ICT to empower themselves, improve productivity, expand democracy to the voiceless and become players in the innovation dynamics of their own cities. This chapter assesses the opportunities presented by the digitisation of a large slum in Nairobi known as Kibera.

In Kibera, however, slum digitisation also has enemies (i.e. those resisting change). These include powerful government officials (Nyamori, 2016) who illegally grab public land to their benefit to build squalid housing for desperate urban migrants. Nyamori demonstrates how their power is being challenged by many well-meaning allies of change that are emerging, including children, to not only challenge ownership but also protest land grabbing. The advent of geographical information systems (GIS) (Amer et al., 2016) shows that technology can help protect such land and improve the livelihood of disadvantaged people.

This chapter aims to establish how we should deal with the problem of making relevant data available through ICT and GIS to assist policymakers to deal with a lack of infrastructure for elements such as water, energy, solid waste, sanitation and general urban planning. Virtually every country in the world wants to make its cities intelligent to improve productivity. This chapter therefore seeks to explain the nexus between data and policymaking to solve the problems associated with growing inequality and the expansion of slums in urban areas.

Definition of Slums

Since the concept of slums first emerged in Europe in the early nineteenth century, its meaning has been progressively changing. 'Slum' once meant a 'room of low repute' or 'low, unfrequented parts of the town', but it is now used to describe other conditions (UN-Habitat, 2003b).

UN-Habitat recently expanded its description of slums as places with squalid living conditions, including physical, spatial, social and behavioural aspects of urban poverty (UN-Habitat, 2003a). This definition can serve as a universal definition with minor modifications.

Operationally, UN-Habitat defines slums as

> one or a group of individuals living under the same roof in an urban area, lacking in one or more of the following five amenities: 1) durable housing that protects inhabitants from extreme weather conditions; 2) sufficient living area; 3) access to clean water that is sufficient, affordable and available; 4) access to sanitation facilities; and 5) secure tenure protection against forced eviction. (UN-Habitat, 2006/2007)

For this study, the definition includes the above UN operational definition and parts of the previous definitions, such as physical and spatial attributes. The emerging concept of 'slum digitisation and spatial planning' offers hope that, with a generation of massive data, methods of measurement can be developed to aid the renewal of neglected neighbourhoods.

Kibera Slum and Map Kibera

Kibera lies in Nairobi's southwestern region around 7 km from the Nairobi Central Business District. It is an informal settlement on public land comprising ten villages that cover approximately 2 m^2. These villages are Lindi, Kisumu Ndogo, Soweto, Makina, Kianda, Mashimoni, Siranga, Gatuikira, Laini Saba and newly founded Raila. Most inhabitants of these villages belong to the Kikuyu and Luo ethnic groups. The original inhabitants, Nubian migrants from Sudan, were squeezed into about 20% of Kibera's space. In Nubi, Kibera means 'forest' or 'jungle'.

Until the youth developed digital maps of Kibera in 2009, there were few verifiable facts about its existence, because different people gave their own accounts depending on their interest. One 'fact' that emerged suggested that Kibera was that largest slum in Africa with a population of 2 million (Karanja, 2010). Some estimates (Mutisya & Yarime, 2011) put the population at just 1 million. However, with the combination of mapping and census, the 'fact' was proven false. It was provided by people seeking more donor funding to 'assist the poor'. The 2009 census confirmed a population of 170,000. More recently, using digital maps, the population is estimated to be below 300,000.

Kibera was a blank spot on the map until November 2009, when young Kiberans, with the help of the US non-governmental organisation (NGO) GroundTruth, created the first free and open digital map of their community. 'Map Kibera' has grown into a complete interactive community information project. Project members work in Kibera, Mathare and Mukuru.

The Problem

The first post-independence census by the National Bureau of Statistics reported that the population in Nairobi was 509,000 in 1969. Per the 2009 census, the population had ballooned to 3.2 million; it continues to grow, especially around Nairobi. This is a result of people migrating out of rural areas in search of non-existent opportunities in urban areas. Government reports show that about 71% of the population in Nairobi lives in slums (GoK, 2013). The problem, however, is likely to continue, owing to planning inconsistencies around housing and other social services, leading to massive social exclusion (Zwanenberg, 2008).

Today, Nairobi has about 135 slum areas. Together, they occupy approximately 1% of the city area, but contain more than 50% of the city's population (Wanjiru & Matsubara, 2017). Many slums have existed since colonial times, when native Africans were forced to stay in temporary residences outside the city. With the introduction of African socialism in 1966, the new post-independence president, Jomo Kenyatta, fearing there would not be enough employment in urban areas, introduced a policy dubbed 'Rudi Mashambani (Go back to the land)', which urged urban residents to move back to their rural homes and adopt agricultural practices (Wafula, 2011). Whereas the government had good intentions, many people did not have land to move back to. Thus, the slums were the only places they could call 'home'.

The Growth of Informal Settlements

Urbanisation has been associated historically with industrialisation and a corresponding growth of employment in the formal sector, and both have spurred and accompanied urbanisation. This is often not the case today (Hendrick-Woong and Angelopulo, 2014), especially in Sub-Saharan Africa, where urbanisation is fast-growing and informal settlements are expanding. A 2015 World Bank report showed that while the percentage of Kenyan people living in urban slums expanded from 55% in 1990 to 56% in 2014, in the Philippines, the numbers declined from 54% to 38% over the same period. Several other African countries have displayed a similar pattern, moving from relatively better to worse. These countries include the Central Africa Republic (88–93%), Comoros (65–70%), Cote d'Ivoire (53–56%), Mozambique (76–80%) and others. This is a worrying trend that comes with many challenges (Amer et al., 2016; UN-Habitat, 2016; Hendrick-Woong and Angelopulo, 2014). It has precipitated internal conflict and has displaced many people to refugee camps.

Rapid urbanisation is an opportunity, as stated by the UN-Habitat (2016) report, but only if developing economies (i.e. in Africa) embrace city planning to provide services and opportunities in enterprise development to create meaningful incomes to all residents.

The Rise of ICTs and GIS for Open Mapping

Kingston (2007) demonstrates how citizens are increasingly using ICTs and GIS to participate in the management of their cities by facilitating 'public engagement in the regeneration of inner city neighbourhoods through more integrated approaches to spatial data management' (p.138). A 2016 United Nations report on world cities concluded that cities have become positive economic platforms 'driving innovation, consumption and investment in both developed and developing countries' (p. 29). The report also acknowledged that the shift to urban life has had profound implications for virtually every aspect of human progress. The report warns that, 'although some of this change is positive, poorly planned urbanisation can potentially generate economic disorder, congestion, pollution and civil unrest' (p. 29). The use of ICTs and GISs has become critically important, especially when it is digitised and openly interactive, generating data that feed into policymaking.

Bangladesh, for example, has been digitising its slums by using GIS techniques to present spatial or geographic data of densely populated clusters of compact settlements (Angeles et al., 2009), for which it was previously considered impossible to develop any form of addressing system. Das, Choudhury, Shobhana, and Vaghela (2014) suggested that the use of geospatial techniques in Rajkot city, India, was instrumental in enabling the government to develop a multi-criteria decision-making system for slum redevelopment and planning. In Dhaka, Bangladesh, Angeles et al. (2009) used GIS to (1) create GIS slum maps, (2) identify slum areas with detailed addresses establishing the number of households and the slum population (slum census) and (3) evaluate environmental conditions, such as water sources and sanitation, or infrastructural issues such as access roads. This information was important for data-driven policymaking decisions on slum upgrades.

Baud, Scott, Pfeffer, Sydenstricker-Neto, and Denis (2013) suggested that digitisation (utilising ICT), spatialisation (i.e. using GIS) and participatory processes can support more inclusive forms of local governance. Their arguments fit into a wider discourse on knowledge construction and circulation in the form of sociopolitical and relational processes (McFarlane, 2011). Some of the emerging outcomes of this discourse include greater competencies in local government, respect for the rights of the people, legitimacy, equity and accountability (McCall & Dunn, 2012).

Even with the promise of inclusivity, there are opponents of digitisation and spatialisation. Bautes, Dupont, and Landy (2013) explored methods of resistance in slum upgrades in Brazil and India and found that slum residents sometimes fight and defend places they live in as theirs to maintain and improve by themselves. In most cases, their defensive behaviour is triggered by civil society organisations that mobilise slum residents to fight for their 'rights', often resulting in direct confrontations with the allies of digitisation and spatialisation seeking to improve their conditions. Bautes et al. (2013) described that some movements arising from the slums are widely acknowledged beyond the slum boundaries.

It is never clear who precipitates resistance when slum renewals are proposed. Pile and Keith (1997) argued that it is imperative to investigate the relation between different types of mobilisation and resistance created to fight change against dominant forces and how resident groups play roles in the complex chemistry between undertakings of individual or communal liberation and strong forces of change.

Giusti de Perez and Perez (2008) demonstrated that open GIS mapping could catalyse quality-of-life improvements in poor urban areas, especially in developing countries. For 30 years, they used the technology in Venezuela to manage urban barrios (i.e. informal settlements) and developed sustainable solutions that went beyond conventional planning programmes. From the literature, providing spatial data openly has the potential to improve informal settlements and make them sustainable.

Potential for Open Data to Improve Slums

The Philippines, with more than 38% (World Bank, 2015) of her people living in slums, plans to use GIS and Global Positioning System (GPS) technologies to conduct the country's 2020 census (Philippines Statistical Authority, 2018). In the past, people living in slums were never accounted for in national statistics (Hagen, 2015), and many slums were neglected, appearing as just blank spaces on national maps. With GIS, this has changed; governments have begun to pay attention to informal settlements.

A 2015 World Bank Policy Note ICT01, 'Open Data for Sustainable Development', demonstrated how Open Data can support development in areas that relate to United Nations Sustainable Development Goals. The report identifies key benefits of Open Data, particularly in developing countries: economic growth and job creation; efficiency improvement; effective and complete public services; transparency; accountability; public participation; and facilitates having better information-sharing within government. The policy note also suggests that trust between governments and citizens is enhanced with Open Data.

Specifically the policy note highlights two case studies where the private sector has leveraged Open Data to create new enterprises. The policy note states,

> *In Ghana, the company Farmerline sends farmers essential information on weather and agriculture by voice and text to their mobile phones. Solapa, an Argentinian company, is building a platform to help farmers analyse their crop strategy and yield by combining market data with sensory and GIS data. In a similar way, Chile has used open U.S. satellite data from Landsat to help estimate water demand under the country's frequent droughts, and help agricultural businesses adjust water use strategically for different crops.* (p. 14)

These three case studies provided motivation for young people to seek opportunities for employment outside urban cities. It is via the provision of Open Data and solutions such as these that more employment opportunities can be created in rural areas to reduce pressure in urban centres while stimulating economic growth at the grassroots level. With so many problems in developing countries, the potential to leverage emerging technologies to create employment is big, as demonstrated in World Bank's 2016 report and the UN-Habitat (2016).

Method

This study was conducted using a qualitative research method. This method (Given, 2008) focuses on human elements of the social and natural sciences. Thomas' (2006) general inductive approach for analysis of qualitative data was applied to 'condense raw textual data into a brief, summary format; establish clear links between the evaluation or research objectives and the summary findings derived from the raw data; and develop a framework of the underlying structure of experiences or processes that are evident in the raw data' (p.237).

Based on Yin's (1984) case study approach, the author developed three semi-structured interview guides (see Appendix I) to collect data from different Kibera slum stakeholders (see Appendix II), including donor organisations, government representatives, residents and service providers (e.g. education providers, healthcare providers, ICT groups, food service providers, landlords, religious groups, community groups). Most stakeholders were also key informants in public discourse on development matters. Marshall (1996) noted that the principle advantages of key informant technique 'relate to the quality of data than can be collected in a limited period of time and the approach is potentially useful as an isolated research technique or in conjunction with other qualitative methods' (p.92).

Three face-to-face rounds of interviews were conducted. Out of 25 targeted interviewees (see Appendix I), I conducted 20 semi-structured interviews to collect primary data from stakeholders, mainly focusing on the challenges they faced before digitisation and the opportunities they found after digitisation.

The second interview process was conducted in debate form, moderated by me, comprising four focus groups created from different stakeholders to dive deeper into issues that hinder the project's progress and to generate insights into the challenges and opportunities of slum digitisation identified by one-on-one interviews. These four focus groups were selected because of their different views towards the project and, as Breen (2006) argued, if well-moderated, it can generate new ideas via debate, allow a deeper understanding of the phenomena, bring new insights when compared with one-on-one interviews and better articulate issues.

For the third round, the author sought to establish what the local and national governments had in their agenda regarding slum upgrades. Two stakeholders were selected because they make policies that impact slum dwellers. The face-to-face interviews were to determine plans and to discuss how Map Kibera would be integrated into regional and national development. Additionally, data from different sources was analysed. These included the 2014 Nairobi Integrated Urban Development Plan, Nairobi City Water and Sewerage Company Limited, Strategic Plan 2014/2015–2018/2019, and plans for infrastructure (i.e. water, energy and housing) development in the coming years. The history of urban planning in Nairobi (from 1927) when the colonial government developed the city's first master plan was also reviewed.

Considering that three methods were used to gather data. The research leveraged data triangulation to identify the relations that emerged from the analysis. This was an 'attempt to map out, or explain more fully, the richness and complexity of human behaviour by studying it from more than one standpoint' (Cohen & Manion, 2000, p. 254). With qualitative research, the main reason for triangulation is to increase the validity and credibility of the evaluation and research findings (Sabina Yeasmin & Rahman, 2012; Yin, 2003).

Case Study: Map Kibera

Map Kibera was started in 2009 with the assistance of a U.S. consultancy firm, GroundTruth Initiative, that provides training for communities in open geographic mapping, citizen journalism and video, collecting and using Open Data, social media, citizen feedback through technology and mobile technology (Hagen, 2017).

With a small grant, GroundTruth trained 13 young men in their 20s to use a handheld GPS, but they did not tell them what to survey (Hagen, 2017). A pattern revealing different facilities, like health, education, entertainment and sanitation, began to emerge to give the team motivation to continue as they discovered that their work was meaningful to the residents. Encouraged, the founders provided visual aids to the mappers 'to give voice to their experiences and slum dwellers'.

Joshua, one of the key members of Map Kibera, said,

> *While it is in the interest of policymakers to know the population size, social and economic needs of the citizenry, it is often difficult to sell data or tools that makes it easier for them to tailor make some of their policies. We are however happy that our data was used to respond to some of the social problems we pointed out. You can see new police posts in areas we classified and black spots for crime, there are six new health centres funded by the First Lady initiative of beyond zero (Campaign to improve maternal and child health outcomes in Kenya). The County Government has responded by giving us street lighting in crime prone areas where many women used to be raped.*

Whereas Map Kibera has not been comprehensively utilised, different groups are prepared to create new digital layers. E-commerce is high on the agenda of map developers and others who see opportunity. Agnes, who has been piloting a new supply chain mechanism between female vegetable vendors and suppliers from the nearby Toi Market, using an existing motorbike logistics firm (with GPS) to make deliveries, stated,

This mapping wasn't just for policy makers to know where we live, we must build new digital layers on these maps. We need e-commerce too.

Sanda, who heads the Safaricom Foundations and works with Shining Hope for Communities to aerially distribute water in Kibera, mentioned (Fig. 6.1),

The mapping of Kibera will create more opportunities than what exists today. This is a place even with good intentions, you work with what you get. You cannot move anyone even when it makes sense to do so. You can see we had to find other means of providing the most precious commodity that everyone needs in the slum—water—but at a very costly method. I am optimistic that the mapping is just the beginning of better planning for services in future. That way the resources will be better utilised to move people out of poverty in line with Sustainable Development Goals.

A senior government official, speaking on the condition of anonymity, stated,

Fig. 6.1 Kibera aerial water distribution: Safaricom Foundation

> *Slums thrive in chaos that any form of order will be resisted either politically or from the invisible power brokers who have no shame using the very disadvantaged people to fight against any good course. Some of the landlords in these slum are rich people, senior public servants who have made poverty their business. The bread and butter of some Non-Governmental Institutions comes out of slums and they too want sustained aid and business.*

Views from all individual stakeholders and focus group converged at one point: slums are not a symptom of a population explosion and demographic shift; they are the outcomes of policy failure at virtually all levels. These include the failure of land use policies, legal frameworks and service delivery systems as well as national, subnational and urban policies. The failure of governance must be addressed if continued expansion of slums is to be halted.

However, the situation could be changing. A senior Ministry of Lands official revealed,

> *We are changing the manual systems that we have used since 1890 when the first land register was created. This system had many loopholes that made land management difficult and subject to abuse leading to land grabbing from rightful owners and converting public land to informal settlements. As a result Lands offices for many years have become veritable dens of corruption. This however is changing as we build a well-defined property regime through the development of a National Spatial Data Infrastructure (NSDI). Already the National Spatial Plan (NSP) is in place and we have built a Computer-Based Land Information System that is supported by geographically referenced data sets. You realise that digitisation of the Ministry records and spatialisation of the country has been in public discourse for a long time but somehow resisted by the powers that be. I am sure that this time round there is no going back and with the NSDI and NSP, planning in urban areas will fall in place.*

Baud et al. (2013) succinctly explained why digitising and spatialising was important and an opportunity to coordinate and exchange knowledge in a more effective local governance. It generally enabled decision-based policymaking. The results have shown that, through mapping, they could understand local problems in Kibera and appropriately respond.

Even as we celebrate the gains from digitisation and spatialisation, results show that more needs to be done to bring hope and optimism to all stakeholders. Results show that, with such transparency, especially after the Ministry of Lands embraced the idea of digitisation and spatialisation, resources can be managed better and with improved governance. This may have never happened without the emergence of open source software, which has enabled ordinary citizens to empower themselves. The change in slums is occurring as citizens take up the role of building data infrastructures and revealing areas for policy interventions, as in Kibera.

While individual interviews did not reveal more than we know, the focus group brought out more details with explicit arguments for and against technology use in slum areas. Some of those opposing were people with deep interests using proxy arguments, such as foreigners using these technologies to incite Kenyans to violence. They have crystallised their arguments around the growing dislike of social media by some key politicians. Allies of this project, however, pointed out that the government through Ministry of Lands has embraced the idea of spatialisation.

In April 2018, the Law Society of Kenya sued the government for embracing technology before there is a legal framework in in place. This was seen as a frivolous delaying mechanism by interested parties fearing loss of revenue from the changes that will make it possible for the poor to afford homes when the government starts to build affordable housing on public land.

Whereas you may seldom get a straight response from government bureaucrats, Kenya has moved too far with technology to reverse the gains. They seemed ambivalent, but that could possibly be that they did not want to contradict top leadership. However, their continued support for digitisation and spatialisation as a strategy to create transparency in government was sufficient to conclude that the country is on its way to fully leverage technology and dealing with many problems coming with unplanned urbanisation. The verdict is that the government has been responsive to issues raised because of Map Kibera. Citizens now engage with government in a more informed manner. The private sector, however, has taken long to overlay their solutions onto Map Kibera. There are different initiatives on the e-commerce front.

Next Steps in Using Open Data to Improve Informal Settlements

By developing GIS mapping, the easy work is done. The difficult work of building new solutions over the maps is yet to be realised fully. There is a knowledge gap between the GIS infrastructure and solution providers. Currently, very few services leverage mapping for services beyond dealing with the government. The little that is happening includes GPS-mounted motorbike deliveries and some piloted e-commerce activity. More services, like water distribution, could also benefit. Some of the problems are structural. However, the percentage of Kenyans with a smartphone has reached 60% (GSMA, 2017). That number is lower in slum areas. Yet, many solutions will require a basic smart phone. There is a need for sensitisation to help residents understand the potential presented them with the mapping exercise. There is a need for policy intervention, especially in the removal of taxes from smart devices, to enable a greater number of people to access and participate in the emerging digital economy.

Conclusion

This chapter demonstrated how use of ICTs and GIS can be used to address growing disparities in urban centres, and that open spatial maps can bring positive change by prompting policy makers to respond. It is not, however, an easy exercise, as there are those who still resist change, requiring the efforts of many allies to prevent opponents from hindering progress. Technology is moving the world from the past. However, policy relies on many factors other than empirical evidence (i.e. observations, benchmarking, instinct or personal experience), to an emerging, resilient data-driven policymaking process. ICT improves not only governance but also productivity, including better resource utilisation, as pointed out in the study findings. It brings many benefits, including better accounting of the citizenry and creates greater opportunities for more socially oriented enterprises and enhanced inclusivity for the most marginalised urban communities. Any form of resistance to the introduction of emerging technologies must be overcome by the allies of change, and the government should support the full development of the NSDI, which will provide the crucial foundation for the development of new GIS applications, particularly the emerging mobile GIS/GPS we now use in difficult areas without addressing systems for efficient e-commerce enterprises.

Appendix I

Questionnaire

1. Respondent Consent

   - Recently some young men asked me to officiate the launch of Map Kibera, which I suspect you know about. I was intrigued by their work and I have decided to look into the project in a more detailed way. My goal in this interview is to capture this important exercise and hope someone else in a similar situation will utilise the knowledge of using GIS Mapping to improve productivity of those living in informal settlements. Please let me know if you are in a position to help me understand, first your role in it and its usefulness to the people. You can withdraw anytime you feel that it is not relevant to your aspirations towards informal settlement. I also want your consent to quote you or else I will keep your contribution anonymous.

2. What role if any did your organisation play in the development of Map Kibera?
3. What direct or indirect assistance did you provide to the project?
4. Did you participate in the process of gathering data for the project? If so, please explain in detail what you did.
5. If you participated in the development of the project, who were your collaborators?
6. When gathering the data, what were key areas of focus?
7. Did you have any challenges? If so, what were they?
8. Do you see any potential opportunities that the project presents?
9. Do you know any other person or organisation that you can recommend to us to talk to?

10. Who financed Map Kibera project?
11. How will this project impact the people of Kibera?
12. What are the take away lessons from this project?
13. Do you have any question for me?
14. Please once more, let me know if you have any concerns to this interview.

Focus Group Discussion Guide

The aims of this focus group (I have selected four of you on the basis of your earlier position taken while conducting individual interviews to anonymously discuss freely on the opposing views) is to dive deeper into Map Kibera Project and discern your differences and the issues that matter to help me to validate earlier information and make generalisations that can be helpful in replicating this study elsewhere. Once more you are free to withdraw from this discussion if you are not comfortable with it.

1. For all intent and purpose, this project could be very useful to the people of Kibera. What in your view is the root cause of the resistance to its implementation?
2. How can we overcome these resistances and enable the residents to benefit from its applications?
3. How do we move forward?

Level 3 Interviews

The purpose of visiting your office today is to ask you to help me understand your role if any with respect to the ongoing digitisation of Kibera slums. I must add that you are at liberty to withdraw from this interview. I have two simple questions with respect to this matter.

I want to know if you are aware of Map Kibera and how it integrates to your wider plans in the city.

Could you share any materials, plans and documents with respect to city planning? I will appreciate it if you can provide any specific plans you have for Kibera.

Appendix II

List of Participants

Table 6.1 List of participants

Name	Role in Kibera Slums	Contribution to Map Kibera
Umande Trust	Water and Sanitation (W&S)	Identification of W&S sites
Carolina for Kibera[a]	Sanitation and health	Partner in the project
Pamoja FM[a]	Broadcast	Communication
Tunapanda Institute	Helping youths to code	Mapathons
Amani Kibera	Peace initiatives	Security mapping
District Peace Committee	Peace initiatives	Identification of safe and unsafe areas
Human Heeds Project	Peace W&S	Collaboration in W&S
Octopizzo Foundation	Peace initiatives	Mapping peace collaborations
Mchanganyiko Women Group	Education	Mapping education
Polycom	Gender	Mapping Gender Violence
Safaricom Foundation	Donor support in many projects	Indirect assistance
Global Health Provider[b]	Health	Supportive but protocol could not permit
Education provider[b]	Education	Neutral
International Academy[c]	Education	Neutral
Transport SACCO[b]	Transportation	Felt it was not relevant
Legal Aid NGO[b]	Legal	Neutral
National Government Administration	Security/peace	Ambivalent
International Development Agency[b]	Donor support in many projects	Indirect support
Kibera Youth Development	Community development	Volunteered services in data collection
Nairobi County Administration	Planning and provision of services	Played no role
Inter-Religious Council	Religious activities	Neutral
Kenya Youth for Christ[a]	Religion/empowerment	Some youth played some role
GroundTruth	Founded the project	Key player
International Sports[b]	Sports	Helped identify some facilities
Kibera Muungano Food vendors Association[a]	Food	Neutral

[a]Interview schedule did not work or withdrew
[b]Chose to remain anonymous
[c]Anonymous and withdrew from the interview

References

African Development Bank. (2012). Inclusive growth. Championing inclusive growth across Africa. A blog by the former Chief Economist and Vice-President. Retrieved May 19, 2018, from https://www.afdb.org/en/blogs/afdb-championing-inclusive-growth-across-africa/post/urbanization-in-africa-10143/.

Amer, A. S., Marcio, C., Bryce, Q., & Philip, S. (2016). Demographic change and development: Looking at challenges and opportunities through a new typology. *Policy Research Working Paper, No. 7893*. World Bank, Washington, DC. © World Bank. Retrieved from https://openknowledge.worldbank.org/handle/10986/25695. License: CC BY 3.0 IGO.

Angeles, G., Lance, P., Barden-O'Fallon, J., Islam, N., Mahbub, A. Q. M., & Nazem, N. I. (2009). The 2005 census and mapping of slums in Bangladesh: Design, select results and application. *International Journal of Health Geographics, 8*(1), 32.

Baud, I., Scott, D., Pfeffer, K., Sydenstricker-Neto, J., & Denis, E. (2013). Spatial knowledge management in urban local: Emerging issues in ICT-GIS-based systems in India, Brazil, South Africa and Peru. *Information, Communication & Society, 16*, 258–285.

Bautes, N., Dupont, V., & Landy, F. (2013). Acting from the slums: Questioning social movement and resistance. In M. Saglio-Yatzimirsky & F. Landy (Eds.), *Megacity slums: Social exclusion, space and urban policies in Brazil and India*. Imperial College Press.

Breen, R. A. (2006). A practical guide to focus-group research. *Journal of Geography in Higher Education, 30*, 463–475.

Cohen, L., & Manion, L. (2000). *Research methods in education* (5th ed., p. 254). Routledge.

Das, S., Choudhury, M. R., Shobhana, B., & Vaghela, B. (2014). Slum redevelopment strategy using GIS based multi-criteria system: A case study of Rajkot, Gujarat, India. *World Journal of Civil Engineering and Construction Technology, 1*(1), 12–41.

Giusti de Perez, R., & Perez, R. (2008). *Analyzing urban poverty: GIS for the developing world*. Retrieved from http://gis.esri.com/esripress/shared/images/139/URBAN_ch01.pdf.

Given, L. M. (2008). *The Sage encyclopaedia of qualitative research methods*. Sage Publications.

GoK. (2013). *National slums and upgrading programme*. Government Printers.

GSMA. (2017). White Paper 2017: Trends from the Kenyan Smartphone and E-Commerce Industry.

Hagen, E. (2015). Mapping the slums TEDxGateway. Mapping the slums. | Erica Hagen | TEDxGateway.

Hagen, E. (2017). Open mapping from the ground up: Learning from Map Kibera. Retrieved from https://opendocs.ids.ac.uk/opendocs/bitstream/handle/123456789/13244/RReport_MapKibera_Online.pdf

Hendrick-Wong, Y. & Angelopulo, G. (2014). 2014 MasterCard African Cities Growth Index: Understanding Inclusive Urbanization. Johannesburg: MasterCard. [Online], retrieved July 2019 from http://newsroom.mastercard.com/mea/files/2014/06/African-Cities-Growth-Index-2014.pdf.

History of Urban Planning of Nairobi. (1927). Retrieved May 19, 2018, from http://www.studio-basel.com/assets/files/029_NRB_ATLAS_11_planning_dr.pdf.

Karanja, M. (2010). Myth shattered: Kibera numbers fail to add up. *Daily Nation.*

Kingston, R. (2007). Public participation in local policy decision-making: The role of web-based mapping. *The Cartographic Journal, 44*, 138–144.

Marshall, M. N. (1996). The key informant techniques. *Family Practice, 13*, 92–97.

McCall, M. K., & Dunn, C. E. (2012). Geo-information tools for participatory spatial planning: Fulfilling the criteria for 'good governance. *Geoforum, 43*, 81–94.

McFarlane, C. (2011). *Learning the city, knowledge and translocal assemblage.* Sussex: Wiley Blackwell.

Mutisya, E., & Yarime, M. (2011). Understanding the grassroots dynamics of slums in Nairobi: The dilemma of Kibera informal settlements. *International Transaction Journal of Engineering, Management, & Applied Sciences & Technologies, 2*, 197–213.

Nairobi City Water and Sewerage Company Limited (2014–2019) Strategic Plan. Retrieved May 19, 2018, from https://www.nairobiwater.co.ke/images/strategic_plan/NCWSC_2014-15_to_2018-19_Strategic_Plan.pdf.

Nairobi Integrated Urban Development Master Plan for the City of Nairobi in the Republic of Kenya. (2014). Retrieved May 19, 2018, from http://citymasterplan.nairobi.go.ke/index.php/downloads/cat_view/4-niuplan-studies/23-draught-masterplan.

Nyamori, M. (2016). Title for Langata primary "still in hands" of land grabbers. *The Standard News Paper*. Retrieved May 19, 2018, from https://www.standardmedia.co.ke/article/2000223497/title-for-langata-primary-still-in-hands-of-land-grabbers

Philippines Statistical Authority. (2018). Use of GIS/GPS in support of the Philippines 2020 Census of Population and Housing. Regional Workshop on the Use of Electronic Data Collection Technology in Population and Housing Censuses UN ESCAP. Bangkok.

Pile, S., & Keith, M. (1997). *Geographies of resistance*. London: Routledge.

Sabina Yeasmin, S., & Rahman, K. F. (2012). Triangulation' research method as the tool of social science research. *BUP Journal, 1*, 154–163.

Thomas, D. R. (2006). A general inductive approach for analysing qualitative evaluation data. *American Journal of Evaluation, 27*, 237–246.

UN-Habitat. (2016). *World cities report: Urbanization and development emerging futures*. United Nations Human Settlements Programme.

UN-Habitat. (2003a). *The challenge of slums: Global report on human settlements*. UK and USA: United Nations Human Settlements Programme.

UN-Habitat. (2003b). *Guide to monitoring target 11: Improving the lives of 100 million slum dwellers*. Nairobi, Kenya: United Nations Human Settlements Programme.

UN-Habitat. (2006/2007). *State of the world's cities 2006/2007*. Nairobi, Kenya. Retrieved June 15, 2019, from https://unhabitat.org/books/state-of-the-worlds-cities-20062007/.

Wafula, N. (2011). Bridging the urban and rural divide: The test case for Kenya. In *FIG Working Week 2011: Bridging the gap between cultures*, Marrakech, Morocco, 18–22 May 2011. Retrieved June 15, 2019, from https://www.fig.net/resources/proceedings/fig_proceedings/fig2011/papers/ts01d/ts01d_nabutola_5253.pdf.

Wanjiru, M. W., & Matsubara, K. (2017). Slum toponymy in Nairobi, Kenya. A case study analysis of Kibera, Mathare and Mukuru. *Urban and Regional Planning Review, 4*, 21–44.

World Bank. (2015). Population living in slums (% of urban population). Retrieved from https://data.worldbank.org/indicator/EN.POP.SLUM.UR.ZS.

Yin, R. K. (1984). *Case study research: Design and methods*. Beverly Hills, CA: Sage Publications.

Yin, R. K. (2003). *Case study research: Design and methods*. Thousand Oaks, CA: Sage Publications.

Zwanenberg, R. (2008). History and theory of urban poverty in Nairobi. The problem of slum development. I.D.S Discussion Paper No.139, University of Nairobi.

Part II

Knowledge Ecosystems and Resilience

7

Mapping Climate Vulnerability with Open Data: A Dashboard for Place-Based Action

Scott Hawken, Komali Yenneti, and Carole Bodilis

Note regarding data references: The editors encourage all authors to deposit their data sets in a data repository or data store—and cite and link to the dataset in your chapter. If your dataset is not currently published, the editors are offering the option to add the dataset(s) to CityData: https://citydata.be.unsw.edu.au. This may assist you if you'd like to publish your data with a digital object identifier—to give it bibliographic identity. You will be referenced when this dataset is used. If you'd like to have your data hosted here, please contact the editors to submit a proposal.

S. Hawken (✉)
Urban Development and Design, Faculty of the Built Environment, University of New South Wales, Sydney, NSW, Australia
e-mail: s.hawken@unsw.edu.au

K. Yenneti
UNSW Built Environment, University of New South Wales, Sydney, NSW, Australia
e-mail: k.yenneti@unsw.edu.au

C. Bodilis
University of Aveiro, Aveiro, Portugal

S. Hawken et al. (eds.), *Open Cities | Open Data*,
https://doi.org/10.1007/978-981-13-6605-5_7

Abbreviations

ABS	Australian Bureau of Statistics
BoM	Bureau of Meteorology
GHG	Greenhouse gas emissions
GSC	Greater Sydney Commission
IPCC	Intergovernmental Panel on Climate Change
NSW	New South Wales
OEH	Office for Environment and Heritage
OLG	Office of Local Governments
USA	United States of America
UNFCCC	United Nations Framework Convention on Climate Change

Highlights The study emphasises the relationship between Open Data and climate change in cities.

- A heat vulnerability index using public data is developed and demonstrated for Greater Sydney.
- Heat vulnerability varies across Greater Sydney due to differentiated socioeconomic and geographical conditions.
- An Open Data-based vulnerability index can support place-based vulnerability planning and management.

Introduction: Open Data, Climate Action and Cities

Our understanding of climate change has been influenced by data like few other global phenomena. The global consensus on climate change has been established using scientific models integrating big data long before the term big data became popular. Big data on temperature, atmospheric carbon dioxide levels, glacier melt, land cover changes and coral dieback has fed scientific models that formed the evidence base for the Intergovernmental Panel on Climate Change (IPCC). The scientific data foundations established by the IPCC (IPCC, 2015) are integral to the decisions that have been made at the global climate agreements from the Kyoto Protocol of 1992 up until the recent Paris Agreement. The agreements that have been reached have called for evidence-based action and emphasised Open Data for meeting climate goals and ensuring resilient communities. However, initiatives leveraging Open Data for climate action have conventionally focused on supranational and nation states rather than urban areas. A new paradigm of city-level Open Data access is needed to ensure that societies can moderate vulnerability to climate change and increase resilience, as more than 50% of the current world's population lives in urban areas and is projected to rapidly increase in number.

This chapter reviews the relationship between Open Data and the climate change movement in cities. Different sources of Open Data are considered for their relevance in empowering both global and local action with specific reference to cities as 'centres of action'. In particular, this research involves the creation of a new heat vulnerability index using accessible public data. We use Greater Sydney's vulnerability to heat waves as a case study to demonstrate the index and its relevance for vulnerability management and resilience building. The index is presented through online mapping and communication technologies designed to promote engagement and action.

Open Data and Climate Change as Global Challenges and Movements

The first wave of climate science has focused on the measurement of climate and biophysical processes. The science for such challenges is well accepted as being data intensive and empirical in nature. However, over the last decade it has become clear that the science that must inform the strategies and approaches to addressing climate change must draw from a markedly different range of sources. Climate change adaptation and the development of global climate resilience require diverse data sets describing the health of populations, demographics, economic investment, infrastructure, political decision-making and public opinion and sentiment. This second wave of climate science has its basis in the social sciences and complements the first wave (Pidgeon & Fischhoff, 2011).

Both the physical and social sciences of climate change have the potential to be transformed by the Open Data movement (Boulton, Rawlins, Vallance, & Walport, 2011). Global scientific consensus on anthropogenic climate change has motivated the world to implement global Open Data policies. It is evident that because global climate change is a profound crisis with drastic worldwide consequences, the data sources for enabling and motivating climate action must be open, transparent and useable so that they can empower different stakeholders around the world in enhancing their efforts to combat climate change at different geographical scales and jurisdiction levels. Climate science must focus more on climate action, and Open Data is essential to achieving this.

The Open Data movement is a worldwide undertaking that can help transcend political rivalry and stimulate collaboration between government, scientific, private sectors and citizens. Open Data is data that is available for anyone to use without restrictions (Opengovdata, 2018; Vetrò et al., 2016). Various groups are using Open Data as a critical resource to tackle climate change. End users include local governments, entrepreneurs, transparency and civil rights advocates, civic leaders, media, businesses, research scientists and the average citizen who can see the opportunities for Open Data to inform or empower individual climate action (Goldstein, Dyson, & Nemani, 2013).

Better access to quality and timely Open Data has the potential to stimulate action on climate change in a number of ways. One of the principles of Open Data as outlined by Emer Coleman, the architect of the London Data Store, is engagement. One of the biggest challenges of public sector reforms is not simply making services more reliable or efficient but to shape them so that they are more communal and inclusive (Goldstein et al., 2013). Addressing this challenge requires initiatives that invite people to participate and engage with the information on climate change in a way that promotes scaling up efforts and replication. Open Data can contribute to climate change action in important ways, including:

- *Make both public and private sector impacts on climate more transparent and accountable.* Open Data can contribute to increased scrutiny on public and private operations. For example, an Open Data programme can result in increased focus on polluting industries, which can further lead to public pressure and new industry pollution standards (World Bank Group, 2015).
- *Stimulate innovation for rapid societal and technological change.* Open Data is a source of potential value for new enterprises and initiatives that can develop products and services with lower carbon footprints or promote lower carbon behaviours related to energy use.
- *Improve public communication,* and the communication of scientific data, approaches and findings.
- *Help politicians communicate and introduce new policies* by making reliable and transparent data available to support new climate sensitive initiatives.

The Greater Sydney-focused case study presented in this chapter attempts to achieve all four of these objectives. The following section introduces the relevance of Open Data initiatives for city-level climate action. This emerging area of knowledge is essential for operationalizing high-level global policy.

Open Data for City-Level Climate Action: Identifying Vulnerabilities

There have been serious obstacles to urban climate action to date. One of the major difficulties in implementing climate-based city initiatives is a disconnect between global and local initiatives. For much of the past several decades, science and action on climate change has focused on the global, supranational, national and regional levels. This is reflected in the emphasis on global- and continental- or national-level biophysical datasets such as continental ecosystems or agriculture-based climate initiatives. It has been largely assumed that the nation state is the most appropriate level to address climate change. However, country- or nation-focused negotiations have been largely ineffective, firstly in negotiating climate agreements, and then, acting upon them. There have also been other barriers to urban climate action.

Following from the first barrier, outlined above, is the assumed adaptability of cities as agile sociotechnical and resilient systems. Researchers and policymakers have traditionally assumed that cities of wealthy developed nations would be resilient to climate change impacts. This viewpoint is no longer tenable with climate-related disasters such as Hurricane Katrina's lasting impact on New Orleans and Hurricane Sandy's devastation in New York in 2005 (Comfort, 2006), highlighting both the vulnerability of the cities and the inadequacy of current urban resilience when the affected cities needed to bounce back (Rosenzweig, Solecki, Hammer, & Mehrotra, 2010).

The third barrier is the failure of imagination to see cities as key actors for climate change mitigation and adaptation. There is a pervasive belief that as cities are responsible for the majority of carbon emissions they cannot be a part of the solution. For several decades cities have been viewed as climate culprits rather than as agents of climate adaptation and mitigation. However, some scholars have challenged the veracity of this assertion, arguing that the urban phenomenon and carbon intensive activities are much more nuanced than a simple division between rural and urban sources of carbon (Satterthwaite, 2008). Scholars have also suggested that we need to be careful in terms of how we measure and attribute data to urban areas to avoid a futile blame game which will neutralise much of the momentum and action that cities are so good at mobilising (Castán Broto & Bulkeley, 2013; Satterthwaite, 2008).

It is this idea of cities as agents of political, environmental and economic change that is so important to the current 'global wave of local initiatives'. Cities, governments and citizenry have realised that if climate impacts are to be addressed, then cities have to act and lead. Much of the current wave of global disillusionment and intransigence on climate can be linked to dysfunctional and divisive federal politics. Wedged by partisan debates and political acrimony, federal politics has seen climate action come to a standing halt in nations such as Australia and the USA. In contrast to the global climate policy stagnation representative of many national policy landscapes, cities are becoming dynamic centres of dramatic change. Katz and Bradley (2013) have emphasised that such dynamism is partly due to the diverse populations that inhabit them. This realisation has also emerged from the revelation that cities are on their own when it comes to climate adaptation and climate resilience. If climate adaptation and mitigation is to be initiated and progressed, it is apparent that this must occur at the urban level.

Cities, home to majority of the world's population and the most dynamic social regions on the planet, have been leading the changes and initiatives necessary for adapting to climate change and mitigating it (Castán Broto & Bulkeley, 2013). In contrast to the uneasy alliances and mistrust between nations and climate agreements, cities have networked to form collaborative organisations for climate resilience and action (Rosenzweig et al., 2010). Organisations such as C40 consist of a global network of cities that promote climate action and share knowledge on innovative approaches to climate adaptation and mitigation. At the global level, the recognition of the importance of cities in addressing the gap between science and action has come late. In 2018 the first IPCC conference focusing on cities was held to establish a global research and policy agenda on cities and climate change (IPCC, 2018).

Risks for cities are varied and relate to a range of factors. For example, cities on the coast may face sea-level rise and storm surges (Leichenko & Thomas, 2012), whilst cities in dry regions will face water shortages (Gober, 2010; Leichenko & Thomas, 2012). Others in hot regions will face heat stress (Nitschke & Tucker, 2007). The diversity of factors and impacts has motivated the formation of urban networks to implement local-based actions and thereby improve resilience and mitigate local climate change.

Among many actions targeting different sectors, cities must actively pursue vulnerability assessment to determine how vulnerable they are to climate change and then act to improve resilience and adaptive capacity. Vulnerability is a complex concept which is altered by both geographic context and scale and is influenced by a range of factors including economy, land use, power, politics, extreme climatic events, demographics, time and duration (Yenneti, Tripathi, Wei, Chen, & Joshi, 2016). The IPCC Fourth Assessment Report has defined vulnerability as 'the degree to which a system is susceptible to and unable to cope with adverse impacts of Climate Change' (Adger, 2007). More precisely, vulnerability depends on the level of exposure of the system to climate change, on the system's sensitivity (the degree to which the system will be affected by climate change) and on its capacity to adapt. While an increased exposure and sensitivity result in a higher vulnerability, better adaptive capacity contributes to a lower one (see Fig. 7.2).

Due to the diversity and complexity of the concept of vulnerability, Open Data has an important role to play to permit and promote access and synthesis of the multiple sources of information required for its definition (Janssen, Charalabidis, & Zuiderwijk, 2012). It is an evolving concept which is defined more through conceptual approaches rather than any discrete formula. A range of Open Data around the world can be useful for defining various forms of vulnerability at various scales. Some of these sources are listed in Table 7.1.

The following section addresses a specific type of vulnerability associated with climate change: that is, heat vulnerability. The section provides an overview of current approaches to address urban heat vulnerability and takes Sydney as a case study to emphasise the need for more in-depth and enduring strategies and policies informed by evidence within the city.

Sydney's Approach to Heat Vulnerability: Going Beyond Current Event-Based Policies

There is global consensus that climate impacts on vulnerable people should be addressed and not overlooked. This is enshrined in the recently signed Paris Agreement (UNFCCC, 2015), which emphasises the vulnerability of developing nations. However, the idea that there are vulnerable people distributed throughout wealthy developed nations and cities, who also stand to suffer through climate change, has not been well acknowledged. Part of the reason for this is a lack of policy focus and the other reason is that there have been limited urban data sets and indexes designed to reveal where vulnerable people reside and the urban geography of climate vulnerability.

This section of the chapter looks at the specific type of vulnerability associated with heat impacts in Greater Sydney. This is known as 'heat

Table 7.1 Open Data for climate change vulnerability assessment and adaptation from global to urban scales

Scale of Open Data for climate change vulnerability definition	Examples of type of data required	Examples of Open Data	
		Source	Type of data available
At global scale, the issue is related to identifying vulnerable global regions and countries to particular climate-related risks and hazards based on the current and past global climate variability	Country-level current and past data on extreme climate-related natural disasters and sea-level rise, economic and population patterns, natural resources availability, government resources, physical and social infrastructure availability	International research organisations and development institutions: IPCC—http://www.ipcc-data.org/ The World Bank— • http://sdwebx.worldbank.org/climateportal/ • https://data.worldbank.org/ EM-DAT—http://www.emdat.be/ NASA—http://beta.sedac.ciesin.columbia.edu/theme/climate/data/sets/browse	Temporal global- and national-level data on climate estimations from observations, global climate models, socioeconomic baseline and future scenarios Temporal national- and subglobal regional-level data on historical and projected climate, socioeconomic, energy, health, natural resources, agriculture, infrastructure, development and government resources Temporal data on occurrence and effects of different types of disasters from 1900 to present day. Spatial and/or temporal data on global urban heat island (UHI), global climate, land cover and land surface temperatures, socioeconomic, health and environmental sustainability indicators.

(*continued*)

Table 7.1 (continued)

Scale of Open Data for climate change vulnerability definition	Examples of type of data required	Examples of Open Data	
		Source	Type of data available
At national and subnational scales, vulnerability analysis involves disaggregated administrative units or small geography areas to capture fine spatial heterogeneity in vulnerability	National and subnational level current and past data on socioeconomic and population patterns, energy and infrastructure, natural resources and green spaces, government resource status	Meteorological departments—http://www.bom.gov.au/ (e.g., Australia)	Spatial and/or temporal data on climate and past weather, floods and hydrometeorology, coastal waters and ocean activities
		Country Census—http://www.abs.gov.au/census (e.g., Australia)	Spatial and/or temporal data on socioeconomics, housing, infrastructure
		Household surveys (living standards, demographic health, multiple indicators)—http://www.abs.gov.au/ (e.g., Australia)	

(*continued*)

Table 7.1 (continued)

Scale of Open Data for climate change vulnerability definition	Examples of type of data required	Examples of Open Data	
		Source	Type of data available
At a regional level, vulnerability assessment is used as a starting point for developing localised adaptation measures and uses aggregated regional data (district level, subregional level and could include both rural and urban areas). This can be effective as it can enable sharing of resources, collective actions and acknowledgement of regional-level social and ecological systems at risk	District level and subregional level current and past (and if available future) data on specific effects of local climate change on regional systems, human, social, natural, biophysical and financial characteristics and regional government service delivery	State/Provincial governments —http://climatechange.environment.nsw.gov.au// (e.g., New South Wales, Australia) Country Census —http://www.abs.gov.au/census (e.g., Australia) Natural resources and environment —http://data.environment.nsw.gov.au/ (e.g., New South Wales, Australia) Planning and Environment —http://www.planning.nsw.gov.au/Research-and-Demography/Demography/Population-projections (e.g., New South Wales, Australia)	Regional spatial and temporal data on climate baselines and projections/models Temporal current data on socioeconomic profile, housing, infrastructure Spatial current data on vegetation distribution, green cover and other biophysical resources Temporal future projections data on socioeconomic profiles

(continued)

Table 7.1 (continued)

Scale of Open Data for climate change vulnerability definition	Examples of type of data required	Examples of Open Data	
		Source	Type of data available
At local level, vulnerability assessment involves fine grain local assessment of climate risks and cross-sectoral vulnerability to such risks and help local governments (urban, small town or community) to improve community resilience and minimise local impacts	Local government level current and past (and, if available, future) data on: climate, socioeconomic profiles, health, natural resources and government finances,	Meteorological departments—http://www.bom.gov.au/ (e.g., Australia) Country Census—http://www.abs.gov.au/census (e.g., Australia) Planning and Environment—http://www.planning.nsw.gov.au/Research-and-Demography/Demography/Population-projections (e.g., New South Wales, Australia)	Current and past local weather data (by weather stations) Temporal current data on socioeconomic profile, housing, infrastructure Temporal future projections data on socioeconomic profiles

Notes:

1. A number of other indicators such as health and socioeconomic development (worker productivity, economy, etc.) also play important roles in vulnerability assessment. As such, relevant data on the impacts of climate risks on health and socioeconomic development should be made available by the national, subnational, regional and local governments.
2. Qualitative data play a significant role in regional and local vulnerability assessment. Regional and local governments should conduct public participation workshops, household surveys and stakeholder engagement workshops to gather relevant data and be included in any vulnerability assessments. Such data should also be made open and accessible.

vulnerability'. In recent years, Sydney has experienced heat waves for extended periods. Extreme temperatures and more frequent and extended heat waves are further projected to increase in Sydney. According to Dean and Green (2018), the number of days over 35 °C is projected to increase from 3.1 days (the average between 1981 and 2001) to between 8.2 and 15 per year by 2090. Heat waves affect the city unevenly due to geographic, social and economic inequalities. For example, the sea breeze cooling the eastern parts of Sydney contrasts with the warmer conditions of the west. The topography of the basin prevents both pollution and warm air from escaping. Instead it is trapped and affects settlements, especially in Western Sydney (Dean & Green, 2018; Santamouris et al., 2017). There are also various social groups that are susceptible or sensitive to heat including the elderly and sufferers of particular health conditions such as diabetes.

Current approach to heat vulnerability in the Greater Sydney area and New South Wales (NSW) has focused on event-based action plans which aim to improve resilience by a series of protocols focused on heatwave events. These include:

1. NSW government's State Heatwave Subplan (NSW Government, 2018), which is triggered when the Bureau of Meteorology (BoM) announces a risk.
2. As a part of the 100 Resilient Cities initiative, a report on Resilient Sydney (City of Sydney, 2016b) has identified key shocks and stresses and a list of actions to heat resilience building in Sydney:

   - Develop and evaluate more integrated emergency response planning,
   - Identify groups that are most vulnerable to extreme events

3. The long term climate adaptation strategy of the local council of City of Sydney (City of Sydney, 2016a) has a Heatwave Response Plan which aims to:

   - Raise awareness and provide an understanding of heatwaves
   - Advocate for a heatwave and extreme event warning system.
   - Align with the NSW State Heatwave Subplan.

These approaches are focused primarily on the symptoms of heatwaves rather than the underlying sensitivities, exposure and enduring adaptive capacities. This emergency response approach is at odds with a climatic impact which will be lasting and is already very much part of Sydney's everyday experience. Although Sydney generally has a comfortable climate with mild winters and warm summers, it is increasingly subject to heatwaves. Official reports suggest that heatwaves will become hotter, last longer and occur more frequently in the future (Climate Council, 2017). In 2017, Sydney recorded the hottest summer in 157 years (BoM, 2017). Temperatures are forecast to reach 50 °C in the future (Lewis, King, & Mitchell, 2017).

The research conducted by the authors of this chapter sought to address the limitations of event-based policy by developing a tool for facilitating strategic planning decisions and for developing approaches that prevent or ameliorate heat-related impacts on the residents and landscapes of the city. It does this by using Open Data to generate a spatial index that communicates vulnerability in relation to sensitivity, adaptation and exposure. It therefore forms an evidence-based tool to assist in making decisions to best address metropolitan heat vulnerability. In scientific literature such a tool is referred to as a decision support system or a planning support system (Geertman and Stillwell, 2004; Pettit et al., 2018). This innovative approach contrasts with current event-based policies in several ways. Using accessible Open Data, our research

- adopts a fine grain approach that seeks to build up a picture of the variation in heat vulnerability across the whole metropolitan area;
- helps to identify places that are either heat-resistant or vulnerable so that governments and communities can act and address the vulnerabilities identified;
- identifies how the various components that contribute to vulnerability have changed over a four-year period;
- assesses the underlying place-based vulnerabilities that make heatwaves dangerous;
- promotes engagement through the development of a heat vulnerability planning support system that uses Open Data to generate a rich range of metrics for each place within metropolitan Sydney; and

- generates an open picture of vulnerability that can be easily explored and examined through a series of interactive and integrated maps. The heat vulnerability of each place can be understood through the hands-on visualisations, promoting engagement and motivational approach that invites the formulation of projects and positive adaptation initiatives.

The following section explores the heat vulnerability planning support system in more detail and puts forward some principles that can be followed by other cities wishing to develop or adopt similar approaches. The experience within this project is complemented by insights from other heat vulnerability projects including one implemented in London.

Open Data and Climate Decision Support Systems for Cities

Since 2016, planning agencies in Sydney have increasingly adopted online visualisation and planning support tools. These have predominantly been to help communicate future development projects. Some agencies such as the Greater Sydney Commission (GSC), charged with coordinating the planning approach in Sydney, have initiated a preliminary data dashboard with some sustainability metrics. However, there is a conspicuous absence of data, visualisations, projects and tools relating to climate. Similarly, the GSC's complementary agency, Planning NSW, does not have an easily accessible, dedicated Open Data webpage but seems to be in a transitional phase with Open Data delivered through other agencies.

Information on climate change is typically difficult to access and understand for policymakers and the general public, as it is produced by scientists for scientists (Boulton et al., 2011). Many data sets, including remote sensing and economic datasets, are not freely available due to the cost in acquiring them and their value for commercial purposes and security issues. In contrast to these difficulties in accessing scientific climate data, there is a burgeoning Open Data movement that seeks to counter this. This research has therefore used Open Data approach to heat vulnerability, using data from open government sources and public-funded university resources to produce high-value open information in the form of easy-to-use research tool that do not require specialist skills to access or understand. The project used the three principles of the Open Data movement, transparency, participation and collaboration (Gonzalez-Zapata & Heeks, 2015), to guide the construction of the planning support system or planning tool.

A planning support system has been defined as a technological tool that presents geographic information to support planners accomplish specific tasks (Geertman, 2006). Open Data repositories such as EM-DAT (CRED, 2018) can help identify the climate-related disaster impacts on health, economy and environment as well as present insights into vulnerability. The tool described above uses a range of metrics constructed from Open Data to generate an index which allows the visualisation of areas that are most vulnerable to climate-related heat impacts.

A series of generic steps outlined in the previous literature (Tapia et al., 2017; Yenneti et al., 2016) were followed to create the Open Data information resources (Fig. 7.1). The specific heat vulnerability index was constructed using the definition in the Intergovernmental Panel on Climate Change (IPCC) fourth assessment report. Drawing on the IPCC recommendations, vulnerability in this research is defined as the propensity of the population of Greater Sydney to be adversely affected by urban heat (Adger, 2007). Based on this review and data availability, a set of indicators were selected to describe heat vulnerability in Sydney (Table 7.2). The data was then gathered from openly available databases and official government reports (Table 7.3). Metadata for the indicators was generated to establish the provenance of the data.

The overall index values were computed by averaging the three components of the vulnerability index. We opted to average the indicators because, as emphasised by previous scholarship (Yenneti et al., 2016), for advocacy purposes, a simple approach is preferable to more complex approaches that weighed variables differently. The individual indicator's attributes are normalised and rescaled to produce values between 0 and 1 using absolute, relative and reverse scaling methods as described in Table 7.2.

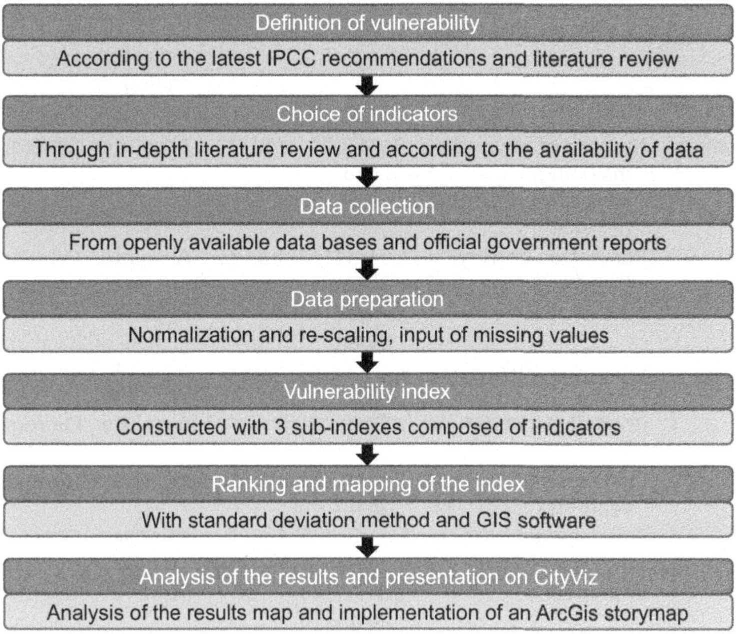

Fig. 7.1 Methodology and workflow to produce Open Data for the heat vulnerability index. Adapted from Tapia et al. (2017)

- $$\text{Absolute}\left(\text{for non percentage values}\right): \frac{\text{Value} - \min\left(\text{Values}\right)}{\max\left(\text{Values}\right)}$$
- $$\text{Relative}\left(\text{for percentage value}\right): \frac{\text{Value}}{\max\left(\text{Values}\right)}$$

$$\text{Reverse scaling}: \frac{1 - \text{Value}}{1 - \min\left(\text{Values}\right)}$$

An Open Data licence was selected to develop the visualisations; in this instance the Open Data Commons Attribution Licence. The range of indicators was then used to generate three sub-indexes that describe vulnerability. Each sub-index was calculated as the average of several indicators selected from previous studies and according to data availability.

Table 7.2 Indicators used to compute the heat vulnerability index

Category	Indicator	Data source (year)	Indicator scaling method	Openness
Exposure	Number of days when maximum temperature >38 °C per year	BoM (2011 and 2016)	Absolute	On request
	Population density (hab/km²)	Census (2011 and 2016)	Absolute	On request
	Poverty rate	Census (2011 and 2016)	Relative	On request
	Inequalities (Gini index)	Census (2011 and 2016)	Absolute	On request
	Road density (km/km²)	NSW Spatial Services (2016)	Absolute	On request
Sensitivity	People older than 65 years (%)	Census (2011 and 2016)	Relative	Open
	People younger than 4 years (%)	Census (2011 and 2016)	Relative	Open
	Women (%)	Census (2011 and 2016)	Relative	Open
	People living alone (%)	Census (2011 and 2016)	Relative	Open
	People needing care (%)	Census (2011 and 2016)	Relative	Open
	Indigenous people (%)	Census (2011 and 2016)	Relative	Open
	Multi-dwellings (%)	Census (2011 and 2016)	Relative	Open
Adaptive capacity	Households having access to internet (%)	Census (2011 and 2016)	Reverse scaling	Open
	Tree canopy (% of area)	NSW OEH (2011)	Reverse scaling	Open
	Parkland (% of area)	Mesh blocks data—ABS (2016)	Reverse scaling	On request
	Adults with a high school degree (%)	Census (2011 and 2016)	Reverse scaling	Open
	Water bodies (% of area)	NSW Spatial Services (2016)	Reverse scaling	On request
	Relevant local governments' expenses (% of total expenses)	OLG (2011 and 2016)	Reverse scaling	Open

Table 7.3 Open Data sources used in the construction of the heat vulnerability index

Data	Data source	Data format
Census, geographical data, park data	The Australian Bureau of Statistics (ABS)—a once-in-a-five-year census with availability of data packs at the mesh block level[a] to download without any restrictions	CSV, Esri Shapefile
Temperature data	The Bureau of Meteorology (BoM) for weather data through the Weather Station Directory[b]	CSV
Road data	NSW Spatial Services (SS) Spatial Information exchange for road data[c]	Esri Shapefile
Tree canopy	The NSW Office for Environment and Heritage (OEH) for tree canopy[d]	Raster
Expenses of LGA	The Office of Local Governments (OLG) for the data on expenses of local government[e]	PDF
Water bodies	NSW Spatial Services (SS) Spatial Information exchange[f]	Esri Shapefile

[a]http://www.abs.gov.au/websitedbs/D3310114.nsf/Home/2016%20DataPacks
[b]http://www.bom.gov.au/climate/data/stations/
[c]https://six.nsw.gov.au/
[d]https://data.nsw.gov.au/data/dataset/tree-canopy-2006
[e]https://www.olg.nsw.gov.au/public/my-local-council/yourcouncil-website
[f]https://six.nsw.gov.au/

The index was then ranked and mapped using the standard deviation method and GIS analytics. Finally, the index was visualised and uploaded to UNSW's Cityviz website using a range of available mapping and statistical software including CARTO, Tableau and ESRI storymap. ESRI storymap has inbuilt capacity to describe the collection and provide analytic and visualisation methods to users of the tool.

The majority of data used was freely available under a Creative Commons licence, while the remainder was available upon request. The processing of the data was made with Excel and ArcGIS, computing a value of vulnerability for each of the statistical areas using the method developed by Weis et al. (2016). This approach was selected for its place-based policy emphasis. In future the index can be updated according to the availability of new datasets or updating the existing datasets.

The index is an example of making sense of Open Data by putting it together visually, with a story to tell. It transforms opaque file formats as

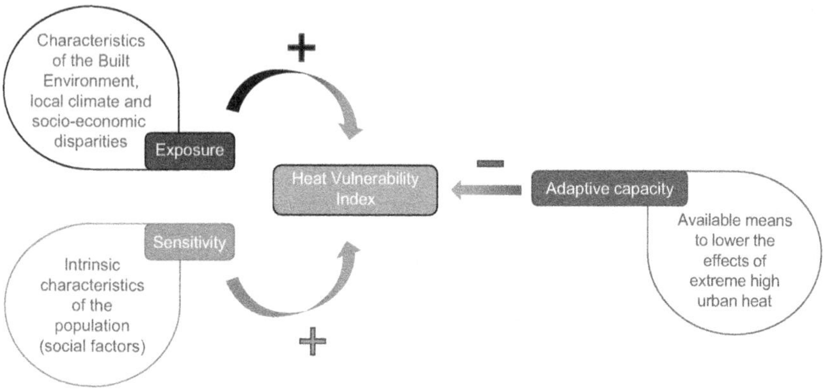

Fig. 7.2 The framework for heat vulnerability index. It is a function of exposure, sensitivity and adaptive capacity

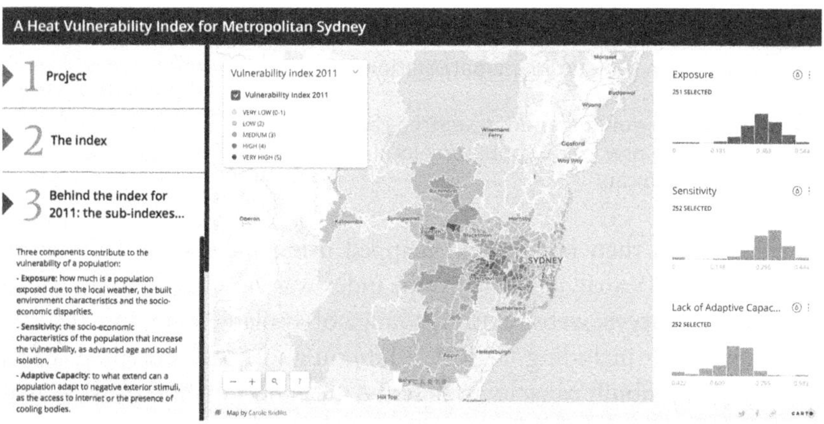

Fig. 7.3 The heat vulnerability planning support system as presented in CityViz. The index can be accessed at the link https://www.arcgis.com/apps/MapSeries/index.html?appid=dd7e39a138fd4449abe758914c6da801

CSV and PDF into meaningful user-friendly visualisations (Fig. 7.3). To accomplish this, an ArcGIS Story Map was used. The visualisation helps a user to visualise the index from an overall vulnerability analysis level up to individual indicator analysis. At each step, the user can play with the data and select to display the areas, for example, with the worst exposure or the best adaptive capacity. This provides transparency at every level.

The advantage of the use of Open Data for sustainable development indexes is well explained by Tiepolo and Bacci (2017)

> with just a few indicators, the [index] can easily be updated and allows central and local governments [...] to orient the measures for adaptation to CC to the most vulnerable sectors, to appreciate the improvements obtained year after year and to assess the impact and sustainability of local projects. (p.121)

Ideas for the usefulness of Open Data include:

- Open Data can be used by small businesses to foster innovation in climate change adaptation.
- Once successful Open Data stories are communicated, it opens the market for even more possibilities.
- Government can make use of it for policy development and climate adaptation strategies.

It is to be well-remembered that all index-based vulnerability assessments are subject to availability of data, choice of indicators and choice of calculation methods. The validation of the results often requires access to confidential, restricted data such as health and well-being data, which the researchers did not have access to. Bao, Li, and Yu (2015) note that to ensure the robustness of the heat vulnerability index, it should be validated with health data. Nevertheless, indexes such as this one contribute to the ongoing debates on Open Data in sustainable development initiatives in a number of ways. First, they provide easily apprehensible methods to identify hotspots and act where the population is more vulnerable. Second, it empowers researchers and provides a way forward for further studies, since the data can be easily reused, and more and improved data can be added to strengthen the index. Third, it provides a tool for decision-makers to address inequalities within urban areas. The online platform makes it possible to map the index, along with the individual indicators and sub-indexes used to build it. Finally, citizens themselves can access the Open Data using the planning support system, which provides evidence and focus for citizen-driven climate advocacy.

Conclusion: Place-Based Approaches to Heat Vulnerability with Open Data

The heat vulnerability index created for Greater Sydney aims to raise awareness about the heat impacts of climate change at the scale of the urban district. This Open Data approach can be strengthened in future by inclusion of 'closed' health data to validate the index. However, it is a myth that a dataset in itself can communicate to a diverse range of people (Janssen et al., 2012). For such Open Data approaches to be effective, it needs to be communicated to researchers, decision-makers and the public in an interactive and engaging way: in this case, by integrating it into a planning support system. When integrated with interactive digital tools, the index functions as a powerful decision support system to guide investment and allocation of resources for focused urban climate adaptation planning. The planning support system reveals not only hotspots in vulnerable areas such as Western Sydney and inner-city locations but also the specific mix of factors and metrics that contribute to this vulnerability.

The geographic nature of the urban heat vulnerability index from the years 2011 to 2016 provides evidence for climate action and place-based policies that build resilience rather than simply address impacts and heat events (Cutter et al., 2008). It demonstrates that even economically developed cities such as Sydney need to focus on climate adaptation for their most vulnerable people and that climate justice can be delivered through policies which build place-based resilience in relation to specific factors and metrics. The planning support system therefore breaks down the monolithic and fearful image of climate change and sets a foundation for a possible range of defensible and evidence-based policy actions.

Acknowledgements The authors would like to acknowledge the Joint European Master in Environmental Studies—Cities & Sustainability (JEMES CiSu) programme and the Smart Cities Cluster at the University of New South Wales who supported Carole Bodilis for a mobility internship during the second year of her Master's degree. The authors would also like to acknowledge the comments from the three anonymous reviewers, which helped to significantly improve the chapter.

References

Adger, W. N. (2007). *Climate change 2007: Impacts, adaptation and vulnerability: Summary for policymakers: Working Group II contribution to the Intergovernmental Panel on Climate Change Fourth Assessment Report*. Geneva: WMO, IPCC Secretariat.

Bao, J., Li, X., & Yu, C. (2015). The Construction and validation of the heat vulnerability index, a review. *International Journal of Environmental Research and Public Health, 12*(7), 7220–7234. https://doi.org/10.3390/ijerph120707220

BoM. (2017). Sydney in January 2017: The warmest month on record. Retrieved from http://www.bom.gov.au/climate/current/month/nsw/archive/201701.sydney.shtml#notes

Boulton, G., Rawlins, M., Vallance, P., & Walport, M. (2011). Science as a public enterprise: The case for open data. *The Lancet, 377*(9778), 1633–1635. https://doi.org/10.1016/S0140-6736(11)60647-8

Castán Broto, V., & Bulkeley, H. (2013). A survey of urban climate change experiments in 100 cities. *Global Environmental Change, 23*(1), 92–102. https://doi.org/10.1016/j.gloenvcha.2012.07.005

City of Sydney. (2016a). Adapting for climate change: A long term strategy for the City of Sydney (p. 72).

City of Sydney. (2016b). Resilient Sydney—Preliminary Resilience Assessment 2016 (p. 26): 100 Resilient Cities.

Climate Council. (2017). *2017: Another record-breaking year for heat and extreme weather*. Retrieved from Australia: https://www.climatecouncil.org.au/resources/2017-heat-report/

Comfort, L. (2006). Cities at risk: Hurricane Katrina and the drowning of New Orleans. *Urban Affairs Review, 41*, 501–516.

Cred. (2018). EM-DAT | The international disasters database. *The international disasters database*.

Cutter, S. L., Barnes, L., Berry, M., Burton, C., Evans, E., Tate, E., & Webb, J. (2008). A place-based model for understanding community resilience to natural disasters. *Global Environmental Change-Human and Policy Dimensions, 18*, 598–606.

Dean, A., & Green, D. (2018). Climate change, air pollution and human health in Sydney, Australia: A review of the literature. *Environmental Research Letters, 13*(053003), 1–15.

Geertman, S. (2006). Potentials for planning support: A planning-conceptual approach. *Environment and planning B: Planning and Design, 33*(6), 863–880.
Geertman, S., & Stillwell, J. (2004). Planning support systems: An inventory of current practice. *Computers, Environment and Urban Systems, 28*(4), 291–310.
Gober, P. (2010). Desert urbanization and the challenges of water sustainability. *Current Opinion in Environmental Sustainability, 2*(3), 144–150.
Goldstein, B., Dyson, L., & Nemani, A. (2013). *Beyond transparency: Open Data and the future of civic innovation.* San Francisco, CA: Code for America Press.
Gonzalez-Zapata, F., & Heeks, R. (2015). The multiple meanings of open government data: Understanding different stakeholders and their perspectives. *Government Information Quarterly, 32*(4), 441–452. https://doi.org/10.1016/j.giq.2015.09.001
IPCC. (2015). *Climate change 2014: Synthesis Report. Contribution of Working Groups I, II and III to the Fifth Assessment Report of the Intergovernmental Panel on Climate Change.* Geneva, Switzerland: IPCC.
IPCC. (2018). Cities IPCC—Cities and climate change science conference [WWW Document]. Retrieved June 14, 2019, from https://citiesipcc.org/.
Janssen, M., Charalabidis, Y., & Zuiderwijk, A. (2012). Benefits, adoption barriers and myths of Open Data and open government. *Information Systems Management, 29*(4), 258–268. https://doi.org/10.1080/10580530.2012.716740
Katz, B., & Bradley, J. (2013). *The metropolitan revolution how cities and metros are fixing our broken politics and fragile economy.* Washington, DC: Brookings Institution Press.
Leichenko, R. M., & Thomas, A. (2012). Coastal cities and regions in a changing climate: Economic impacts, risks and vulnerabilities. *Geography Compass, 6*(6), 327–339.
Lewis, S. C., King, A. D., & Mitchell, D. M. (2017). Australia's unprecedented future temperature extremes under Paris limits to warming. *Geophysical Research Letters, 44*(19), 9947–9956.
Nitschke, M., & Tucker, G. (2007). Morbidity and mortality during heatwaves in metropolitan Adelaide. *Medical Journal of Australia, 187*, 662–665.
NSW Government. (2018). *State Heatwave Subplan—A Subplan of the NSW State Emergency Management Plan.* Retrieved from Surry Hills, Sydney https://www.emergency.nsw.gov.au/Documents/plans/sub-plans/SubPlan_HeatWave.pdf
Opengovdata. (2018). The annotated 8 Principles of Open Government Data. Retrieved from https://opengovdata.org/

Pettit, C., Bakelmun, A., Lieske, S. N., Glackin, S., Hargroves, K., Thomson, G., … Newman, P. (2018). Planning support systems for smart cities. *City, Culture and Society, 12*, 13–24. https://doi.org/10.1016/j.ccs.2017.10.002

Pidgeon, N., & Fischhoff, B. (2011). The role of social and decision sciences in communicating uncertain climate risks. *Nature Climate Change, 1*(1), 35–41. https://doi.org/10.1038/nclimate1080

Rosenzweig, C., Solecki, W., Hammer, S. A., & Mehrotra, S. (2010). Cities lead the way in climate-change action. *Nature, 467*(7318), 909–911. http://dx.doi.org.wwwproxy1.library.unsw.edu.au/10.1038/467909a

Santamouris, M., Haddad, S., Fiorito, F., Osmond, P., Ding, L., Prasad, D., … Wang, R. (2017). Urban heat island and overheating characteristics in Sydney, Australia. An analysis of multiyear measurements. *Sustainability, 9*(5), 712.

Satterthwaite, D. (2008). Cities' contribution to global warming: Notes on the allocation of greenhouse gas emissions. *Environment and Urbanization, 20*(2), 539–549. https://doi.org/10.1177/0956247808096127

Tapia, C., Feliu, E., Mendizabal, M., Antonio, J., Fernández, J. G., Laburu, T., & Lejarazu, A. (2017). Profiling urban vulnerabilities to climate change: An indicator-based vulnerability assessment for European cities. *Ecological Indicators, 78*, 142–155. https://doi.org/10.1016/j.ecolind.2017.02.040

Tiepolo, M., & Bacci, M. (2017). Tracking climate change vulnerability at municipal level in rural Haiti using open data. In *Renewing local planning to face climate change in the tropics* (pp. 103–131). Springer.

UNFCCC. (2015). *Climate change 2014. Synthesis report: Contribution of Working Groups I, II and III to the Fifth Assessment Report of the Intergovernmental Panel on Climate Change*. IPCC.

Vetrò, A., Canova, L., Torchiano, M., Minotas, C. O., Iemma, R., & Morando, F. (2016). Open data quality measurement framework: Definition and application to Open Government Data. *Government Information Quarterly, 33*(2), 325–337. https://doi.org/10.1016/j.giq.2016.02.001

Weis, S. W. M., Agostini, V. N., Roth, L. M., Gilmer, B., Schill, S. R., Knowles, J. E., & Blyther, R. (2016). Assessing vulnerability: An integrated approach for mapping adaptive capacity, sensitivity, and exposure. *Climatic Change, 136*(3–4), 615–629. https://doi.org/10.1007/s10584-016-1642-0

World Bank Group. (2015). *The World Bank annual report*. Washington, DC: The World Bank.

Yenneti, K., Tripathi, S., Wei, Y. D., Chen, W., & Joshi, G. (2016). The truly disadvantaged? Assessing social vulnerability to climate change in urban India. *Habitat International, 56*, 124–135. https://doi.org/10.1016/j.habitatint.2016.05.001

8

Urban Metabolism and Open Data: Opportunities and Challenges for Urban Resource Efficiency

Aristide Athanassiadis

Abbreviations

AURIN	Australian Urban Research Infrastructure Network
BCR	Brussels Capital Region
GHG	Greenhouse gas
IOA	Input Output Analysis
LCA	Life Cycle Analysis
LGA	Local government areas
MFA	Material Flow Analysis
WESC	Water and Energy Supply and Consumption

A. Athanassiadis (✉)
Department of Building, Architecture and Town Planning (BATir), Circular Economy and Urban Metabolism, Université Libre de Bruxelles, Brussels, Belgium
e-mail: arisatha@ulb.ac.be

S. Hawken et al. (eds.), *Open Cities | Open Data*,
https://doi.org/10.1007/978-981-13-6605-5_8

Highlights

- Open Data is essential to develop comprehensive urban environmental policies and evaluate urban metabolism.
- Open metabolic data at small territorial scales can enhance the identification of resource use and pollution emission drivers.
- To compare urban metabolism studies, data standards and accounting methodologies are needed.
- Platforms such as the global urban metabolism database that collect and curate metabolic data will be essential in the future.

Introduction

To meet Paris Agreement goals of keeping the increase in global average temperature to well below 2° C above pre-industrial levels, cities will have to play a major role. In fact, cities are responsible for approximately three-quarters of greenhouse gas (GHG) emissions and energy use (Seto et al., 2014). Therefore, there is a pressing challenge to reduce urban environmental effects.

Urban metabolism is a research field that compares cities to urban systems where resource flows enter, are transformed/used/stocked and ultimately exit as waste or emissions (Athanassiadis et al., 2017b). Through this working metaphor, the complex relationship between an urban system on the one hand and its resource needs and waste/emissions generation on the other is questioned in order to ultimately propose policies and strategies that will reduce urban environmental impacts (Hendriks et al., 2000). To better understand this relationship, most studies go through a relatively important data collection and data analysis process of energy, water, material, waste and GHG emissions flows (Kennedy & Hoornweg, 2012). Accessing reliable and recent data is one of the main challenges for urban metabolism studies as it can heavily affect their accuracy and comparability to other cities. In fact, data are often available (when not confidential) through statistical yearbooks, websites, urban administrations, grid operators and other providers (Athanassiadis et al., 2017b). This variety of data sources makes data collection highly time consuming and hinders a more regular update and replicability of studies (Hoekman & von Blottnitz, 2017).

The emergence and generalisation of Open Data at an urban level is an exceptional opportunity for urban metabolism studies. New sources of data now become available opening new avenues of research. For instance, it now becomes easier to gather machine-readable data into a single database in order to provide comparative analysis between metabolic flows, year or cities. In addition, environmental and urban data are now provided at smaller spatial scales enabling, through a bigger sample, a more accurate and reliable analysis of their relationship. Nevertheless, Open Data also present a number of challenges before they are fully and more easily integrated in urban metabolism studies. Metadata are often missing or not detailed enough, making data almost unusable as researchers are unaware of what is exactly measured. Moreover, very few international standards exist on environmental data collection making comparative analysis between cities within the same country or internationally very difficult. This chapter will thus discuss the opportunities and challenges of using Open Data in urban metabolism studies and consequently to urban environmental policymaking.

Urban Metabolism

The metaphor and concept of urban metabolism has been used by a number of researchers to accomplish different objectives. These objectives include amongst others learning from natural ecosystems to inform urban design, quantify entering and exiting flows cities to optimise their processes and identify drivers to mitigate environmental impacts, to discuss the relationship between human activities and their resulting socioecological regimes and so on (Broto, Allen, & Rapoport, 2012; Beloin-Saint-Pierre et al. 2017; Zhang, 2013; Newell & Cousins, 2015). The breadth of objectives of urban metabolism as well as its extensive use of (Open) Data could clearly provide valuable insights for Open Cities. In fact, through the interdisciplinary scope of urban metabolism that wishes to better understand and model the functioning of urban systems, in regards of resource requirements and environmental impacts, it becomes possible to inform and co-create pathways to more socially just and less environmental harmful cities.

In this chapter, urban metabolism will be used to specifically refer to the analysis of material and energy flows entering and exiting for the purpose of urban environmental monitoring and assessment. From this perspective, perhaps one of the pioneering studies was the one of Wolman (1965), where he studied the flows of a theoretical American city of one million inhabitants. Since then, a number of case studies were examined by researchers studying resource flows entering in urban area, being transformed/used/stocked and ultimately exiting as pollution flows. Numerous standardised accounting methodologies exist that are used to study the metabolism of cities including Material Flow Analysis (MFA), Life Cycle Analysis (LCA) and Input Output Analysis (IOA) (Loiseau, Junqua, Roux, & Bellon-Maurel, 2012). Nevertheless, these methodologies were originally designed for different spatial scales (mostly national scale for MFA and IOA, and at product or industry level for LCA). Therefore, their application to urban systems is less straightforward, due to a mismatch between the data requirements of these methodologies and the data availability at an urban level (Goldstein, Birkved, Quitzau, & Hauschild, 2013; Athanassiadis, Bouillard, Crawford, & Khan, 2017a; Voskamp et al., 2017). In practice, studying the metabolism of a city can result in a large data collection exercise (Kennedy, 2012), scraping data through statistical yearbooks, websites, urban administrations, grid operators and other providers (Athanassiadis et al., 2017b). The variety of data sources makes data collection highly time consuming and hinders a more regular update and comparability of urban metabolism studies. Therefore, it becomes hard to compare resource use and pollution flows across cities and to identify patterns that could in turn propose mitigation strategies. In fact, one of the major criticisms for current urban metabolism studies is how descriptive they are, merely showcasing a table of numbers (Athanassiadis et al., 2017b), instead of looking at the intricate interrelationships between flows and local socioeconomic and territorial factors. At this stage, it can be argued that current data quality and accessibility crucially hinders the quality of analysis and therefore better understanding of resource use patterns (Pincetl & Newell, 2017).

The following section will briefly showcase some of the current challenges in the data collection process that are common in urban metabolism studies through three different case studies covered by the author in the last years.

(Open) Data Collection in Urban Metabolism: The Case of Brussels, Melbourne and the Creation of a Global Urban Metabolism Database

In the framework of the urban metabolism definition used in this Chapter, data collection and processing is perhaps one of the most important and time-consuming tasks. Researchers need to seek data about energy and water use, material imports and exports, waste composition, production and treatment, pollution emissions and so on to better understand the resource use demands of cities in order to develop mitigation strategies. As it will be presented here below, some data are readily available, while others require numerous queries to urban administrations and private companies holding or producing the data.

The Case of Brussels: Stitching Reports and Excel Sheets

The data collection process to study the metabolism of Brussels took over a three to four-month period of full-time research, which spanned over one and a half years. Brussels metabolism was described by creating a patchwork of tables collecting energy, water, materials and pollution data for each spatial scale and each time interval available. Metabolic data were not available through one main data source that is continuous across time and consistent across space. Instead, metabolic data are gathered by a number of public and private providers such as official reports originating from regional and national environmental, economic and mobility administrations, and are not only more scattered but also not harmonised. Building a metabolic profile is therefore time consuming and requires particular attention to the metadata for each flow to ensure compatibility with other flows, spatial scales and temporal intervals. In this chapter, only the flow of energy will be discussed as it covers a range of challenges that are generally encountered in urban metabolism studies.

In the case of Brussels, data on energy flows could be subdivided into different energy sources (natural gas, electricity, petroleum products and others) and sectors of consumption (industry, tertiary, residential and transport). All data sources are summarised in Table 8.1 (in most cases data were available through annual reports for a number of years, but only one year was mentioned).

Data on energy use were collected from the annual reports of grid operators and the national federations of electricity (*Fédération des Professionnels du secteur de l'Electricité*, currently Synegrid) and of natural gas (*Fédération des Industries du Gaz*, currently Synegrid). These federations reported delivered energy values for Brussels until the early 2000s. The values reported in this research were obtained by physically going through the archives of these federations. In some cases, and for some years, natural gas use could be disaggregated into low and medium pressure and electricity use into low and high voltage, indicating usage (residential and small tertiary activities vs. large tertiary activities and industrial activities).

Another possibility for obtaining energy data for Brussels was to use data from the energy grid operators of Brussels (Interelec, Interga and Sibelgaz until 2003, and Sibelga since then). An additional source of information that was used was the annual reports of the energy balance carried out by the environmental administration of Brussels Capital Region (BCR), starting from 1990 until the current day (IBGE, 2012). These annual reports document the energy use by source and sector in a consistent way for the last 25 years. These were used for the values after 1990. When comparing the data between the federation of grid operators and the environmental administration reports, there is a slight difference in metrics as the former provides the amount of delivered energy, whereas the latter provides the final energy use of households and the rest of the sectors. While no additional elements are provided to characterise what was measured, the discrepancy between the two sources create a break in a time-series.

A final potential source of data is the energy retailers of Brussels. However, since 2004 (for non-residential users) and 2007 (for all residen-

Table 8.1 Data availability of energy use and energy sources in Brussels

Scale	Energy source	Energy use	Time interval	Source
Region	Electricity	High/low voltage	2003–2014	(SIBELGA, 2011)
Region	Electricity	High/low voltage Several industrial and tertiary activities	1972–1995 1982–1995	(FPE, 1972)–(FPE, 2000)
Region	Electricity	Residential, tertiary, industry, transport (each category is sometimes further disaggregated)	1990–2012	(IBGE, 2012)
Region	Natural gas	Medium/low pressure	2003–2014	(SIBELGA, 2011)
Region	Natural gas	Residential, non-residential	1975–1989	(FIGAZ, 1980)
Region	Petroleum products	Residential, tertiary, industry, transport (each category is sometimes further disaggregated)	1990–2012	(IBGE, 2012)
Region	Coal, wood, heat/ steam	Residential, tertiary, industry, transport (each category is sometimes further disaggregated)	1990–2012	(IBGE, 2012)
Municipality	Electricity	Unknown	2003–2014	(SIBELGA, 2011)
		Per type of meter	2014	(SIBELGA, 2014)
Municipality	Electricity	Unknown	1976; 1978; 1983; 1986	(AB, 1976, 1978, 1983, 1986)
Municipality	Natural gas	Unknown	2003–2014	(SIBELGA, 2011)
		Per type of meter	2014	(SIBELGA, 2014)

Source: Author

tial users), the Belgian market of energy has become liberal and *Electrabel* is no longer the unique producer and retailer of electricity and natural gas. Instead a large number of companies have emerged allowing households to choose their electricity and natural gas providers (it is possible for a household to choose from a single retailer for both electricity and natural gas supply). While retail data allows a great level of detail both from a temporal and from a spatial perspective, in practice it would result into a "patchwork" of data that would be very difficult to aggregate at spatial boundaries for which other data exist.

To identify the drivers of resource use (in this case energy use), this type of data needs to be complemented by information characterising the local specificities of a city such as climate, demographics, socioeconomic characteristics, territorial organisation aspects and so on. As opposed to the energy flows, these types of data are far more centralised and easy to access through the regional statistical administration (either through excel/csv documents or through .shp files). Yet, as most of the energy data are at a regional scale, one of the only ways to identify drivers is by contrasting them with urban characteristics data over a longer time period (Athanassiadis, Crawford, & Bouillard, 2015b). When performing a 40-year temporal analysis, it underlines how difficult it is to find any correlations between some urban indicators and metabolic flows and how counter-intuitive some relationships can be. For instance, during the 40-year period population (and population density) and resource flows had very different trends. The difficulty to characterise this relationship underlines the need to use other disciplinary vantage points but also to perform spatial and comparative analyses.

The case of Brussels shows a relative common process of data collection in urban metabolism studies and highlights some common issues encountered. As shown, data are scattered around different administrations and available in different formats (Excel tables, PDF documents, printed copies of archives, etc.). While all these data can be characterised as open as they are available and can be used by anyone with no specific copyright, they are not easily accessible as they require to go through different archives and one needs to figure out the entire ecosystem of urban administrations and energy providers. In addition, some figures come from relatively old and printed reports with no description of the methodology used to measure them hindering their comparison with more recent data.

The Case of Melbourne: Juxtaposing Datasets

This second example will briefly explain the data collection process to study the metabolism of Melbourne (Athanassiadis, Crawford, & Bouillard, 2015a). Unlike Brussels, far fewer data about metabolic flows were available for Melbourne and therefore the data collection time period was dramatically reduced. Moreover, in the case of Melbourne, data on metabolic flows were limited to water and centralised in one place, namely the Australian Urban Research Infrastructure Network (AURIN). Within AURIN, three promising datasets about water use were found as of December 2015 (see Table 8.2), namely:

- Postcode Water Use 2008–2009 for Melbourne Water in Melbourne
- Postcode Water Consumption 2010–2014 for City West Water Postcode Regions in Melbourne
- Postcode Water Consumption 2011–2014 for Yarra Valley Water Regions in Melbourne

As visible in Table 8.2, data for water flows in Melbourne were disaggregated at the postcode level. Such level of disaggregation is highly useful

Table 8.2 Source of disaggregated data for Melbourne's water use

	Melbourne water	Yarra valley	City West water
Year	2008 and 2009	2011–2014 (quarterly readings)	2010–2014 (quarterly readings)
Unit	Litres per household per day	Kilo litres	kilo litres
Use	Total use	Commercial, industrial and residential uses	Residential and non-residential uses
Area covered	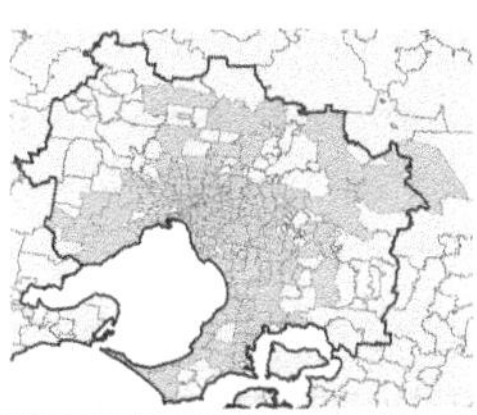	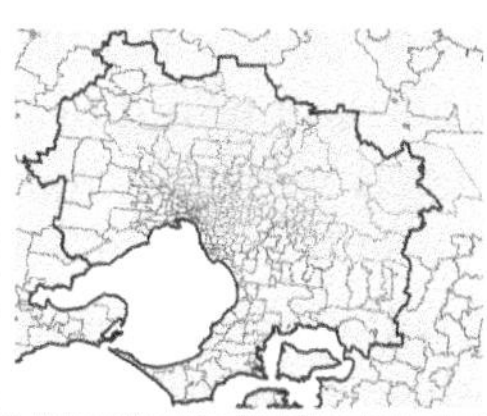	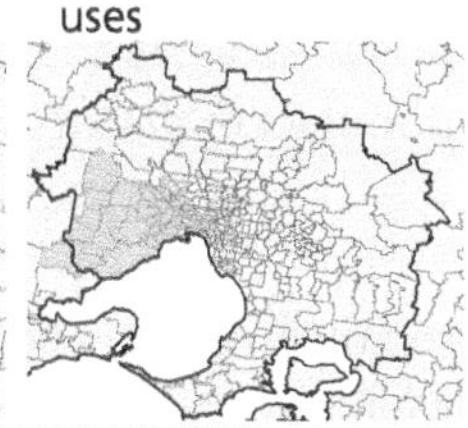

Source: Athanassiadis et al. (2015a)

when it comes to looking for the relationship between water use and other socioeconomic and territorial characteristics of Melbourne. Nevertheless, the disaggregated data coming from the source covering the most important area of Melbourne (i.e. Melbourne Water) is expressed in litres per household per day. This composite indicator aggregates residential and non-residential use, which limits the identification of water use drivers. Furthermore, the dataset of Melbourne Water did not offer more details as to what exactly was measured or what source was used to measure the households per postcode water use.

Although some aspects (such as putting together data from different companies) still need to be addressed before identifying resource use drivers, the case of Melbourne presents how the emergence of Open Data could develop an entire new path for urban metabolism. Finally, another significant element that the case of Melbourne showcased was the use of the WESC (Water and Energy Supply and Consumption) data standard for Yarra Valley and City West Water water usage. The use of a standardised format for spatially disaggregated resource use is a prerequisite for comparing resource use patterns and drivers across case studies.

With disaggregated data, the identification of drivers is obtained through spatial analysis (if the sample of spatial units is sufficiently large to perform it) (Athanassiadis et al., 2015b). To perform this type of analysis however it is necessary that data characterising a city are also available at the same spatial scale as the metabolic data. In the case of Melbourne, the difference between the spatial boundaries of metabolic and urban data was very small (postcode vs. postal area). Nevertheless, in other cases, disaggregated metabolic data could be provided under an arbitrary spatial level that does not offer urban data. As an example, when using the Open Data of Melbourne to perform a spatial analysis, it was possible to observe that water use was negatively correlated with population density and the amount of flats or apartment units, while it was positively correlated with the amount of households owning 4+ motor vehicles per dwelling. While the correlation coefficients were relatively weak, they still provide a more evidence-based understanding Melbourne's metabolism and offer a first potential avenue of mitigation policies.

Creating a Global Urban Metabolism Dataset

The two previous examples highlighted some challenges in data collection and urban metabolism studies to identify resource use patterns and drivers. It underlined that in the case of Brussels, a temporal approach should be considered to identify drivers, whereas in the case of Melbourne a spatial approach should be considered. Nevertheless, both these cases offer individual insights about resource use patterns but do not enable us to have a better understanding on how urbanisation and urban areas affect the environment as a whole. To do so, it would be necessary to compare the relationship between resource and urban characteristics across different cities.

Metabolism of Cities is an open-source platform that promotes urban metabolism to policymakers and a wider audience by centralising publications, data, research as well as creating free tools and learning materials in one central place. The global urban metabolism database created by Metabolism of Cities attempts to do exactly that (Fig. 8.1). The general

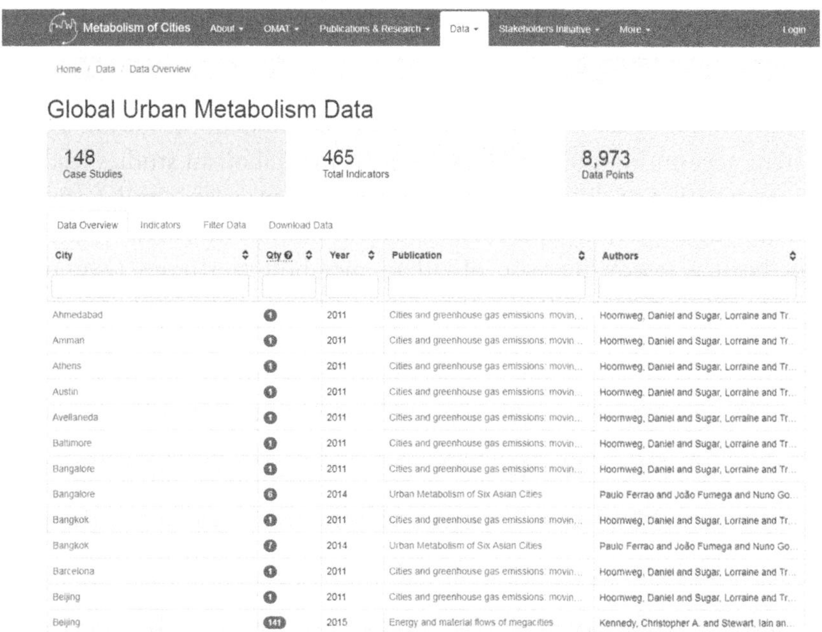

Fig. 8.1 The global urban metabolism database. Source https://archive.metabolismofcities.org/page/casestudies

philosophy of this dataset was to gather data from existing urban metabolism publications in one single place. Currently this database has collected data from 12 publications, providing metabolic 8777 data points classified in 419 indicators for a total of 45 cities. The indicators are loosely structured in six main categories (named areas in the database) including emissions, energy, materials, urban characteristics, waste and water. These main categories are then further subdivided in subcategories. For instance, the energy category is subdivided into electricity consumption, electricity sources, mobile, stationary and total. Finally, each subcategory gathers a number of indicators for which data are listed. In the case of the electricity source sub-indicator, there are 11 indicators (e.g. coal, hydropower, natural gas, nuclear and oil).

While this database is attempting to form a valuable tool for urban metabolism researchers and policymakers by proposing an open dataset that can filter and download data by flow and city, it still presents shortcomings. To maximise the reuse of data from extracted by publications, it was internally decided that data will be uncurated, that is not classified following a specific indicator list that has already been proposed (C. Kennedy, Stewart, Ibrahim, Facchini, & Mele, 2014) nor create a new one. As such the data would remain as close to the initial description of data. Nevertheless, as aforementioned, the lack of a consistent accounting methodology for urban metabolism studies leads to a juxtaposition of different indicators that measure very similar aspects, yet are not directly comparable. On top of that, in the recent year there has been a steady increase of urban metabolism studies (Musango, Currie, & Robinson, 2017). Therefore, the task of updating this global urban metabolism database is not only highly time-consuming but also not automated. As a conclusion, identifying drivers through a comparative analysis is not entirely possible at this stage through the global urban metabolism database as data from one city to another are not necessarily comparable. At this stage, the urban metabolism database helps to raise awareness amongst researchers as to the urgency to develop more consistent and comparable indicators to assess the environmental footprint of cities as well as enable transferable lessons and policies from one city to another. In the future, it would become essential to develop a common agreement between researchers, policymakers and urban officials on at least the major categories and subcategories of indicators and perhaps enable a certain flexibility to accommodate for context-specificity and local data (un)availability. In addition, in order to make this database useful to a wider range of stakeholders, data collection should probably shift from scientific publications (as it is currently doing) to official Open Data portals and develop an API to facilitate the integration of different data sources, a regular semi-automated update, as well as ease up the use of this database to create other applications.

Open Data and Urban Metabolism Studies

As shown from the previous cases, quality data is essential to understand what resource use demands of cities are and what causes them in order to propose mitigation strategies. As previously shown, the advent of Open Data has already provided a glimpse of a new research pathway in urban metabolism studies. Yet, the previous examples also pinpointed that Open Data does not necessarily equate to easily usable and comparable data. This section will provide a synthesis of the opportunities and challenges using Open Data in urban metabolism studies.

Opportunities of Using Open Data in Urban Metabolism Studies

Thanks to a wider movement advocating for cities to open their data to their citizens or researchers, urban data are becoming more and more available either by one single access portal (e.g. the UN-Habitat urban data platform[1]) or by individual urban Open Data portals.

The advent of urban Open Data presents numerous opportunities for urban metabolism studies. Firstly, as more cities open up their data, the

[1] http://urbandata.unhabitat.org/

size of the sample of comparable studies ever increases. Very recently, OpenDataSoft created Open Data portals for 500+ US cities[2] having populations between 65,000 and 375,000 inhabitants. The portals do not yet include metabolic flows data, but such initiatives become more common across different countries which would enable the comparison of cities through consistent databases.

Other promising initiatives include the creation of accounting protocols and standards that provide consistent results for numerous cities. For instance, C40 provides comparable greenhouse gas emissions for a number of their city members that are freely available and accessible[3]. Another example is the International Standard ISO 37120 which has established a number of indicators (profile, core and supporting), some of which could be used in urban metabolism studies, yet only partially and superficially covers resource and emission flows. Cities can receive different levels of certification depending on the amount of indicators for which they can provide data. As of October 2017, this website offers results for 30 cities across the globe (although Melbourne is measured both at the local government area [LGA] and at the Greater Melbourne scale), following the ISO 37120 indicators[4]. Data are open and easily comparable through the website (between cities but also for different years), although at this stage it is not possible to download the entire database. In both cases data collection and encoding is verified by a third party, ensuring the reliability and consistence of data. Therefore, it can be expected that comparing the metabolism of urban areas at a city-scale will become easier and studies identifying drivers at city level (Facchini, Kennedy, Stewart, & Mele, 2017) will help to develop international urban environmental policies.

While increasing aforementioned initiatives could increase the sample size of metabolic studies enabling a more detailed and more accurate analysis of metabolic drivers (at a city level), the most important opportunity of using Open Data in urban metabolism is the availability of metabolic data at a disaggregate level (Stephanie Pincetl & Newell, 2017). Disaggregate level data depict metabolic flows at an infra-urban territorial

[2] https://opendataamerica.com

[3] http://www.c40.org/research/open_data/5

[4] http://open.dataforcities.org/

scale such as municipalities, zip codes, blocks, neighbourhoods and so on. By breaking down metabolic data into smaller territorial units, it becomes possible to take into account all the complex internal mechanisms that are driving resource use and pollution emission and are unique to each city (Athanassiadis et al., 2017b; C. A. Kennedy et al., 2015; S. Pincetl, Graham, Murphy, & Sivaraman, 2015). In addition, in some cases the number of infra-urban territorial units for which metabolic data are available can be very significant (over 66,000 districts for Chicago[5]). In the future even larger samples could become available if utility companies make their individual metres openly available ever (yet keeping the anonymity of their customers). For instance, if energy use data were available every hour of a year for all Brussels' customers, this would produce over 10 billion data points.

The availability of such data could open new avenues of research for the Urban Metabolism field, ranging from predicting future resource use, to building robust models that would investigate different sustainable pathways, to proposing urban environmental policies and so on. Finally, data could be automatically updated and access through APIs minimising the efforts of researchers to update their results.

Challenges of Using Open Data in Urban Metabolism Studies

Due to the relative novelty of Open Data, a number of challenges still remain before they can be streamlined in urban metabolism studies. Some of these challenges were already discussed earlier in this Chapter.

Similarly to urban metabolism data, Open Data portals are scattered across a number of websites and are difficult to be found, let alone be comparable. Some initiatives have attempted to centralise Open Data portals,[6,7] yet they do not differentiate portals that specifically target urban areas nor metabolic flows from others. As an ever-increasing

[5] https://data.cityofchicago.org/Environment-Sustainable-Development/Energy-Usage-2010/8yq3-m6wp

[6] http://opendatainception.io

[7] http://www.europeandataportal.eu

amount data will become available through an ever-increasing number of sources, it will become extremely challenging to properly curate, categorise or even store these data. Yet, without a unique place where metabolic Open Data are centralised, consistent and comparable future metabolism studies would suffer from the same shortcomings as current ones.

As mentioned, the current development of global inventories, protocols and standards seem to offer a satisfactory solution into centralising and comparing data. Yet the increasing development of such initiatives every year makes the choice of city officials and researchers relatively difficult (as to how they will format their data). Additionally, most of these initiatives are developed for city level and do not have sufficient descriptors or criteria for large amounts of disaggregated data,[8] which greatly hampers the usability and comparability of data.

Another major challenge that urban metabolism studies will face when identifying drivers is the territorial units and the temporal intervals for which urban and metabolic data are available. Very often socioeconomic data (that are usually used to be correlated with metabolic data to identify drivers) are only available at different administrative boundaries, which can be rather arbitrary. The arbitrary definition of territorial units over which urban data are available can limit the possibility of crossing them with metabolic data (which are provided at spatial levels that are specific to utilities). Finally, as mentioned before, Open Data on metabolic flow are now already being provided by month (or even by hour), but all socioeconomic data are available (at best) on a yearly basis (but other socioeconomic data are available at best every year). Therefore, to better understand the real-time demand of urban resource use and its drivers, new ways to retrieve socioeconomic data need to be developed.

Conclusion

Urban metabolism is a relatively new concept with a loose definition that attempts to characterise the interactions between society, economy, technology and the environment. The advent of Open Data could signifi-

[8] A first attempt to provide guideline on Open Data is the WESC data standard of the CSIRO (http://wescml.org/)

cantly impact and shape this field by developing and strengthening some research pathways. More and better metabolic data will profoundly help to develop urban environmental policies needed to meet the Paris Agreement targets by better understanding the drivers of resource use and pollution emissions. The examples of Brussels and Melbourne show that at this stage urban metabolism could offer an environmental monitoring, detect temporal and spatial trends of resource use and pollution emissions and provide broad mitigation policy guidelines. It is perhaps still early to provide specific spatially explicit mitigation actions and transferable policies based on similar urban characteristics, but this could become possible in the future through a more coherent and consistent database and analytical framework. Policymakers and city officials could then for instance specifically target parts of the city to decrease their resource needs by planning or transport policies. Specific socioeconomic groups could also be targeted to alter consumption patterns or propose new lifestyles.

This chapter highlighted the main opportunities and challenges urban metabolism studies will face when using Open Data through three different cases. All three pointed out that without enhancing comparability and accessibility of data, the use of Open Data will not help in overcoming current shortcoming of metabolic studies. In the future, new data standards and protocols as well as centralising platforms would need to emerge. The Global Urban Metabolism Database provides a first step towards an integration of dispersed efforts in the field as well as making explicit data harmonisation challenges. Of course, the challenge remains as to how and who will define an urban metabolism indicator that will be used both by researchers and by policymakers. To do so, the indicator set should be a compromise between being as comprehensive and as accurate as possible. The indicator set should be the outcome of close collaboration between researchers, city officials, statistical offices, utility companies and international organisations in order to align interests and develop linkages and collaborations (Bai et al., 2016). Most probably different versions of the indicator set would need to exist to cater to different needs, yet they should use the same structure to facilitate the transfer of knowledge between all stakeholders. An iterative and incremental collaborative process to build an urban metabolism data structure, indicator set and ultimately to study the findings coming from analysing the gathered data could contribute in a significant manner to current knowledge gaps (Bai et al., 2018).

Finally, it is important to underline that the use of data alone will not necessarily be sufficient to depict the whole complexity of the relationship between an urban system and its flows. Without adding sufficient understanding of what a city is, how it was developed, how its citizens interact (amongst them or with their environment) and how local policies were shaped over the years, it becomes very hard to interpret the results from data analysis and models. To do so, urban metabolism studies should complement and interpret the analysis of data with theories and models developed by disciplines such as social sciences, history and geography and others.

Acknowledgement The author would like to thank the reviewers for their valuable comments that helped improve the quality of this Chapter.

References

Agglomération de Bruxelles. (1976). Radioscopie de Bruxelles. Brussels.

Agglomération de Bruxelles (AB). (1978). Radioscopie de Bruxelles. Evolution 1976–1978. Brussels.

Agglomération de Bruxelles (AB). (1983). Radioscopie de Bruxelles. Brussels.

Agglomération de Bruxelles (AB). (1986). Radioscopie de Bruxelles. Brussels.

Athanassiadis, A., Crawford, R. H., & Bouillard, P. (2015a). *Exploring the relationship between Melbourne's water metabolism and urban characteristics.* Paper presented at the State of Australian Cities 2015.

Athanassiadis, A., Crawford, R. H., & Bouillard, P. (2015b). *Overcoming the "black box" approach of urban metabolism.* Paper presented at the Living and Learning: Research for a Better Built Environment, 49th International Conference of the Architectural Science Association, Melbourne, Australia. http://anzasca.net/2015-conference-papers/

Athanassiadis, A., Bouillard, P., Crawford, R. H., & Khan, A. Z. (2017a). Towards a dynamic approach to urban metabolism: Tracing the temporal evolution of Brussels' urban metabolism from 1970 to 2010. *Journal of Industrial Ecology, 21*(2), 307–319.

Athanassiadis, A., Fernandez, G., Meirelles, J., Meinherz, F., Hoekman, P., & Cari, Y. B. (2017b). Exploring the energy use drivers of 10 cities at microscale level. *Energy Procedia, 122*, 709–714. https://doi.org/10.1016/j.egypro.2017.07.374

Bai, X., Dawson, R. J., Ürge-Vorsatz, D., Delgado, G. C., Barau, A. S., Dhakal, S., ... Roberts, D. (2018). Six research priorities for cities and climate change. *Nature, 555*(7694), 23–25.

Bai, X., Surveyer, A., Elmqvist, T., Gatzweiler, F. W., Güneralp, B., Parnell, S., ... Webb, R. (2016). Defining and advancing a systems approach for sustainable cities. *Current opinion in Environmental Sustainability, 23*(Suppl. C), 69–78. https://doi.org/10.1016/j.cosust.2016.11.010

Beloin-Saint-Pierre, D., Rugani, B., Lasvaux, S., Mailhac, A., Popovici, E., Sibiude, G., ... Schiopu, N. (2017). A review of urban metabolism studies to identify key methodological choices for future harmonization and implementation. *Journal of Cleaner Production, 163*, S223–S240.

Broto, V. C., Allen, A., & Rapoport, E. (2012). Interdisciplinary perspectives on urban metabolism. *Journal of Industrial Ecology, 16*(6), 851–861.

Facchini, A., Kennedy, C., Stewart, I., & Mele, R. (2017). The energy metabolism of megacities. *Applied Energy, 186, Part 2*, 86–95. https://doi.org/10.1016/j.apenergy.2016.09.025

Fédération des Industries du Gaz (FIGAZ). (1980). Rapport d'activités. Brussels.

Fédération Professionnelle du secteur Electrique (FPE). (1972). Rapport d'activités. Brussels.

Fédération Professionnelle du secteur Electrique (FPE). (2000). Rapport d'activités. Brussels.

Goldstein, B., Birkved, M., Quitzau, M. B., & Hauschild, M. (2013). Quantification of urban metabolism through coupling with the life cycle assessment framework: Concept development and case study. *Environmental Research Letters, 8*(3), 035024.

Hendriks, C., Obernosterer, R., Müller, D., Kytzia, S., Baccini, P., & Brunner, P. H. (2000). Material flow analysis: A tool to support environmental policy decision making. Case-studies on the city of Vienna and the Swiss lowlands. *Local Environment, 5*(3), 311–328.

Hoekman, P., & von Blottnitz, H. (2017). Cape Town's metabolism: Insights from a material flow analysis. *Journal of Industrial Ecology, 21*(5), 1237–1249.

Institut Bruxellois pour la Gestion de l'Environnement (IBGE). (2012). Bilan énergétique de la Région de Bruxelles-Capitale 2010. Brussels.

Kennedy, C. (2012). A mathematical description of urban metabolism. In *Sustainability science* (pp. 275–291). New York, NY: Springer.

Kennedy, C., & Hoornweg, D. (2012). Mainstreaming urban metabolism. *Journal of Industrial Ecology, 16*(6), 780–782.

Kennedy, C., Stewart, I. D., Ibrahim, N., Facchini, A., & Mele, R. (2014). Developing a multi-layered indicator set for urban metabolism studies in megacities. *Ecological Indicators, 47*, 7–15. https://doi.org/10.1016/j.ecolind.2014.07.039

Kennedy, C. A., Stewart, I., Facchini, A., Cersosimo, I., Mele, R., Chen, B., … Sahin, A. D. (2015). Energy and material flows of megacities. *Proceedings of the National Academy of Sciences, 112*(19), 5985–5990. https://doi.org/10.1073/pnas.1504315112

Loiseau, E., Junqua, G., Roux, P., & Bellon-Maurel, V. (2012). Environmental assessment of a territory: An overview of existing tools and methods. *Journal of Environmental Management, 112*, 213–225.

Musango, J. K., Currie, P., & Robinson, B. (2017). *Urban metabolism for resource efficient cities: From theory to implementation*. Paris: UN Environment.

Newell, J. P., & Cousins, J. J. (2015). The boundaries of urban metabolism: Towards a political–industrial ecology. *Progress in Human Geography, 39*(6), 702–728.

Pincetl, S., Graham, R., Murphy, S., & Sivaraman, D. (2015). Analysis of high-resolution utility data for understanding energy use in urban systems. The case of Los Angeles, California. *Journal of Industrial Ecology, 20*(1), 166–178.

Pincetl, S., & Newell, J. P. (2017). Why data for a political-industrial ecology of cities? *Geoforum, 85*(Suppl. C), 381–391. https://doi.org/10.1016/j.geoforum.2017.03.002

Seto, K. C., Dhakal, S., Bigio, A., Blanco, H., Delgado, G. C., Dewar, D., … Ramaswami, A. (2014). Chapter 12—Human settlements, infrastructure and spatial planning. In *Climate change 2014: Mitigation of climate change. IPCC Working Group III contribution to AR5*. Cambridge University Press.

SIBELGA. (2011). *Rapport d'activités 2011*. Brussels: SIBELGA.

SIBELGA. (2014). [Personal Communication on Energy data].

Voskamp, I. M., Stremke, S., Spiller, M., Perrotti, D., van der Hoek, J. P., & Rijnaarts, H. H. (2017). Enhanced performance of the Eurostat method for comprehensive assessment of urban metabolism: A material flow analysis of Amsterdam. *Journal of Industrial Ecology, 21*(4), 887–902.

Wolman, A. (1965). The metabolism of cities. *Scientific American, 213*(3), 178–193.

Zhang, Y. (2013). Urban metabolism: A review of research methodologies. *Environmental Pollution, 178*, 463–473.

9

Tackling the Challenge of Growing Cities: An Informed Urbanisation Approach

Christopher Pettit, Elizabeth Wentz, Bill Randolph, David Sanderson, Frank Kelly, Sean Beevers, and Jonathan Reades

C. Pettit (✉)
Urban Science, Faculty of the Built Environment, University of New South Wales, Sydney, NSW, Australia
e-mail: c.pettit@unsw.edu.au

E. Wentz
Department of Social Sciences, College of Liberal Arts and Sciences, Gainesville, FL, USA

School of Geographical Sciences and Urban Planning, Arizona State University, Tempe, AZ, USA
e-mail: wentz@asu.edu

B. Randolph
City Futures Research Centre, Faculty of the Built Environment, University of New South Wales, Sydney, NSW, Australia
e-mail: b.randolph@unsw.edu.au

S. Hawken et al. (eds.), *Open Cities | Open Data*,
https://doi.org/10.1007/978-981-13-6605-5_9

Abbreviations

ASU	Arizona State University
ADRRN	Asia Disaster Reduction and Response Network
CRED	Centre for the Research of the Epidemiology of Disasters
CAZs	Clean Air Zones
CCS	Congestion Charging Scheme
DFID	UK Department for International Development
KCL	King's College London
LAEI	London Atmospheric Emissions Inventory
LAQN	London Air Quality Network
LEZ	Low Emission Zone
SDG	Sustainable Development Goal
uLEZ	ultra Low Emission Zone
UNSW	University of New South Wales

Highlights

- This chapter presents an Informed Urbanisation framework for policy-makers, planners and citizens to work together in response to the grand challenges facing our cities.
- Examples of informed urbanisation activities illustrate the principles of open cities and Open Data in the context of London, Sydney and Phoenix. Each of these cities has followed different developmental paths, but between them they offer scope for identifying different 'baskets' of stresses and emergent sets of policy solutions.

D. Sanderson
Faculty of the Built Environment, University of New South Wales, Sydney, NSW, Australia
e-mail: david.sanderson@unsw.edu.au

F. Kelly • S. Beevers
Environmental Research Group, King's College London, London, UK
e-mail: frank.kelly@kcl.ac.uk; Sean.Beevers@kcl.ac.uk

J. Reades
Department of Geography, King's College London, London, UK
e-mail: jonathan.reades@kcl.ac.uk

- The examples employ quantitative data to highlight the interconnected systems of cities, and more consensual, co-design approaches to devise 'solutions'—however provisional—to the challenges so identified.
- Frameworks such as Geodesign enable domain experts and scientists to turn Open Data into information and evidence to force a greater accountability in city shaping and citizen involvement.

Introduction

We are living in the 'urban century'. It is well known that urban areas are now home to more than 50% of the global population, and today's urban population of 3.5 billion will nearly double to 6 billion by 2030 (United Nations, 2014). Of course, the definition of 'urban' varies dramatically from continent to continent, and country to country, but these forecasts nonetheless suggest that urbanised areas are expanding at an unprecedented rate and in often uncontrolled ways to accommodate this growing population. This expansion has potentially severe consequences for citizens and their quality of life, the environment and sustainability and infrastructure and energy usage. While urban areas have undergone several phases of rapid growth in recent history, it is the global scale at which this process is now occurring that marks this era as being part of a new phase of the urbanisation process.

But much of what is happening in cities that are facing rates of unprecedented change is highly contested. It is a truly 'wicked problem' and city governments face the increasingly difficult task of responding to the global economic forces and increasingly footloose financial flows that drive and shape our cities (Kourtit, Nijkamp, & Geyer, 2014). While populations are being drawn to our cities in record numbers, we are still struggling to understand the long-term implications of these changes on the economic and social well-being of these new urban populations. Key issues, such as mounting urban inequalities, housing affordability, resilience to extreme weather events, employment change and transportation and congestion, defy simple solutions. Public reaction to these pressures has re-emerged in many cities—from the global 'Reclaim the City' movement to local protests about perceived overdevelopment or poor service delivery.

This disruptive urban revolution is taking place in the context of another disruptive revolution, the rapid rise of the digital economy. This digital revolution is unleashing a major economic and social shift in the way we interact at a global scale and has generated an exponentially expanding range of data at an unprecedented scale on almost every human transaction and activity—the so-called data deluge (Anderson, 2008). The longer-term impacts of this revolution are unknown, but we can be sure the outcomes for citizens will be unprecedented, not least on the restructuring of global labour markets and employment opportunities, movement patterns and social cohesion that will have comparable effects to previous historical technological revolutions.

These two major transformative changes—rapid urbanisation and mass digital disruption—are brought together in the concept of 'Informed Urbanisation', which underpins the contributions discussed in this chapter. While there has been much excitement about the concept of 'Smart Cites', much of this has focused either on the technicalities of the 'smarts'—the technologies and applications that are being developed and deployed to monitor and display urban activities—or on the issues concerning the gathering of 'big data' and its analysis and processing. However, in order for these technologies and data to be effective in assisting in the better governance and management of our cities, to the greater benefit of the citizens who are now living there, they must be accompanied by the development of a better understanding of how these new digital tools and information can be applied to solving real-world urban problems (Kleinman, 2017). Data, and the 'smarts' to analyse them, are not enough. We not only need to know *what* is happening but *why* (Townsend, 2014). The notion that big data has done away with a need to understand processes and to develop concepts and theories to account for them by simply relying on correlation to 'explain' events is clearly deficient. If the emergence of 'city science' is to have any lasting value, then it requires the complementary development of new theories and explanations of cause and effect of the processes that are driving the urbanisation process.

In this context, 'Informed Urbanisation' therefore encapsulates both the 'smarts' of big data and the analytics to explore them, as well as the conceptual developments in understanding how cities work as supported through these new data and tools. The new knowledge generated by developing and testing new hypotheses about the urban process using new methods of city science will, in turn, inform better decision-making to support policies that will seek to tackle the mounting challenges of an urbanised humanity. This approach stands in contrast with the more common and more problematic 'accidental urbanisation' that is unsustainable, responsive growth due to pressures of population demand and economic development. Informed urbanisation can be seen as (i) incorporating aspects of 'smart cities' research, whilst also (ii) drawing on the long experience of planners and urbanists that sees technology and data science as providing only part of the solution, and (iii) calling for the involvement of wider expertise and insight through, for example, the principles and methods of Geodesign, which is at the intersection of disciplines of design, planning and geography (Steinitz, 2012). At the same time, the emergence of both urban big data and advances in geospatial analytics and visualisation prompt critical questions about how the data is gathered and from whom, as well as how it is used, by whom and for what purposes (Goodspeed, Pelzer, & Pettit, 2018). Informed urbanisation encompasses and interlinks three domains of knowledge: research, scholarship and policymaking.

Our premise is that cities cannot be seen as simply places in space but should be analysed and planned as an integral system of networks and flows generated by and responding to the behaviours of urban citizens, governments and businesses (Batty, 2013). The role of human agency and relationships that underpin these networks and flows is therefore a central concern in how we utilise the data and analytics to better understand these actions. Informed urbanisation offers the means to decipher such a system through rigorous and comprehensive analysis of the multitude of data on housing, transport, city resilience, city migration and other aspects of urban change; it is in this domain that international collaboration can make the most effective contribution, bringing in a multidisciplinary team and expertise across different cultural settings that can cover the diverse and complex nature of urban change.

In this chapter, we introduce an Informed Urbanisation framework (see Fig. 9.1), which has been developed through the PLuS Alliance (http://www.plusalliance.org/), a multi-institutional initiative involving Arizona State University (ASU), King's College London (KCL) and University of

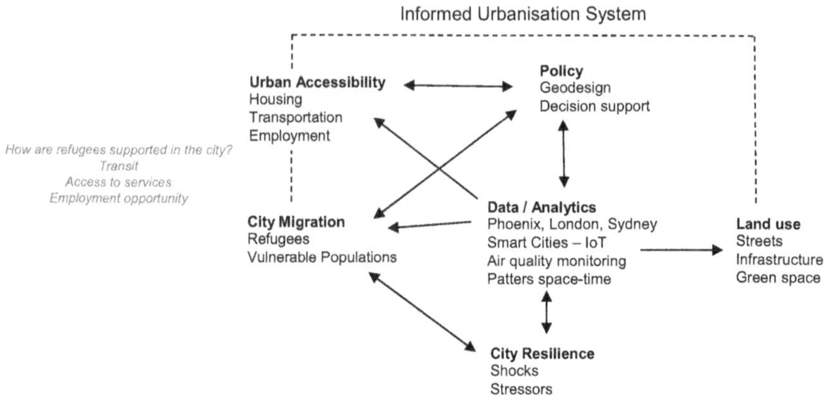

Fig. 9.1 Informed Urbanisation framework (Source: authors)

New South Wales (UNSW), which aims to deal with the 'grand challenges' facing our planet. In this chapter we will present some case studies on how the Informed Urbanisation framework is being implemented and activated in the respective cities of Phoenix, London and Sydney.

The PLuS Alliance

The Informed Urbanisation project is one of several research activities that have emerged from the Research Network of the PLuS Alliance (https://www.plusalliance.org/). An overarching goal of the PLuS Alliance is to tackle global challenges through an integrated university partnership approach. The PLuS Alliance consists of a Research Network and a Global Learning Network formed through this tri-university partnership. The goal of the Global Learning Network is to increase accessibility to the vast number of courses at the three institutions. The education objective is met by university courses being taught by faculty at all three institutions (some jointly taught) with students enrolled from all three institutions. The Research Network aims to increase the research opportunities and impact global challenges of Health, Social Justice, Sustainability, and Technology and Innovation. The partnership aims to integrate the strengths of these three outstanding universities to more effectively and efficiently address global problems.

Informed Urbanisation in the PLuS Alliance

Informed urbanisation seeks to use new and existing forms of data to better understand the local dynamics of urbanisation: how do changes in demographics, mobility, land use and policy interact with one another to promote the emergence of more resilient cities and mitigate their exposure to environmental and man-made shocks? The use of 'big data' and 'data exhaust' to better understand urban environments is common within the 'smart cities' movement; however, it is our feeling that insufficient attention has been paid to the unique policy context of real cities and the need for co-creation in the design of what are, ultimately, political responses to a range of challenges.

So, although this framework is, in principle, equally applicable in low, middle and high-income country contexts, we have chosen to focus on Phoenix, London and Sydney: not only are they cities with which our Alliance members are already familiar and for which significant amounts of data are accessible, but they also exist in very different political and policymaking contexts. These cities therefore provide a suitable test bed as they differ substantially in terms of (i) urban form (monocentric to polycentric, and compact to sprawling); (ii) population size, distribution and density; (iii) air quality and modes of transportation; (iv) social and cultural context; and (v) policy and legal frameworks. Each of these cities has followed different developmental paths, but between them they offer scope for identifying different 'baskets' of stresses and emergent sets of policy solutions.

It is the contrast between these contexts that 'informs' our thinking on how to begin responding to the very significant challenges that each will face in the coming decades. Two initial areas have been selected for this research given there is systematic impact across cities which hold global relevance. These are:

1. The balance of travel modalities within cities—particularly the scope for 'active travel' use—and the cumulative impact that these modes have on urban health through their impact on air quality and pollution. Understanding the drivers of mode choice requires an understanding of land use, mobility, environment and policy.

2. The rise of both voluntary and involuntary international migration flows, and their impact on cities and urban policy. At a time when politics is increasingly inflected by an 'us-and-them' narrative, it is important to understand how international migrants impact on—and are impacted by—the cities where they settle through housing and employment, as well as the unequal health outcomes likely to result from poor-quality provision of both.

Although our three cities all face these challenges, they are far from alone in doing so. Consequently, it is our aspiration that the Informed Urbanisation initiative will lead to new, transferable and evidence-led insights into the dynamics of cities in real policy contexts, which support both top-down (policy and planning) and bottom-up (local community) action. Ultimately, cities as noted by Batty (2013) are highly complex systems, in which it is impossible to isolate any one component from the components with which it interacts. Our approach seeks to employ quantitative data to highlight these interconnections, and more consensual, co-design approaches to devise 'solutions'—however provisional—to the challenges so identified.

In the ensuing sections of this book chapter, we will present a number of examples of informed urbanisation activities being undertaken by the three PLuS Alliance partners/co-authors of this chapter. These examples illustrate the principles of open cities and Open Data in the context of the three cities of Phoenix, London and Sydney—which also serve as urban living laboratories.

Informed Urbanisation Examples

Air Quality and Health

Poor air quality is a significant public health issue especially in cities where traffic is the major source of pollution. KCL's research based on the London Air Quality Network (LAQN), emissions modelling and vehicle profiling indicated that improvements in air quality could be achieved by restricting entrance of specific vehicle classes into London.

These research outputs were utilised by the Mayor of London to introduce the Congestion Charging Scheme (CCS) in 2003 and Low Emission Zone (LEZ) in 2008. This research, together with KCL's ongoing evaluation of air quality in London, has created interest in this method of pollution control, resulting in the adoption of similar schemes across Europe.

In 1995, KCL established the LAQN, which has evolved into the largest urban air quality network in Europe, consisting of over 120 fixed measurement sites for a range of pollutants including oxides of nitrogen (NO_x) (notably nitrogen dioxide [NO_2], ozone [O_3] and particulate matter [PM2.5 and PM10]). The LAQN has been the predominant source of air pollution information and health advice in London, benefiting the UK, regional and local government, private sector companies and the public. Monitoring data from the LAQN is available via the LAQN API (http://www.londonair.org.uk/Londonair/API/), giving access to any of the historic monitoring data beginning in 1993. The data is available in raw form, hourly measurements that have undergone QA/QC checks or summarised by relevant air quality EU limit and Target values (http://www.londonair.org.uk/LondonAir/LATools/SiteSpeciesComparison.aspx). A unique 'nowcast' system (http://www.londonair.org.uk/LondonAir/nowcast.aspx) also provides current air quality in map form, down to individual streets across the city, air pollution forecasts for the days ahead and advanced infographic information (see Fig. 9.2).

Another area of activity has been our focus on the provision of real-time data and advanced warnings of potentially health-damaging events, in the form of national air quality indices and proactive alert services to the public (Kelly, Fuller, Walton, & Fussell, 2012). The aim is to empower people to modify behaviour—for example, when to increase medication, which route/mode of transport to take to school or work or the appropriate time to pursue outdoor activities—in a way that protects their health as well as the quality of the air they breathe (http://www.londonair.org.uk/Londonair/MobileApps/).

Research has also been directed at improving exposure assessment—the Achilles heel of epidemiological research (Elliot & Warternberg, 2004). Although the most accurate exposure assessment equates to a scenario where each individual is equipped with a fast-response, specific personal measurement device, it is unlikely in the near future to be applicable on

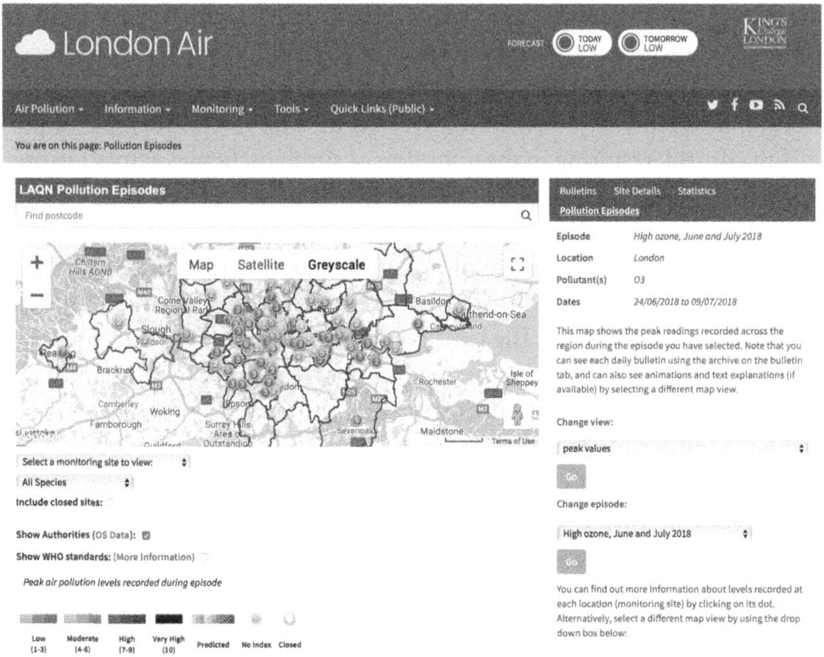

Fig. 9.2 The London Air website which is updated hourly with the concentration of all major health-related air pollutants. (Source: ERG, King's College London, 2018)

the large scale required by epidemiological studies of long- or short-term exposure (Beevers et al., 2013). Consequently, appreciating that hybrid models come closest to a "best" estimate of human exposure, we combined our dispersion with its detailed emissions' estimates from London Atmospheric Emissions Inventory (LAEI) to produce a dynamic hybrid exposure model for London (Smith et al., 2016).

London is now preparing for the introduction of an ultra Low Emission Zone (uLEZ) in April 2019. The uLEZ will initially cover London's central Congestion Charging Zone, an area bounded by the London Inner Ring Road, which includes the City of London plus adjacent areas of eight radially adjoining Boroughs. It is projected to deliver for central London by 2020 emission reductions of −49% for NO_x; −47% for NO_2; and −11% for PM10. Subject to consultation, the uLEZ will subsequently extend to include an inner London area (within London's North and South Circular roads), and a London-wide area, for heavy vehicles only. With the UK Government preparing a new air quality strategy, which will include the introduction of Clean Air Zones (CAZs) in cities across the UK, assessing the successes/failures of London's uLEZ will help other cities and towns in deciding whether an uLEZ will help solve their air quality problems. The work planned by KCL in monitoring London's uLEZ will help provide this evidence base to support these decisions as linking improvements in air quality with improvements in health is key to the development of effective policies and, importantly, their acceptance by the public.

Assessing the impact of the uLEZ will be undertaken through both the examination of the routine air quality monitoring data that is collected by the LAQN and a bespoke children's health study funded by the National Institute of Health Research (NIHR) (https://youtu.be/o8j6J-3F3M3g). By undertaking this work, we aim to demonstrate to the public that London Government's decision to introduce further measures to control traffic emissions is of direct benefit to their family's health. Furthermore, these collective projects involving traffic composition, vehicle emissions, penalty costs, health statistics, travel patterns and community engagement illustrate the dynamic nature of modern city life and the possible negative and positive influences on urban life.

Community Resilience

Resilience is a concept that aims to determine how well an urban area or community can respond to a system-wide shock or a long-term stress on the physical infrastructure or social and economic functions (Sanderson & Sharma, 2016). Community resilience in the PLuS Informed Urbanisation group is a strategy for supporting municipalities, non-governmental organisations, industry and universities to work together to (i) assess relative vulnerabilities and existing coping practices; (ii) allocate limited resources more efficiently; and (iii) decipher rapidly changing contextual factors affecting vulnerable populations. Successfully achieving these objectives requires collection, analysis and visualisation of vast and diverse data; collaboration with representative stakeholders; and prioritisation and implementation of action-oriented plans.

Built on a theme of collective and shared data, we are building a digital data hub, to be known as PLuSData (see Fig. 9.3). PLuSData has been developed on the open source GeoNode software stack and has proven be useful for hosting city data such as active transport trajectory data (Leao et al., 2017). The data portal is to support the release of open city data across three city living laboratories and other cities from across the globe. In supporting community resilience, PLuSData will serve as home to Sustainable Development Goal (SDG)-related projects, which are producing data, information and knowledge pertaining to the 17 SDG targets and supporting indicators. A number of PLuS Alliance projects are contributing to the SDG agenda and creating data products which will reside on PLuSData. The data on PLuSData is publicly available through the URL: http://data.plusalliance.org/.

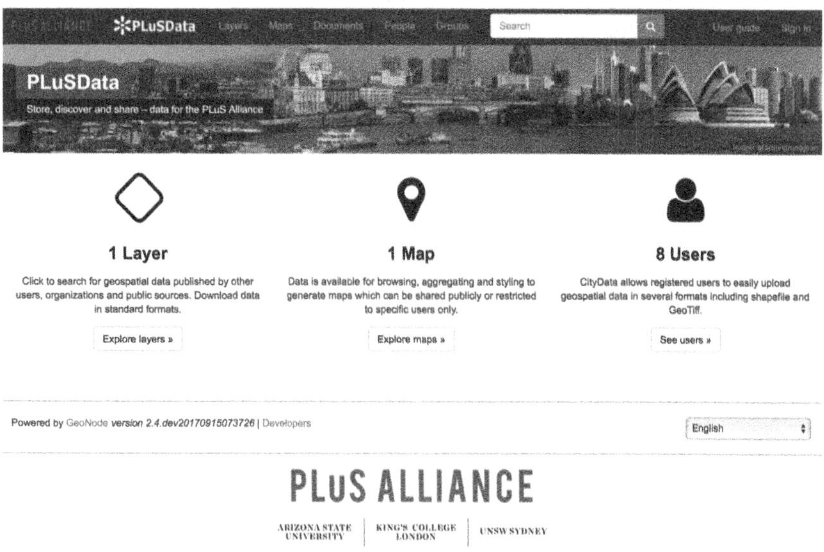

Fig. 9.3 PLuSData—a data infrastructure supporting open cities, the SDGs and other PLuS Alliance data-sharing activities (Source: University of New South Wales 2016)

Active Transport

Active Transport refers to self-powered human mobility, such as walking or bicycling, for commuting, personal transport and recreation. The Informed Urbanisation group is analysing the effectiveness of crowdsourced data on bicycling activity (e.g., mobile apps such as Riderlog and Strava) to understand the spatial patterns of bicycling activity and the factors that influence where and when people use active transportation. This is illustrated in CityViz Sydney online cycling map (see Fig. 9.4) (Leao et al., 2017; Pettit, Lieske, & Leao, 2016). The advantages of active transport are widely known to improve individual health outcomes (e.g., obesity reduction), to reduce automobile congestion, to improve air quality and to provide alternative transportation modalities. Informed urbanisation tools such as CityViz can ultimately assist policymakers, planners and communities in shaping more sustainable futures cities. Through the provision of such digital tools, such groups and individuals explore

Fig. 9.4 CityViz Sydney cycle map (Source: UNSW Built Environment, 2015a). Anonymised versions of the data underpinning CityViz are available via https://citydata.be.unsw.edu.au/

data-driven strategies that can improve the conditions and infrastructure that encourage or discourage individuals from using active transport. Some of the conditions are culturally driven—an attitude that active transport, either alone or in combination with other modalities, is an acceptable alternative to strictly single-occupancy automobile transportation. Other conditions are more physical, which could include safe infrastructure and access to bike share programs through mobile apps. In the context of open cities and Open Data, there is a need for online interactive visualisations to support citizen science in building evidence to support policymaking. Applications, such as www.Bikemaps.org, enable citizens to enter in information such as reporting of cyclist safety incidents (see Fig. 9.5). Such online mapping tools are valuable to underpin policies and planning of safer and more active transport-ready cities.

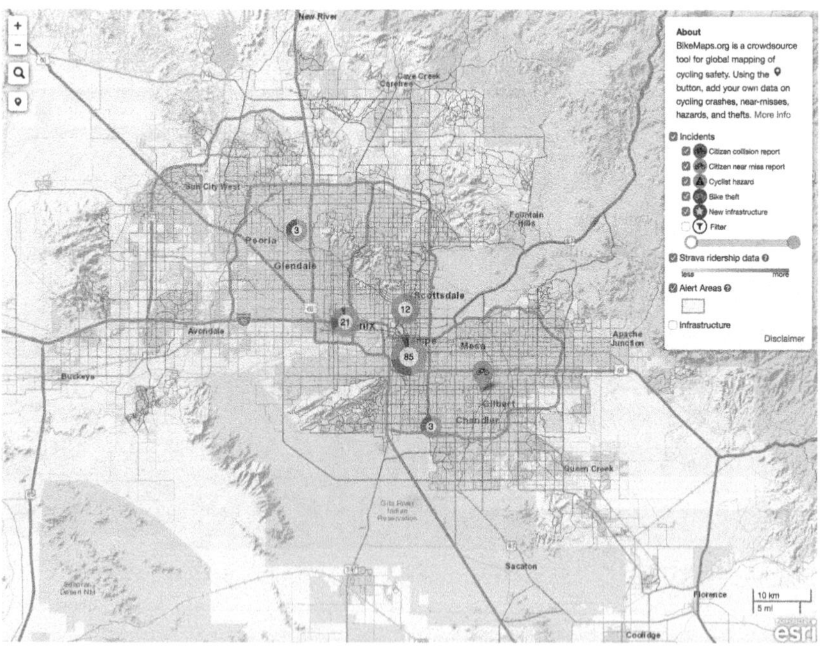

Fig. 9.5 A screenshot of Phoenix from Bikemaps.org, a crowdsource tool for cycling safety (Source: Nelson, Denouden, Jestico, Laberee, & Winters, 2015)

Geodesigning Future Cities

One of the distinct components of informed urbanisation is the use of systems frameworks such as Geodesign (Steinitz, 2012) for supporting a multi-agency and multidisciplinary approach for data-driven strategic planning and urban design. Geodesign is a planning and design approach that uses the strength of digital geographic technologies (e.g., Geographic Information Systems and remote sensing) to empower the community to come together to solve problems and design future land uses. Geodesign has been applied in many localities. However, in the context of this chapter it has been piloted both in Phoenix and in Sydney (see Fig. 9.6). Also, Geodesign methodologies are forming the basis for a co-developed course between ASU and UNSW. Upper division undergraduate and graduate students in programmes in geography, urban planning, urban design and city analytics will have the opportunity to learn the techniques of Geodesign. Using a multi-day workshop format, municipal planners, stakeholder groups and community members join together to address a specific planning programme and generate land use plans. One of the critical elements of the workshop is the facilitator, who provides the group with focus and requires that they negotiate and make compromises throughout the process. In the end, a land use plan emerges that reflects the collective wisdom of the workshop participants.

Fig. 9.6 Geodesign workshop for planning South East Sydney 2050 (Source: Pettit et al., 2019)

ASU conducted a Geodesign workshop to integrate the ASU Thunderbird campus with the ASU West campus and the Banner Health medical facility on the west side of the Phoenix metropolitan area (see Fig. 9.7). The project was challenging because much of the land is already in use and infill was needed to create vibrant connections between these hubs of activities. The workshop therefore focused on connective land uses such as transportation, green spaces and commercial districts that would integrate these spaces and mitigate urban heat (see Fig. 9.8). Like the workshop in Sydney, a strength of the Geodesign approach was the ability to harness the collective wisdom, use of sophisticated spatial analysis tools and a trained facilitator to reach a consensus for future precinct planning (Fig. 9.9).

Fig. 9.7 The Geodesign workshop in Phoenix south to create connections between the ASU West campus and Banner Health medical facility (Source: authors)

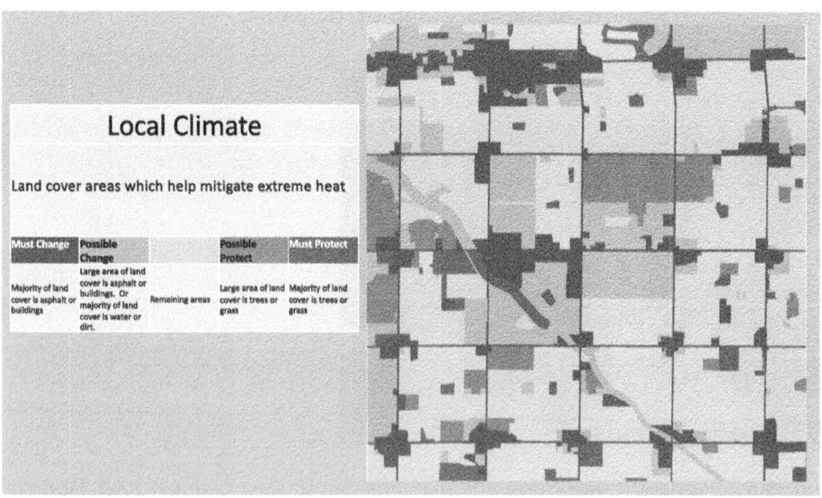

Fig. 9.8 Goal statement for Geodesign workshop, with map showing where land use changes are possible (Source: authors)

Fig. 9.9 Facilitator working with workshop participants, answering questions about software tools (Source: authors)

Housing Affordability

In recent years, housing affordability has become the focus of intense political and media scrutiny, much of it speculative or ill-informed. The financialisation of housing as an investment asset on a global scale in the last 30 years has transformed the way housing markets work, especially in major cities (Forrest & Hirayama, 2014). In Australia, pressures of growth and migration have been compounded by the impact of financial liberalisation to create intense pressure on property prices, both in the rental and sales markets (Daley & Coates, 2018). At the same time, market volatility has become more pronounced in the form of speculative booms and busts. The three cities of Phoenix, London and Sydney have all witnessed periods of escalating price growth as well as downturns. But overall, the price of housing in these cities, as elsewhere, has increasingly outstripped household incomes for many on lower incomes. Changing accessibility to affordable housing has been largely responsible for the suburbanisation of poverty and lower income communities in recent years (Randolph, 2017).

The Informed Urbanisation PLuS Alliance team across the three institutions have been working on joint projects to develop comparable analyses of changing property price data and interpreting the dynamics of housing affordability. The prevalence of publicly accessible and regularly updated address-level property price data is a key component of this. Geocoding such data has now become commonplace, making the analysis of time series spatialised data at a range of geographical scales much more feasible. Analytical approaches to calculating housing affordability for key groups (e.g., young families, the low paid and elderly renters) can be coupled with regularly updated price data to explore affordability outcomes across a range of urban housing market areas. These kinds of data also offer excellent opportunities for more advanced visualisation and analysis techniques.

Work at UNSW has generated several initial analyses that we foresee adapting to the London and Phoenix contexts. These include census tract-level analyses of affordability for low-income workers (van den Nouwelant, Crommelin, Herath, & Randolph, 2016) and address-level assessments of the affordability of multi-unit residential redevelopment schemes (Troy, Randolph, Easthope, & Pinnegar, 2015). Figure 9.10 illustrates the results of housing affordability as calculated and mapped for Sydney. Incorporating affordability analyses into scheme- and precinct-level development viabil-

Fig. 9.10 Sydney Housing affordability index (Source: UNSW Built Environment, 2015b)

ity assessments has also been undertaken, using spatially enabled point level sales datasets as a key input (Leao, Troy, Lieske, Randolph, & Pettit, 2018). Next steps in the research will be to extend and evaluate these methods for both London and Phoenix.

Marginalised Urban Communities: 'Informal Dwellers' and Forced Migrants

Matched with rapid urban growth is the increased risk of poverty and vulnerability. Globally, just under one billion people live in low-income settlements (often called slums); a figure that may well grow to 1.8 billion by 2025 (United Nations, 2014). Those living in such conditions often lack access to basic services, security of tenure, and are more vulnerable to shocks and stresses. At the same time, forced migration (those fleeing in fear of their lives, usually from conflict) is at its highest level since World War II, with over 65 million people fleeing their country to become refugees, or having shifted within their own country, becoming internally displaced people (UNHCR, 2016). The majority of forced migrants move to cities and towns: globally it is estimated that some 60% of refugees are urban, while in countries such as Jordan, the figure is close to 90% of Syrian refugees (Sanderson & Sharma, 2016). The majority of forced migrants—those who fled without goods or access to savings—find themselves often living in poor-quality housing, also with limited access to services and are often isolated from the everyday life of their host neighbourhood. Traditional humanitarian aid approaches, based largely on providing services to people in refugee camps, are not geared to this new urban reality of dispersed populations.

To these ends, the question the PLuS Alliance is intending to grapple with is, how can informed urbanisation help improve security and build opportunity for forced migrants? And, in the wider picture of low-income settlements, how does informed urbanisation provide alternatives to traditional top-down planning that has largely ignored the voices of those living in low-income settlements in the past? As a first step, the Informed Urbanisation PLuS Alliance team convened a round-table discussion on how big data can be harnessed to address such seemingly intractable issues, looking at a range of issues, including systems thinking and applying the concept of resilience.

What was clear from this event, and remains the case, is that any approach developed that will meaningfully address these two challenges needs to be based on partnerships between different disciplines and needs to embrace the opportunities (and not just the problems) inherent in these areas. The Informed Urbanisation team will also look to learn from current initiatives underway. For example, UNSW is currently engaged in a project[1] in Bangladesh's capital Dhaka, identifying and supporting neighbourhood-level innovations in Korail, one of the city's largest slums, which will improve health outcomes (see Fig. 9.11). The project, a collaboration between health-care providers, NGO workers and built environment professionals, is based on the premise that lasting improvements are based on local skills, knowledge and above all local ownership of any interventions.

Conclusion

The challenges presented through a rapidly urbanising world and the pressing need to better understand the dynamics of the processes in different urban contexts can only benefit from an informed approach. Big data, city analytics, modelling and simulation and systems-thinking frameworks need to be used to ensure we use the best available evidence and expertise to plan for our future cities. In this chapter we have presented an informed urbanisation approach for support to policymakers, planners and citizens to work together in responding to some of the grand challenges facing our cities. A number of vignettes have been discussed which outline existing and future-planned informed urbanisation research. City governance needs to also embrace an informed approach to monitoring and responding to rapid urbanisation and thus there is a need for Open Data platforms such as London Air, CityViz, Citydata, Bikemaps and PLuSData to support the democratisation of data and increased transparency in decision-making. To harness the power of this data revolution, which has seen rise to an array of Open Data platforms, is the need for systems-thinking frameworks such as Geodesign to enable

[1] The project consortium comprises Dhaka Community Health Trust, the Indian NGO SEEDS, the Asia Disaster Reduction and Response Network (ADRRN), the Centre for the Research of the Epidemiology of Disasters (CRED) at the University of Louvain and UNSW. It is funded by the Start Network, with funding from the UK's Department for International Development (DFID).

Fig. 9.11 A typical street in Korail (Source: authors)

domain experts and scientists to turn this data into information and evidence to force a greater accountability in city shaping and citizen involvement, rather than social control. An informed urbanisation approach is urgently needed to breakdown the silos of government to enable sustainable, productive, liveable and resilient cities to prosper.

References

Anderson, C. (2008, June 23). The end of theory: The data deluge makes the scientific method obsolete. *Wired.*

Arizona State University. (2016). PLuS Alliance. Retrieved February 15, 2018, from https://www.plusalliance.org/terms-use

Batty, M. (2013). *The new science of cities*. Cambridge, MA: MIT Press.

Beevers, S. D., Kitwiroon, N., Williams, M. L., Kelly, F. J., Anderson, H. R., & Carslaw, D. C. (2013). On the use of air pollution dispersion models for human exposure predictions in London. *Journal of Exposure Science and Environmental Epidemiology, 23*(6), 647–653.

Daley, J., & Coates, B. (2018). *Housing affordability: Re-imagining the Australian dream*. Melbourne: Grattan Institute.

Elliot, P., & Warternberg, D. (2004). Spatial epidemiology: Current approaches and future challenges. *Environmental Health Perspectives, 112*(9), 998–1006. https://doi.org/10.1289/ehp.6735

ERG, King's College London. (2018). London Air. Retrieved February 15, 2018, from www.londonair.org.uk

Forrest, R., & Hirayama, Y. (2014). The financialisaton of the social project: Embedded liberalism, neoliberalism and home ownership. *Urban Studies, 52*(2), 233–244.

Goodspeed, R., Pelzer, P., & Pettit, C. (2018). Planning our future cities: The Role Computer Technologies can Play. In T. W. Sanchez (Ed.), *Planning knowledge and research*. Abingdon: Routledge.

Kelly, F. J., Fuller, G. W., Walton, H. A., & Fussell, J. C. (2012). Monitoring air pollution: Use of early warning systems for public health. *Respirology, 17*, 7–19.

Kleinman, M. (2017) *Cities, data and digital innovation*. IMFG Papers on Municipal Finance and Governance, 24. Institute on Municipal Finance and Governance and Innovation Policy Lab, University of Toronto.

Kourtit, K., Nijkamp, P., & Geyer H.S. (2014) Managing the urban century. *International Planning Studies, 20*, 1–2, 1–3.

Leao, S. Z., Lieske, S. N., Conrow, L., Doig, J., Mann, V., & Pettit, C. J. (2017). Building a national-longitudinal geospatial bicycling data collection from crowdsourcing. *Urban Science, 1*(3), 23.

Leao, S. Z., Troy, L., Lieske, S. N., Randolph, B., & Pettit, C. (2018). A GIS based planning support system for assessing financial feasibility of urban redevelopment. *GeoJournal.* https://doi.org/10.1007/s10708-017-9843-2

Nelson, T. A., Denouden, T., Jestico, B., Laberee, K., & Winters, M. (2015). Bikemaps.org: A global tool for collision and near miss mapping. *Frontiers in Public Health, 3*, 53. Retrieved February 15, 2018, from https://bikemaps.org/

Pettit, C., Hawken, S., Ticzon, C., Steinitz, C., Ballal, H. T., Cranfield Leao, S., Afrooz, A., & Lieske, S. (2019). Breaking down the silos through Geodesign—Envisioning Sydney's urban future. *Environment and Planning B: Urban Analytics and City Science*. https://doi.org/10.1177/2399808318812887

Pettit, C., Lieske, S. N., & Leao, S. Z. (2016). Big bicycle data processing: From personal data to urban applications. In *ISPRS Annals of the Photogrammetry, Remote Sensing and Spatial Information Sciences, Volume III-2*. Paper presented at the ISPRS XXIII Congress (2016), Prague, Czech Republic, 12–19 July.

Randolph, B. (2017). Emerging geographies of suburban disadvantage. In J. Hannigan & G. Richards (Eds.), *The handbook of new urban studies*. Thousand Oaks: Sage.

Sanderson, D., & Sharma, A. (2016). *World disasters report, resilience: Saving lives today, investing for tomorrow*. Geneva: IFRC.

Smith, J. D., Mitsakou, C., Kitwiroon, N., Barratt, B. M., Taylor, J. G., Anderson, H. R., … Beevers, S. D. (2016). The London Hybrid Exposure Model (LHEM): Improving human exposure estimates to NO2 and PM2.5 in an urban setting. *Environmental Science and Technology, 50*(21), 11760–11768.

Steinitz, C. (2012). *A framework for Geodesign: Changing geography by design*. ESRI Press.

Townsend, A. M. (2014). *Smart Cities: Big Data, Civic Hackers, and the Quest for a New Utopia*. New York: W.W. Norton & Company.

Troy, L., Randolph, B., Easthope, H., & Pinnegar, S. (2015). *Renewing the Compact City—Final Report*. Sydney: City Futures Research Centre, UNSW Australia.

UNHCR. (2016). *Global forced displacement hits record high*. Retrieved from: http://www.unhcr.org/en-au/news/latest/2016/6/5763b65a4/global-forced-displacement-hits-record-high.html

United Nations. (2014). *World Urbanization Prospects: The 2014 revision, highlights*. Population Division, United Nations: Department of Economic and Social Affairs.

UNSW Built Environment. (2015a). Bicycling Movement (Capital Cities). City Futures Research Centre. Retrieved February 15, 2018, from https://cityfutures.be.unsw.edu.au/cityviz/bicycle-trips-time-day-capital-cities/

UNSW Built Environment. (2015b). Sydney Housing Affordability Index. City Futures Research Centre. Retrieved February 15, 2018, from https://cityfutures.be.unsw.edu.au/cityviz/affordability-index/

van den Nouwelant, R., Crommelin, L. M., Herath, S., & Randolph, B. (2016). *Housing affordability, central city economic productivity and the lower income labour market* (AHURI Final Report No. 261). Melbourne: Australian Housing and Urban Research Institute.

10

Linking Complex Urban Systems: Insights from Cross-Domain Urban Data Analysis

Lelin Zhang, Bang Zhang, Ting Guo, Fang Chen, Peter Runcie, Bronwyn Cameron, and Roger Rooney

Highlights While Open Data unlocks limitless opportunities for innovation, the potential could only be fully realised by a comprehensive understanding of data from multiple urban sub-systems.

L. Zhang (✉) • T. Guo • F. Chen
Data61 CSIRO, Eveleigh, NSW, Australia

University of Technology Sydney, Ultimo, NSW, Australia
e-mail: lelin.zhang@uts.edu.au; ting.guo@data61.csiro.au; fang.chen@uts.edu.au

B. Zhang • P. Runcie
Data61 CSIRO, Eveleigh, NSW, Australia
e-mail: matt.zhang@data61.csiro.au; peter.runcie@data61.csiro.au

B. Cameron
Sydney Water, Parramatta, NSW, Australia
e-mail: bronwyn.cameron@sydneywater.com.au

R. Rooney
ACT Government, Canberra, ACT, Australia
e-mail: roger.rooney@act.gov.au

S. Hawken et al. (eds.), *Open Cities | Open Data*,
https://doi.org/10.1007/978-981-13-6605-5_10

- Various machine learning and data analytics techniques have been used to address the unprecedented complexity of cross-domain urban data.
- Individual industries could expand the horizon of their data analytics projects by adopting inter-domain data.
- Policymakers can make more informed and comprehensive decisions with a holistic view of urban systems.
- There are still significant challenges in making organisation-owned data open to public. But potential solutions exist and the related techniques are fast developing.

Introduction

Urbanisation is a global trend that has resulted in a fast-growing demand for infrastructure, transportation, energy, dwelling, education, healthcare, entertainment, communication, banking and financing and other urban services. Cities are complex socioeconomic systems that must provide all these services. Urban systems or cities consist of a large number of sub-systems which must support individual demands and also interact closely with each other. A comprehensive understanding of the whole urban system can help us make efficient urban asset maintenance, accurate forecasts and informed plans for future demand.

However, in the past, data about different urban sub-systems are often locked up in different government agencies and organisations. Studies of a sub-system usually can only access isolated and fragmented datasets, and therefore treat the sub-system in an isolated and fragmented way, rather than considering it holistically. Fortunately, following the recent Open Data advocacy, more and more public data about urban sub-systems have been released by government agencies and organisations. This unlocks limitless opportunities for all parties in the urban systems to collaborate and innovate, not only for individual sub-systems but also for the urban system as a whole.

Along with the Open Data movement, studies are now aiming to "move from data to information to knowledge, and, ultimately, to action for urban sustainability and human well-being." (Ramaswami, Russell, Culligan, & Karnamadakala Rahul Sharma, 2016). To derive insights from the data collected in disparate areas, we need cross-domain data-driven solutions. In this chapter, we will shed light on three successful cases on water utility, parking and urban planning projects. This showcases how cross-domain urban data analysis techniques, combining a mixture of public urban data and private sub-system-specific data, can reshape the decision-making processes of businesses, governments and societies.

While current projects are often driven by a public/private organisation of a closed nature, we believe the general public and the organisations themselves could benefit from a more open environment. Following the case studies, we conclude this chapter with discussions on possible directions to increase the openness of urban sub-system data analytic projects, and how such movements could be a win-win for all parties.

Urban Wastewater Pipe Blockage Prediction

Background

For water utilities, wastewater pipe (sewer) blockages pose a great challenge to the daily operation of wastewater pipe networks. Not only do they result in high economic costs in emergency repairs and clean up, but they also have social impacts, such as service disruption, environmental pollution and road and amenity closure. Traditional forecasting and prediction techniques are currently not capable of accurately predicting sewer chokes. Sydney Water, one of the largest water utilities in Australia, sponsored a study with us to address this challenge. The aim was to investigate and understand the key factors that impact the sewer blockage patterns and develop a prediction model to predict future blockage probability. The outcome of this project provides data-driven decision support to water utilities that could lead to a more efficient and predictive maintenance strategy.

If current field validation of the model is successful, a preventative maintenance programme may reduce the social, reputational and environmental impact of sewer chokes, enable more efficient allocation of resources, improve regulatory compliance and reduce disruptions, rebates and property damage costs.

Interestingly, the data from vegetation coverage, climate, soil and demography plays a significant role in the solution, which demonstrates the value of cross-domain data analysis.

Datasets

Sewer blockages can be caused by many things, including tree roots, grease, debris and foreign objects (e.g., soft wipes). There are many factors which may contribute to blockage incidents, including the pipe's intrinsic characteristics (diameter, length, material, etc.), the pipe's external environment (vegetation coverage, climate and soil condition, property type and demographical characteristics) and historical blockage events. In order to achieve accurate prediction, we investigated the following datasets (Table 10.1).

Table 10.1 Datasets used for urban wastewater pipe blockage prediction

Dataset	Description	Source
Sewer network	Characteristics of pipes, including laid year, diameter, length, material, location, etc.	Private—water utility
Blockage records	Historical blockage records, including blocked pipe, date, type of blockage, blockage location, etc.	Private—water utility
Vegetation coverage	Shapes (polygons) of tree canopies for more than 4 million trees, obtained from satellite imagery.	Private/Public—third party
Climate and soil	Rainfall, temperature, evaporation and soil moisture.	Private/Public—Bureau of Meteorology and Office of Environment and Heritage[a]
Demographic	Property types and densities, population, etc.	Public—Australian Bureau of Statistics[b]

[a]http://www.bom.gov.au/climate/data/
[b]http://www.abs.gov.au/websitedbs/censushome.nsf/home/datapacks

Factor Analysis

Based on the historical data of sewers and blockages, we conducted statistical analysis (Li, Zhang, Wang, Chen, & Vitanage, 2014; Li et al., 2015) to discover the influences of different factors. This provides useful insights towards the prediction model, such as:

Pipe characteristics: using only the sewer and blockage data, we could get the insights about the performance of pipes with different characteristics. Figure 10.1 shows the blockage rates of pipes laid in different years. Figure 10.2 shows the blockage rates for a few common materials, which suggests that pipes made of PVC (polyvinyl chloride) are less prone to blockage than other materials. When local knowledge of the number and types of pipes laid in different years is taken into account, these confirm the domain knowledge held by experts in water utility.

Tree coverage: tree root intrusion is one of the most common causes of sewer blockages in the Sydney Water network. For a tree, the extent of its root system is dependent on the species, age, nutrient availability and physical limitations of surrounding soil. However, the tree canopy often provides a good indication of the size of the root system. Sydney Water, in collaboration with Jacobs, developed a tree canopy polygon layer in order to understand the relationships between vegetation and blockages caused

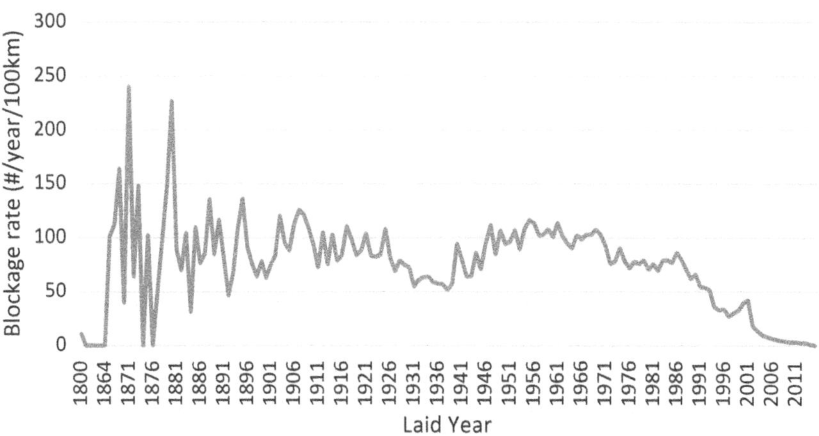

Fig. 10.1 Blockage rates of wastewater pipes laid in different years

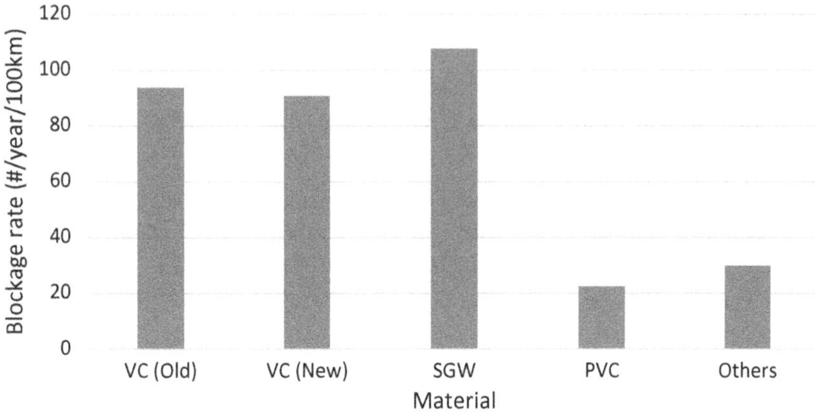

Fig. 10.2 Blockage rates for wastewater pipes made of different materials

Fig. 10.3 Tree canopy polygons

by tree roots. The pipe network is overlaid on the tree canopy polygons (refer to Fig. 10.3) to compute the tree canopy coverage percentage of each pipe. As shown in Fig. 10.4, a clear positive correlation between tree canopy coverage and blockages caused by tree roots can be observed.

Climate and soil conditions: to understand the impact of climate and soil condition to sewer blockages, we matched the temperature, rainfall, evaporation and soil moisture data to each pipe in the whole network. A strong correlation between the climate and soil conditions

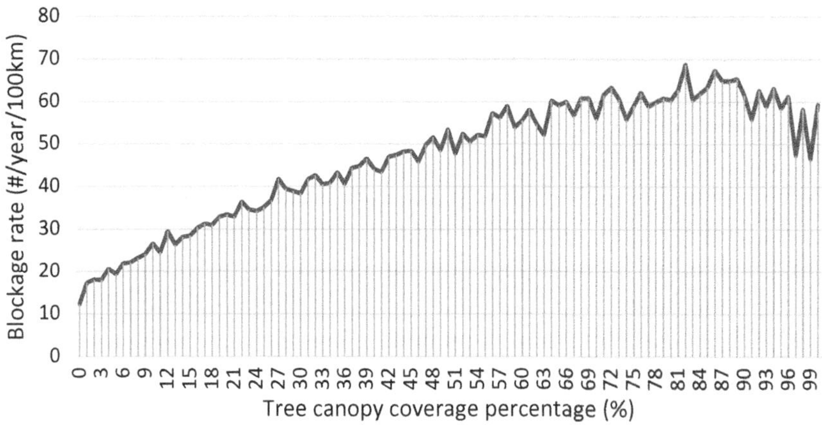

Fig. 10.4 Tree root caused blockage rates under different tree canopy coverage percentages

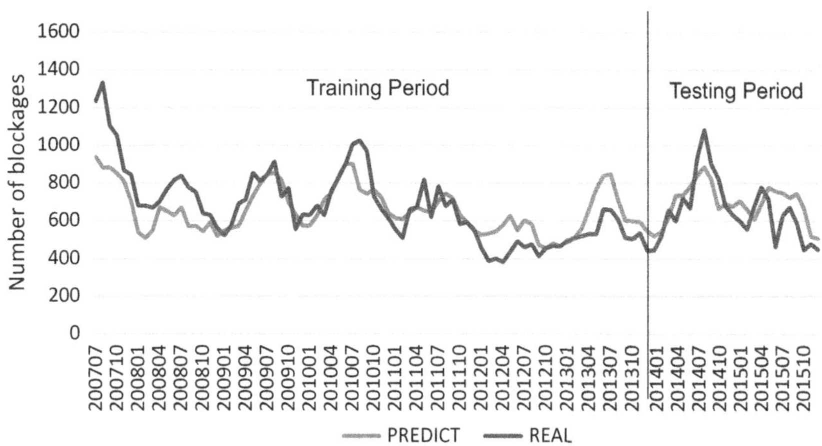

Fig. 10.5 Predictive power of climate factors (temperature, rainfall, evaporation and soil moisture) for predicting tree root caused blockages six months later

and the blockages occurred 6 months later is observed. One theory is that the climate in summer affects the growth of tree roots, which in turn affects the number of blockages for the following winter peak. Figure 10.5 demonstrates the predictive power of the climate and soil conditions on the blockages.

Methodology and Outcome

We proposed a new variant of Hawkes process (Hawkes, 1971) for short-term (one year) sewer blockage prediction (Lin et al., 2015, 2016a). Hawkes process is a stochastic point process (Daley & Vere-Jones, 2002) based statistical model. In this study, each blockage event is treated as a point, and the point intensities (number of blockages in the future) are predicted using both background intensity and trigger intensity. The background intensity models the impact of a pipe's intrinsic characteristics and external environment on blockage behaviours, while the trigger intensity models the contribution of a pipe's past blockage history on future blockage events. All the key influential factors discovered in the factor analysis phase are utilised in the prediction model.

To evaluate the performance of our predictive model, we split the 15-year blockage data into two parts: the first 14 years' data are used to train the model, and the model is blind-tested using the last year's data. As shown in Fig. 10.6, the prediction model could accurately predict risk of blockages in testing year: the top 1% pipes with high-predicted blockage risk contribute to about 8% of the blockages in the testing year. In other words, if top 1% high-risk pipes according to prediction were fixed, about 8% of the blockages would have been prevented. Similarly, fixing 10% of the pipes according to

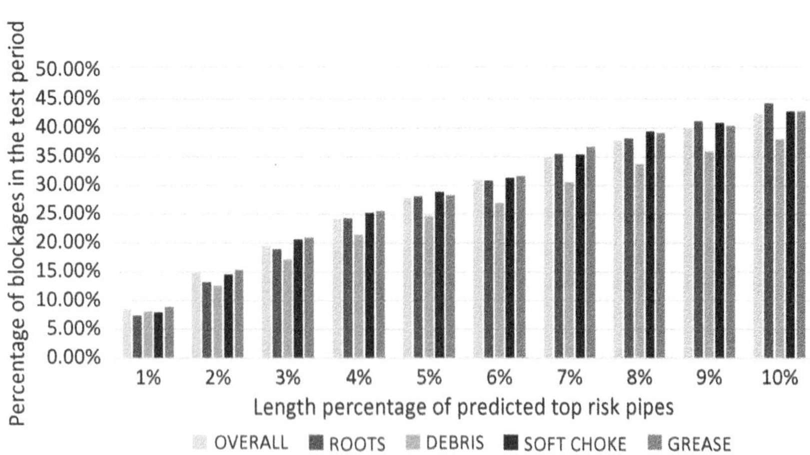

Fig. 10.6 Blockage prediction curve, showing the contributions of blockages in the test period for predicted top risk pipes

the predicted risk would have prevented more than 40% of blockages. Sydney Water is conducting a field validation to confirm the performance of the model, and if passed, the predictive model could improve the efficiency and reduce the cost of the future maintenance schedules.

This project illustrates how water utilities could improve their maintenance schedules by bringing in open urban data. With the predictive model driven by cross-domain data, water utilities could reduce their operation cost and service interruptions, providing a better service with lower cost to the public. Modern cities are a complex system with sub-systems interacting with each other. With the movement of Open Data, datasets of different sub-systems will be empowering similar techniques in other asset-intensive urban sub-systems, bringing better services to the citizens.

Smart Parking Occupancy Pattern Analysis

Background

With the increasing demand and expectation for public parking spaces, smart parking systems are being adopted in many parking lots. They utilise sensing devices to monitor occupancies in real-time, and distribute the information via various channels, such as signboards, Internet websites and mobile applications. They help reduce parking time, ease traffic load burden and better utilise the parking spaces.

In this study, we collaborated with a local government agent on a trial of smart parking system. The aim is to understand the parking patterns from the sensor data, and develop a prediction model for the future occupancy rates. The outcomes of the data analysis and prediction model provide decision support for both policymakers and motorists.

Datasets

The parking lots in this trial are located at a local central business area, surrounded by shops, restaurants, a cinema and a cricket/oval ground hosting Australian Football League (AFL) games. Each of the facilities

Table 10.2 Datasets used for smart parking occupancy pattern analysis

Dataset	Description	Source
Parking records	Records of more than 4 million parking events for more than 400 bays, including start time, end time, bay, lot type (1P Free, 2P Voucher, etc.), location, etc.	Public—Open Data Portal (ACT Government, 2017)[a]
Public Wi-Fi	Number of users connected to nearby public Wi-Fi access points.	Private—Local government
Payment transactions	Payment records for paid parking lots.	Private—Local government
Special days	Dates of public holidays and events (Father's day, Mother's day, AFL game).	Public—Public
Weather	Temperature, rainfall and solar.	Public—Bureau of Meteorology[b]

[a]https://www.data.act.gov.au/Transport/Smart-Parking-Stays/3vsj-zpk7
[b]http://www.bom.gov.au/climate/data/

has different attendance patterns, which generates different parking demand. In order to understand the impacting factors for the parking patterns, we investigated the following datasets (Table 10.2).

Factor Analysis

In order to identify influential factors for predicting parking patterns, we conduct statistical analysis on different factors. We analysed the parking patterns in three metrics:

- Occupancy rate: the utility ratio of a parking bay, defined as the percentage of duration that the bay is occupied by vehicles.
- Arrival rate: the turn-over of a parking bay, defined as the number vehicles arrived per hour per bay.
- Overstay rate: the percentage of vehicles that stayed more than 10 minutes over the parking time limit.

A few useful insights discovered include:

Hour of day: as one would expect, the demand for parking spaces would have its peak and non-peak time during a day. Table 10.3 summarises the

Table 10.3 Occupancy rate for different hour of day and different lot types

Hour	Overall	5Min Free	1/4P Free	1/2P Free	1/2P Voucher	1P Free	1P Voucher	2P Free	2P Voucher	4P	8P	Disabled	Dropoff	EV	LZ
0	0.053	0.119	0.043	0.076	0.079	0.094	0.114	0.126	0.042	0.022	0.030	0.016	0.067	0.013	0.025
1	0.043	0.095	0.036	0.052	0.050	0.081	0.081	0.122	0.032	0.019	0.028	0.012	0.038	0.008	0.016
2	0.042	0.073	0.039	0.039	0.038	0.078	0.082	0.148	0.029	0.018	0.028	0.010	0.022	0.007	0.016
3	0.043	0.064	0.034	0.032	0.034	0.074	0.104	0.166	0.027	0.018	0.028	0.010	0.021	0.007	0.014
4	0.045	0.060	0.037	0.029	0.035	0.069	0.120	0.175	0.027	0.019	0.028	0.010	0.015	0.010	0.014
5	0.062	0.069	0.046	0.029	0.063	0.076	0.199	0.205	0.058	0.019	0.027	0.018	0.016	0.016	0.015
6	0.107	0.224	0.067	0.033	0.148	0.157	0.270	0.232	0.158	0.021	0.028	0.065	0.023	0.016	0.029
7	0.200	0.634	0.339	0.070	0.284	0.337	0.367	0.308	0.357	0.028	0.039	0.120	0.036	0.017	0.067
8	0.308	0.720	0.457	0.452	0.317	0.521	0.402	0.451	0.573	0.050	0.105	0.219	0.043	0.043	0.137
9	0.430	0.776	0.552	0.776	0.419	0.620	0.489	0.653	0.760	0.121	0.202	0.398	0.090	0.097	0.272
10	0.567	0.790	0.625	0.858	0.606	0.769	0.612	0.774	0.897	0.308	0.246	0.691	0.116	0.296	0.422
11	0.634	0.827	0.680	0.885	0.714	0.833	0.671	0.773	0.912	0.450	0.293	0.811	0.162	0.409	0.468
12	0.725	0.863	0.734	0.926	0.874	0.934	0.742	0.784	0.946	0.646	0.334	0.869	0.327	0.521	0.492
13	0.729	0.853	0.735	0.910	0.846	0.897	0.733	0.770	0.929	0.706	0.357	0.788	0.339	0.516	0.446
14	0.646	0.837	0.728	0.865	0.739	0.805	0.676	0.723	0.867	0.536	0.339	0.717	0.236	0.443	0.360
15	0.579	0.841	0.739	0.867	0.689	0.735	0.675	0.679	0.831	0.393	0.289	0.625	0.243	0.371	0.286
16	0.539	0.844	0.760	0.854	0.725	0.731	0.695	0.640	0.770	0.335	0.237	0.508	0.237	0.345	0.289
17	0.543	0.859	0.764	0.831	0.873	0.804	0.762	0.540	0.739	0.383	0.148	0.444	0.376	0.419	0.478
18	0.653	0.870	0.735	0.883	0.917	0.913	0.811	0.530	0.888	0.643	0.108	0.550	0.501	0.596	0.691
19	0.694	0.870	0.678	0.885	0.913	0.920	0.792	0.523	0.902	0.761	0.160	0.592	0.540	0.625	0.723
20	0.605	0.847	0.574	0.792	0.868	0.830	0.739	0.451	0.759	0.646	0.147	0.455	0.483	0.550	0.647
21	0.404	0.764	0.414	0.538	0.729	0.607	0.601	0.316	0.466	0.371	0.094	0.243	0.352	0.375	0.510
22	0.232	0.594	0.212	0.306	0.476	0.399	0.431	0.231	0.221	0.181	0.059	0.121	0.218	0.197	0.221
23	0.113	0.224	0.067	0.146	0.219	0.203	0.262	0.179	0.090	0.072	0.039	0.045	0.115	0.053	0.069

occupancy rate for different lot types. Two peak times for lunch (11:00–13:00) and dinner (18:00–20:00) are observed for most lot types.

Special days: we matched the parking records with the dates of public holidays and events to see if there are any differences in parking patterns. As depicted in Fig. 10.7, such days do exhibit different parking patterns compared to a normal day, and each day has its unique pattern. For instance, the occupancy rate is higher on Easter Saturday, AFL game day and Mother's Day, but much lower on Christmas and Boxing Day.

Weather: to investigate the parking behaviours in different weather, we matched the parking records with the local temperature, rainfall and solar records. However, no significant difference in parking pattern is observed. Figure 10.8 shows the results for rainfalls, no matter a day is sunny or rainy, the occupancy, arrival and overstay rate stay in same levels.

Methodology and Outcome

In order to help motorists plan their trips, we build an ensemble-based regression model (Polikar, 2006) that takes all the factors as input to predict next hour occupancy rate for every parking lot. The model is chosen

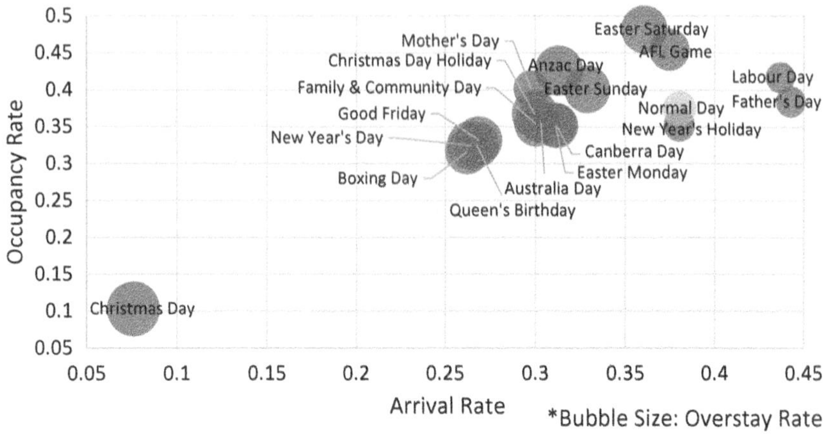

Fig. 10.7 The occupancy rate, arrival rate and overstay rate for different public holidays/events

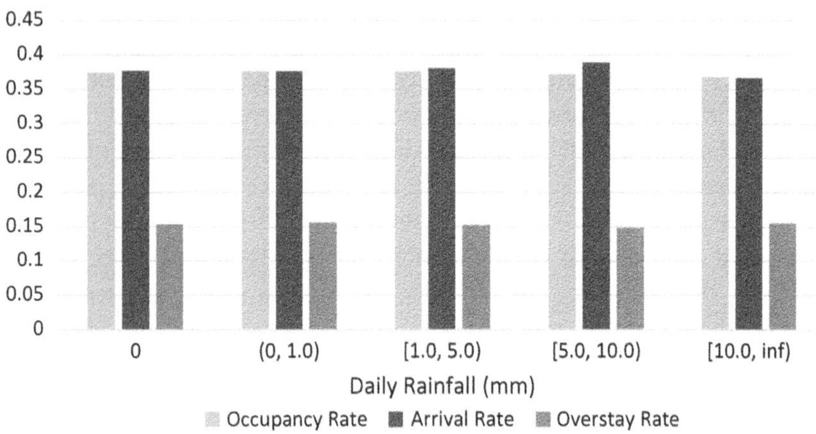

Fig. 10.8 The occupancy rate, arrival rate and overstay rate on days with different rainfall

because it is able to handle heterogeneous data types from different data sources, and it offers competitive prediction performance.

The predictive model is tested by using the standard ten-fold cross-validation process. That is, the data are split into ten folds, and the model is trained and tested for ten rounds, where in each round, nine folds of data are used to train the model, and one fold is left to test the results.

The overall R^2 score is 89.9%, which means the model could predict the occupancy rate with high accuracy, explaining about 90% of the variation. According to a survey for the trial smart parking app, most users find the occupancy rate prediction function to be useful in planning their trips and reducing parking time.

The project illustrates how Open Data could be contributing to the modelling of parking availability. Like other urban sub-systems, the traffic condition of a city is impacted by many factors. By providing a better guide, the parking time is reduced, and the traffic condition improved. To bring greater good to the community, and to inspire further research, the parking data is made available on the Open Data Portal (ACT Government, 2017).

Urban Function and Region Popularity Analysis

Background

In recent years, the fast-growing urban population and shifting urban functions pose a major challenge for urban planners. In a joint study with the planning department of a regional government, we aim to understand various factors involved in these processes, and answer the question: "where, when and why a region develops?" (Zhang, Zhang, Guo, Wang, & Chen, 2018)

Datasets

There are many factors influencing the developments of urban systems. In this study, we investigated the following datasets (Table 10.4).

Methodology and Outcome

We performed three different analyses on the collected cross-domain datasets:

Dwelling productions: dwelling production is an important indicator of urban development. Using the macroeconomic indicators as input factors,

Table 10.4 Datasets used for urban function and region popularity analysis

Dataset	Description	Source
Development application	Time, location, type, scale of new developments.	Public—Planning department[a]
Macroeconomic indicators	GDP, inflation rate, labour markets, housing, population, market/consumer performance, etc.	Public—Australian Bureau of Statistics and Reserve Bank of Australia[b]
Urban functions	More than 100 categories of business and public facilities, including location, type and size.	Public—Google Places API[c]
Properties	Type (house/apartment/unit), configurations (number of bedrooms), historical price.	Public—real estate websites[d]

[a]https://data.gov.au/dataset/sydney-region-dwellings/resource/cd918b27-05a8-426e-9d9e-e54d5b963c53
[b]https://www.rba.gov.au/statistics/tables/
[c]https://developers.google.com/places/web-service/intro
[d]https://www.realestate.com.au/

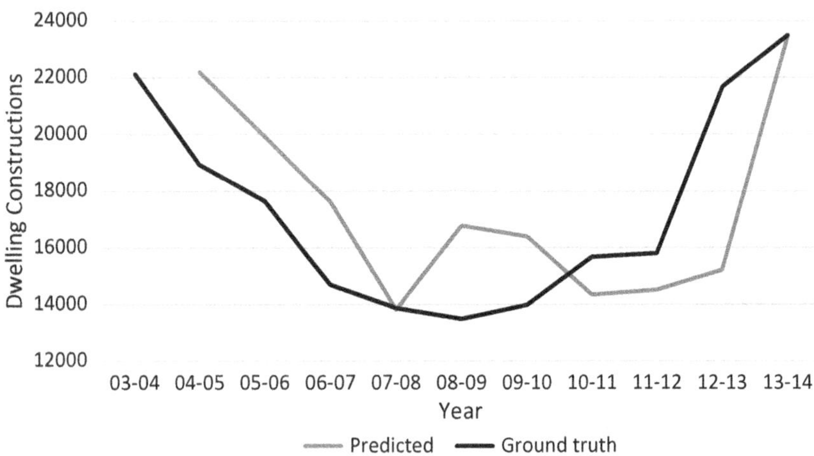

Fig. 10.9 Predictive power of macroeconomic indicators on dwelling constructions (DC)

we aim to predict when and where new dwelling will appear. As shown in Fig. 10.9, macroeconomic indicators are informative predictors for the dwelling production, yielding an accurate prediction.

Fig. 10.10 Examples of urban functions for the greater Sydney area

Urban functions: we try to understand the similarities and differences of different urban regions in terms of urban functions. To achieve this, we first compute the function intensities for each region from the business and public facility location data. The function intensity for a particular type of business and public function (e.g., bank, restaurant or school) measures the density of facilities with that function. Some examples of urban functions are shown in Fig. 10.10.

With the function intensities for each region, we then cluster the regions into different groups using a Bayesian nonparametric mixture model with spatial constraints (Lin et al., 2016b). The algorithm offers a tool to discover the regional patterns of urban functions and the similarities among regions.

Region popularity: lastly, we combine the urban functions and property data to analyse the region popularity. A self and mutually excited stochastic interaction point process (Lin et al., 2016a, 2016b) is used to

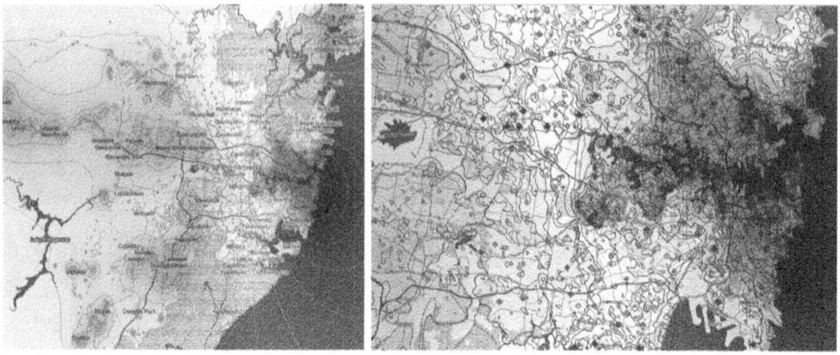

Fig. 10.11 Region popularity (median property price) heat map and evaluation contour line for the greater Sydney area

estimate the region popularity based on both the historical property price and region functions. The estimated popularity heat map and price contour lines are shown in Fig. 10.11. The model (Zhang et al., 2018) provides a useful tool to understand the interactions between urban functions and regional popularities.

Conclusion and Future Direction

The advocacy of Open Data helps make a huge amount of valuable urban data publicly available. The potential of such data, however, can only be fully realised by adopting a holistic cross-domain view of them. In many cases, the open urban data needs to be exploited with the support of organisation-owned private data for maximising its value. In this chapter, we showcased how industries and government agencies can utilise a combination of cross-domain data from both public and private sources to improve their capabilities in planning, operating and maintenance. These successful projects demonstrate the demand, methodology and benefit of the cross-domain urban data analysis. In the following, we share some of our observations and considerations on how to fully release the power of open urban data and how to gradually make more privately owned urban data publicly open for the society to use.

- Thanks to the advanced data analytic techniques, especially the recent advance in artificial intelligence and machine learning, the value of open urban data can be greatly released and help generate valuable insights for supporting efficient urbanisation.
- Synergy exists between open and private urban data. Open urban data can help leverage the potential value of organisation-owned private data. Both open and private urban data can work together to empower the efficient operation and maintenance of urban systems.
- Urban system has many sub-systems, for example, urban infrastructure, urban transportation and so on, which are working closely with each other. Making one sub-system's data public open can benefit other sub-systems via cross-domain urban data analysis.
- Due to the data security and privacy concerns, it is difficult for many industrial organisations to make their datasets public open. However, the value of such private urban data is significant. Hence, efforts need to be made to help industrial organisations to gradually release their data and make it publicly available for other parties' utilisation.
- Research community has also paid attention to help make private data safely open for sharing. For instance, privacy preserving data analysis and machine learning on encrypted data techniques (Aldeen, Salleh, & Razzaque, 2015; Graepel, Lauter, & Naehrig, 2013) have attracted increasing attention and have been widely adopted in many different areas for making private data safely sharable.

Performing cross-domain data analysis for a single urban sub-system is just the first step. Looking into the future, more efficient optimisation of urban systems could be achieved if we could jointly optimise multiple sub-systems together. For example, by considering the current demographics and new developments of shops and dwellings, local governments could obtain an accurate forecast of how the urban functions will change in the future, and make collaborated policies to facilitate such transition. Such information could also be passed to utility and telecommunication companies to ensure the infrastructures could be built to match the growing demand. We believe that Open Data, combined with cross-domain data analysis, could transform future cities into smart, inclusive and responsive open cities.

Acknowledgements The authors would like to thank our Sydney Water collaborators, Bronwyn Cameron, Mark McGowan, Craig Mitchell, Judith Winder and Rod Kerr, for the wastewater pipe blockage prediction work.

References

ACT Government. (2017). *Smart parking stays | Open data portal.* [Online]. Retrieved May 21, 2018, from https://www.data.act.gov.au/Transport/Smart-Parking-Stays/3vsj-zpk7

Aldeen, Y. A. A. S., Salleh, M., &. Razzaque, M. A.. (2015). *A comprehensive review on privacy preserving data mining.* s.l.: SpringerPlus.

Daley, D., & Vere-Jones, D. (2002). *An introduction to the theory of point processes.* s.l.: Springer.

Graepel, T., Lauter, K., & Naehrig, M. (2013). *ML confidential: Machine learning on encrypted data.* s.l.: Information Security and Cryptology—ICISC 2012. ICISC 2012. Lecture Notes in Computer Science, vol. 7839. Springer, Berlin, Heidelberg.

Hawkes, A. G. (1971). Spectra of some self-exciting and mutually exciting point processes. *Biometrika, 58*(1), 83–90.

Li, B., Zhang, B., Li, Z., Wang, Y., Chen, F., & Vitanage, D. (2015). *Prioritising water pipes for condition assessment with data analytics.* s.l.: OzWater.

Li, Z., Zhang, B., Wang, Y., Chen, F., & Vitanage, D. (2014). *Water pipe condition assessment: A hierarchical Beta process approach for sparse incident data.* s.l.: Machine Learning.

Lin, P., Zhang, B., Wang, Y., Li, Z., Li, B., Wang, Y., & Chen, F. (2015). *Data driven water pipe failure prediction: A Bayesian nonparametric approach* (pp. 193–202). New York, NY: ACM International Conference on Information and Knowledge Management.

Lin, P., Zhang, B., Guo, T., Wang, Y., & Chen, F. (2016a). *Infinite hidden semi-Markov modulated interaction point process* (pp. 3900–3908). Barcelona, Spain: Advances in Neural Information Processing Systems.

Lin, P., Zhang, B., Guo, T., Wang, Y., & Chen, F. (2016b). *Interaction point processes via infinite branching model.* Phoenix, AZ: Thirtieth AAAI Conference on Artificial Intelligence.

Polikar, R. (2006). Ensemble based systems in decision making. *IEEE Circuits and Systems Magazine, 6*(3), 21–45.

Ramaswami, A., Russell, A. G., Culligan, P. J., & Karnamadakala Rahul Sharma, E. K. (2016). Meta-principles for developing smart, sustainable, and healthy cities. *Science*, 940–943.

Zhang, B., Zhang, L., Guo, T., Wang, Y., & Chen, F. (2018). *Simultaneous urban region function discovery and popularity estimation via an infinite urbanization process model.* London, UK: ACM SIGKDD Conference on Knowledge Discovery and Data Mining.

11

Interfacing the City: Mixed Reality as a Form of Open Data

Jeremy Harkins and Christopher Heard

Abbreviations

AECO	Architecture, Engineering, Construction, and Operations
AHU	Air Handling Unit
AR	Augmented Reality
BIM	Building Information Modelling
BMS	Building Management System
FLOPS	Floating Point Operations Per Second
GB	Gigabyte
GUI	Graphical User Interface
HVAC	Heating, Ventilation, and Air Conditioning
IBMS	Integrated Building Management System
IoT	Internet of Things
OBSI	Open Building Systems Integration

J. Harkins (✉) • C. Heard
ineni Realtime, Surry Hills, NSW, Australia
e-mail: jeremy@ineni.co; chrish@ineni.co

S. Hawken et al. (eds.), *Open Cities | Open Data*,
https://doi.org/10.1007/978-981-13-6605-5_11

PoC	Proof of Concept
VDC	Virtual Design and Construct
VR	Virtual Reality

Highlights

- Interface design
- Gamification
- Virtual worlds
- Integrated systems

Introduction

Interface is everything.

How we interact with physical, ephemeral, and virtual objects always involves some form of interface.

The boundary between surfaces, language, writing, imagery, and tools are all forms of interfaces, though the focus of this chapter will be on the interface between humans and our emerging smart cities.

By applying learning that has been driven by the video game industry, in the areas of interface and interaction design, we can make existing industry processes more engaging and valuable. Planning, design, construction, and operation of our buildings, infrastructure, and cities can all benefit from more intuitive interfaces through 'gamification'.

Game engine technology provides a unique method of interfacing with huge data sets, enabling contextualised access to information that would otherwise be inaccessible, confusing, masked, or hidden.

Data sets can be either open or proprietary, and while the generally accepted definition of Open Data refers to publicly available data, for this chapter, the concept of Open Data is taking a slightly different interpretation. The interpretation being referred to here is how proprietary data sets can be shared among commercial entities freely, for efficiency and mutually beneficial outcomes. In this instance, this is a project-based industry move towards more Open Data sharing that is tending towards the concept of Open Data.

By integrating our design and construction technologies, such as Building Information Modelling (BIM), with other activities and systems throughout the lifecycle of a development, including marketing, leasing, training, asset management, facilities management, and many other areas of building activity, there is an opportunity to develop a virtual representation of the real world that can be used as an intuitive interface to our built environment.

Currently, the primary challenge presented by the increasing amount of data being generated by our cities is in how we digest it and subsequently make informed decisions based upon it. The solution to this lies partly in creating interfaces that can access and represent this data in a more human readable form.

A case study of the Barangaroo South Development, a city precinct in Sydney, Australia, will be examined exploring how BIM and integrated systems work together to provide a 3D interface to building data. The business case around the distributed creation of the virtual asset across the Architecture, Engineering, Construction, and Operations (AECO) industries will also be examined, and how a common virtual model for all activities is helping disparate disciplines to collaborate more effectively, providing better building outcomes, and helping to create our smart cities from the inside out.

With the majority of smart city exploration looking from the outside, concentrating on infrastructure, city services, and big data, a building-first approach may be key to making more human-relevant decisions that inform commercial and social interactions.

A Picture Is Worth a Thousand Words

Visual information comprises 80–90% of information that enters the brain, according to some researchers (Hyerle, 2008), so visual interfaces will likely be key in how we as people deal with the information explosion (Zakon, n.d.) related to smart cities, and exposing the underlying interactions of our future cities.

When we think about visual information, one of the industries with the most prevalent use of visuals that comes to mind is the entertainment industry, specifically television and moving pictures. From the time of the first electronic television broadcasts in the 1930s and 1940s, anecdotally we have been absorbing information in the form of images and moving pictures through screens at an ever-increasing rate.

A natural progression that evolved from television and motion pictures was a desire to interact with the media, which led to video or computer games. Video games take the concept of immersion in moving images a step further than passive watching of screens, allowing users to engage interactively with the visual information being displayed, and have a direct effect over the use and outcome of this information.

The age of information has seen an explosion of data related to people and places, with the internet and modern search engines putting a growing amount of information and concepts on an almost endless number of topics only a few keystrokes away. Connected devices such as smart phones further decrease the barriers to accessing this growing resource of data.

Beyond the ability to read and engage with text and images on our handheld mini computers (mobile phones), these portable visual devices are becoming a portal to access an emerging virtual world and the inherent data that comprises this digital landscape. As the infrastructure is developed to keep us all connected no matter where we are, our appetite for information is only growing, and the exploration of this alternate virtual reality is in its infancy.

Data: At the Core

At the core of this influx of accessibility to information is the idea of data, and as an indication of the rate of data generation, we can compare how much information has ever been created to how much we are creating every day in the modern world.

> *Between the dawn of civilization and 2003, we only created five exabytes; now we're creating that amount every two days. By 2020, that figure is predicted to sit at 53 zettabytes (53 trillion gigabytes)—an increase of 50 times. Hal Varian, Chief Economist at Google* (Rizzatti, 2016)

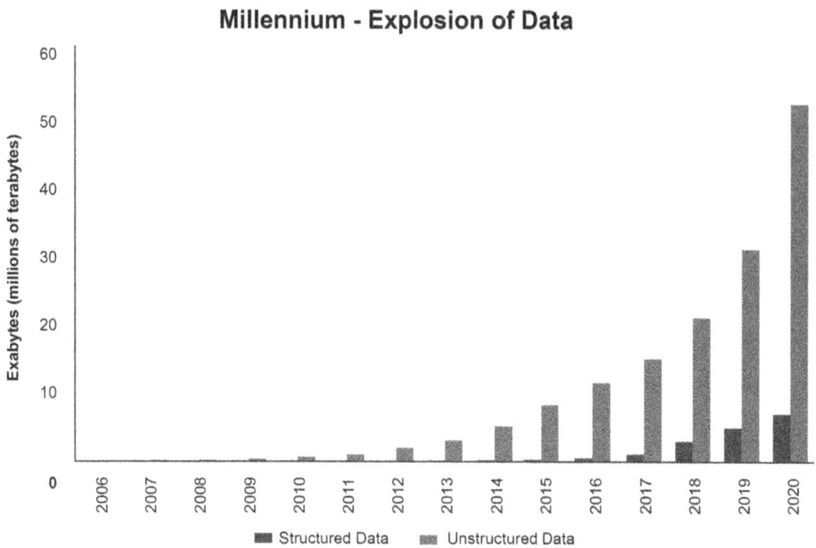

Fig. 11.1 The growth of structured versus unstructured data over the past decade shows that unstructured data accounts for more than 90% of all data. Image adapted from (Rizzatti, 2016)

This is more data than a single person will ever be able to meaningfully consume, yet the propagation of this data within our communities, and by extension our urban areas and cities, is opening our cities to new paradigms, such as the smart city (Fig. 11.1).

The Smart City

Many of our most well-known cities have historically grown organically over time, forging their physical and cultural identity through a slow process of human development and interaction. The physical aspect of our cities is fairly well understood and defined by physics and construction technologies. The cultural and social identity of a city is much more fluid and ambiguous, evolving with trends and technologies in an ever-changing landscape of human activity.

There are many differences between a traditional city and the emerging concept of the "smart city". From how they were planned and built, to how inhabitants live and interact within them. One of the most significant differences however is the integration technology, specifically digital technologies, and one of the defining aspects of digital technologies is the creation and retention of data. We have always collected and stored data, even before computers, in the form of literature and ledgers (among other mediums). The digital age has brought us the ability to index, search, and relate our data with much greater efficiency. To contextualise the magnitude in the change of speed of data access, in the past it may have taken a clerk minutes to find a particular piece of data in a filing cabinet or ledger, effectively achieving one operation. By comparison, the contemporary supercomputers are performing quadrillions of calculations per second, commonly known as Floating Point Operations Per Second (FLOPS) (Feldman, 2016), which is comparable to a million 2018 planet Earths worth of people, each performing a task every second. This has led to the ability to interact with and analyse massive datasets, gleaning insights about complex systems that were previously impossible to articulate, or even know existed.

The question arises of how can we make our cities safer, healthier, cleaner, and more efficient through the application of this ocean of data?

The answer is in the ideal of the smart city, where the very fabric of our cities responds to our needs and adjusts itself to optimise comfort for the individual and the group.

Just as the smart phone has given us a powerful set of tools in our hands, removing the need to remember phone numbers or emails and enabling us to capture high-quality photos and movies at a whim, so too will the smart city smooth our occupancy through tasks like guiding us to where we want to go, providing on demand transport, ensuring that there are no traffic jams, or managing our waste and the cities' cleanliness in the most efficient ways.

The Interface

A significant concept that computers have introduced into the zeitgeist is the idea of the interface; the mechanism or boundary allowing for interactions between the user and the components of a task. The study of UI

(User Interface) and UX (User Experience) didn't even exist before the advent of computers and a need to manipulate virtual space.

It can be argued that any physical object has some type of "interface", for instance the handle on a hammer could be seen as the interface to the hammerhead, but with physical interfaces, the use and function of the object we are interfacing with are inherently linked. In the abstract world of computing, the interface becomes a function unto itself. Without interface there is no way to interact with the invisible 0s and 1s that make up the data being held within our devices.

One of the earliest human-computer interface techniques was text on a screen in the form of written words and syntax to enact the underlying operation of the computers programming, with one of the most well-known examples being Microsoft's DOS prompt. This was a clunky medium and was probably responsible for the stigma of complexity that grew around computers in the early decades. An early step away from this text-based interaction was the concept of the Graphical User Interface (GUI), replacing the text-based interaction with a visual cue-based interaction. This was enhanced by the concept of the mouse that allowed for a "point and click" paradigm. In these early GUIs, designers would borrow from real-world examples, such as the concept of a folder, which was utilised by early adopters such as Xerox, Macintosh, and Microsoft: a virtual representation of the filing cabinet.

This representative medium to interact with computer code has been taken to the extreme in the video game world, where almost anything you can imagine has been turned into some type of game since the early 1970s; from controlling a dot around a screen in early games, such as the classic game "Pong" (Steven, 2010) to the fully immersive experiences of today such as the "Call of Duty" franchise, developed by Activision since 2003.

It was the advent of computer game consoles for consumers in the early 1980s that saw the first step to engaging a wider audience in the use of computers, outside of the Personal Computer that was largely used by professionals at the time, and the GUI is in part what helped to engage non-technical users.

The development of both software and hardware have gone hand in hand, incrementally being refined to give the user the best experience, leading to our modern-day interface techniques such as touch screens and gesture-based controls.

IoT: The Interconnectedness of Things

With the advent of the digital age and the internet, there has been an increase of interconnected people and processes across the world which is having a profound effect on the very nature of what makes a city, a city. Our cities are no longer just a collection of physical spaces; there is now an underlying layer of information and data that helps to operate our environments and buildings, increasingly becoming more autonomous. This layer of information has increased operational efficiencies and allowed for the concept of the smart building and smart city to come into being.

The underlying data being generated by our built environment and being used to operate our cities goes largely unnoticed by most people utilising our spaces, which has been traditionally fine for the majority of subsystems. However, as more and more people are becoming aware of this underlying data, there is a growing desire to access and use this information to our benefit as a population.

To date there is no universal way to easily interact with this information, and where there are ways to gain access, the data is rarely, if ever, contextualised in a meaningful way for non-skilled users to understand.

By creating accessible virtual environments of our buildings and cities, using design processes such as BIM as the basis for these virtual worlds, and enhanced with rich metadata, we are not only democratising the valuable data created during the design, construct, and operation of a development, we are also providing a visual platform for everyday users to better understand the context and availability of the data woven inextricably into the fabric of our cities.

Because of this location-based and context driven approach to our data, the way we are interacting with information is become more contextualised and less abstract, and it is possible that the thing that will change most about the internet in the coming decade will be how we interact with it.

The internet was instrumental in changing the way we deal with data, but since its inception, it hasn't really changed that much. The basics of communicating over long distance and interacting with text, images, videos, and systems through a screen display have been fairly constant.

This ability to access the underlying data behind our physical objects through interfaces on our handheld devices could be a major paradigm shift in computing interaction and seems to be a growing movement evidenced by the global interest in the Internet of Things (IoT).

Amazingly Powerful Software

Gaming as an industry has seen exponential growth in recent years, with 2016 estimated to reach over US $90 billion globally (van der Meulen, n.d.). The user base at the core of the industry, the majority coming from the younger generations, are engaging with cutting edge programming techniques and hardware solutions, such as Virtual Reality (VR) and Augmented Reality (AR). Due to the size of the industry and a need to engage large audiences, developers in the games industry utilise new technology earlier than other industries to gain an edge.

The growth of this industry has attracted interest from many business sectors, especially businesses related to the built environment. This scrutiny has led to the realisation of the power and diversity of the tools used to make video games, and subsequent investment to repurpose these tools to solve complex problems outside of gaming, is gaining traction.

Games have traditionally been about interaction and design, where the goal is to create applications that are intuitively easy to use, have a high level of interactivity, and are visually engaging to make people want to play them.

However, the concepts and toolsets used to create video games are useful for software and tasks outside of the imaginary world of video games and the visual nature can reduce the learning curve to understanding complex datasets.

Leveraging the significant ongoing development occurring in the game industry, innovative, interactive interfaces to real-world data can be developed, which are forming the basis of contextualised interfaces for our smart buildings and cities. This is called gamification.

Case Study: Barangaroo South Precinct, Sydney, Australia

Virtual Barangaroo and the Open Building Systems Integration (OBSI)

Barangaroo South is the business district precinct of the Sydney waterside suburb of Barangaroo. Originally a territory of the Cadigal People, the native Australian traditional land owners of the Sydney region, the land has seen a rich history both before and after colonial settlement. For the Cadigal people it was an area of fishing and hunting for up to 14,500 years, after settlement it became one of the key wharf districts of Sydney starting in about the 1850s (Discover Barangaroo: History, 2011).

After decades of little use of the area, in 2003 the NSW government determined that the entire Barangaroo suburb would be redeveloped from being a shipping and stevedoring area into a mixed-use precinct, consisting of commercial and residential developments, public space, and parklands. After an initial controversial design competition in 2005 that saw the winning entry from Hill Thalis Architects replaced with international Architect Richard Rogers, from the firm Rogers Stirk Harbour + Partners, the design was eventually awarded to Lendlease (as sole developer of the Barangaroo South precinct).

The government's brief for Barangaroo called for "a visionary plan that would position Sydney as Australia's truly global city and would establish a framework for our urban growth" (Rapley, 2014). Based on this, there was a desire by the developers to assure a truly world class development utilising the best-in-breed technology available at the time and provide infrastructure that would make the precinct as "smart" and future-proofed as possible (Fig. 11.2).

In 2005, when Lendlease was awarded the development rights for Barangaroo, "smart" was a term coming into use to describe technology that had an automated aspect, allowing for our technology to perform tasks with little to no supervision. For instance, there were early "smart phones" that were packed with many functions that could help with efficiency in our daily tasks, yet the lack of accessible infrastructure and

Fig. 11.2 Image capture from Virtual Barangaroo, showing an artist's representation of the Barangaroo precinct. A boundary line has been added to show the site extents of Barangaroo South, the area of interest for the Virtual Barangaroo Project

simple to use interfaces on these devices made many of these functions barely usable, and far from what we understand to be "smart".

It was not until the release of the first iPhone in 2007 that we started to see changes in how we interacted with the increasing data being made available to us through an expanding internet.

In just over a decade, we have seen a boom in "smart" technology, with a plethora of products and processes claiming to be "smart", from buildings and phones, all the way down to light bulbs, all the way up to entire cities; and everything in between.

With an 18-year development plan for Barangaroo, aiming to be completed in 2023, and with the mandate of creating a "visionary" development, the site was destined to be a testing ground for many emerging technologies, many aimed at what was to be coined "smart cities".

As with the term "smart", the concept of the Smart Building was also emerging in 2005, with early examples of building automation such as networked systems for lighting, air conditioning, and security being implemented to make buildings more "liveable".

These systems were and still are largely invisible to the regular building occupant, being embedded in the building fabric, with only some of the functions immediately apparent, such as the lights turning on automatically when entering a space, or doors locking, and unlocking based on the occupant's personal credentials in the underlying systems, usually using an access card reader as the touch point.

While the underlying data involved in such systems is likely owned or managed by private entities, the access and interaction with this data can be used by the public, and in that sense the data is "open" for the purposes of individual activities, even if the entire data set is still closed or proprietary. As we become more comfortable as a society with the concept of "access to data", less sensitive information that may still be proprietary could have a tendency to be made open for the public good.

The overall vision of smart buildings is to have many of the tedious tasks involved in interacting with a building become automated for occupants, letting them live more seamlessly without having to worry about repetitive tasks.

Barangaroo South is an interesting development for many reasons (history, location, controversy, technology, size, and budget, among other aspects), though one particular point of interest to this case study is that the development period of 18 years has and will see many changes in technologies and processes throughout its development's duration.

To implement best-in-breed practices across the site and aim to deliver as future-proofed a precinct as possible, Lendlease engaged as many new technologies in the AECO industries as was feasible. Relevant to Virtual Barangaroo and the OBSI platform, they made a commitment to implement Virtual Design and Construct (VDC) practices using BIM for the precinct, which at the time was a relatively new methodology being applied in industry, particularly for large developments.

Beyond the final delivery of Barangaroo South, the project offered Lendlease an opportunity to learn from the experience of developing Barangaroo and embed the learnings across their entire business with the goal of producing global efficiencies for the business.

As Barangaroo South is intended to be a new business, residential, and cultural hub of Sydney, plans to make the development both engaging on completion and future-proofed for at least the next 20 years were forefront in Lendlease's scope.

One initiative in the planning phase was an open tender to Building Service Providers (such as Honeywell, Siemens, Schneider Electric, and Alerton, among others) called the Open Building Systems Integration (OBSI) platform. The scope of the OBSI was for a provider to deliver an Integrated Building Management System (IBMS). The vision was to integrate all of the systems and data generated from these systems into a common data environment allowing for control of all embedded sensors and services from a single 2D Dashboard Interface point. Systems originally planned to be included in the integration, but not limited to, were Interior Lighting, Exterior Lighting, Heating, Ventilation, and Air Conditioning (HVAC), Elevators, Fire Alarms, Doors and Window sensors, Card Access points, Hot-Desking Systems, Room Booking software, Energy Monitoring, and Concierge Services.

Having already made a commitment to utilise BIM and VDC methodologies for the development of Barangaroo South, the question was asked during the tender process, "Can the BIM be used in any way to enhance the OBSI?"

This seemingly innocent question led to a significant addition to the requirements for the OBSI tender; the use of the BIM as the basis for a graphical interface to enhance the traditional 2D building services data dashboards. At this point in time, when they were in the process of planning how the system may work, the OBSI and its potential uses were still just a concept, with no obvious precedents available as an example.

In 2012 when the OBSI tender was proposed, there were very few commercial examples, if any, of the BIM being used in such an innovative way. The concept was talked about in academic circles, but there was no known out-of-the box software to achieve the desired outcome; additionally, there was no business model around the delivery and implementation of such a system.

By chance, one of the tenderers (Alerton) came across a young company (ineni Realtime) at a trade show, who were showing off Realtime Environments created in the UDK (Unreal Development Kit) game engine for interactive architectural visualisation.

After a short development phase of around a week, the two companies' teams integrated BIM's from Barangaroo's proposed International Towers with live data from a BMS called iVivaCloud. A Proof of Concept (PoC) was created where data from the BMS was linked to data and the geometry from the BIM in the game engine environment (Fig. 11.3). Utilising internet protocols, the virtual assets in the BIM were live linked back to the BMS data control dashboard.

Effectively, there were three computers, all communicating through the cloud for the PoC. One computer was controlling a temperature sensor, monitoring the status and output of the device, and sending the data to the BMS system in the cloud. The second computer had an instance of the BMS dashboard showing the data being received from the device in realtime and communicating to the third computer that was running the 3D graphic representation of the building, with the located virtual temperature sensor mimicking the physical devices real-world status.

This proof of concept demonstrated an optimised BIM linked to live data for the purpose of monitoring and controlling sensors in a building environment and eventuated to the Alerton-led consortium being awarded the OBSI contract.

Fig. 11.3 Image capture of the original Proof of Concept (PoC) 3D interface for Lendlease's OBSI platform. This view is showing the early concepts of visual hot-desking and temperature sensors in the ceiling

The difficulty in achieving Lendlease's vision, while not an unprecedented concept in academia, centred around the utilisation of a 3D interface for controls. The integration of many data sources into a common dashboard environment had been achieved by several companies to varying degrees of sophistication by 2012; however, the use of a 3D real-time interactive environment for displaying the dashboard data in context to the assets was still largely untried in industry.

This case study is mainly concerned with the visual interface created from the BIM rather than the complex integration of the back-end systems.

One of the early realisations, once the contract had been awarded, was that there were varying levels of quality and detail available for the 3D visual interface. These different levels of visual information were useful for different activities. The original scope of the OBSI 3D interface was contracted at a low level of detail and quality (Fig. 11.4).

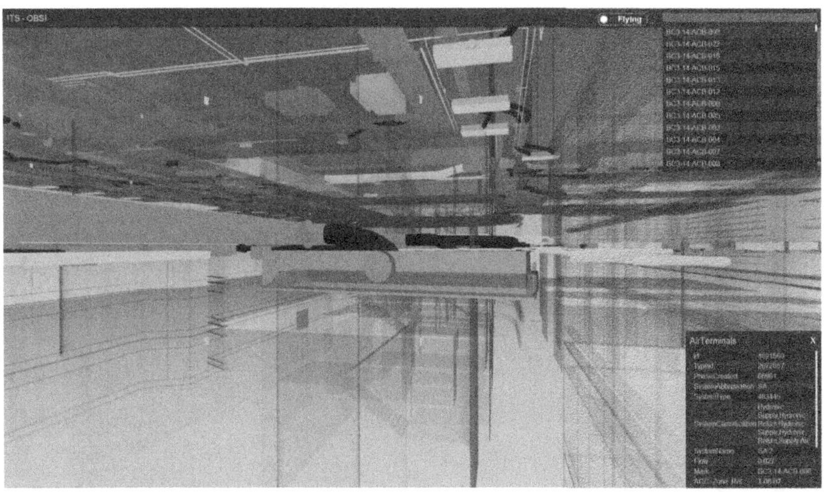

Fig. 11.4 Image capture of an early version of the 3D interface for Lendlease's Open Building Systems Integration (OBSI) platform, produced at a low level of detail and quality. This view of the virtual model is showing a virtual representation of an Air Handling Unit (AHU). The data being displayed is the embedded BIM data associated with the asset, and live operational data being pulled from the Building Management System (BMS)

This low level of model fidelity was suitable for operations; however, there was a desire from the Lendlease project team to utilise high-level detail and quality for purposes outside of building operations.

This led to a secondary contract directly with Lendlease to create Virtual Barangaroo; a high-quality visual model used for visual activities related to the development of the precinct (Fig. 11.5).

This highly detailed visual model was used for many non-operational activities, including, but not limited to, stakeholder and community engagement, design review and testing, operational readiness planning, training and inductions, still renders and fly-throughs, leasing activities, and change management activities.

Virtual Barangaroo became a living representation of the site as it was being developed, being extended area by area through over 35 contracted engagements (Fig. 11.6).

Each contract served a particular purpose, delivering the contracted deliverable, yet allowed for a "greater than the sum of its parts" resultant

Fig. 11.5 Image capture from Virtual Barangaroo showing a representational view from the proposed ferry route to the new Barangaroo South Ferry Terminal. This view was part of the virtual "Day in the Life" tour of a typical Barangaroo inhabitant, produced for stakeholder and community engagement during the development of the precinct

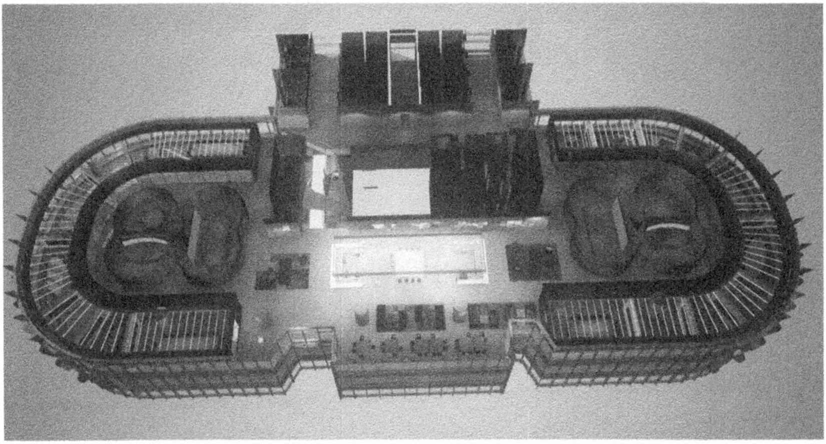

Fig. 11.6 Image capture showing a "Dollhouse" view of the proposed Level 14 fitout in Tower 2 of the International Towers, Barangaroo South. This third-person viewpoint is one potential viewing technique contained within Virtual Barangaroo application, allowing for stakeholder engagement and visual design review to assure a consistent delivery of the proposed design from the Architects

virtual environment form the combination of the progressive development approach (Figs. 11.7 and 11.8).

Virtual Barangaroo and the 3D interface to the OBSI share a symbiotic relationship, where the content produced for each stream contributed to the overall development of the two in tandem. For instance, the "ghosted" geometry that can be seen in Fig. 11.9 is the same geometry developed for the Virtual Barangaroo model from Fig. 11.5, just optimised and treated different visually to provide the best view of the operational OBSI model.

Virtual Barangaroo and the OBSI 3D Interface have been collectively worked on for over four years, slowly developing into the robust and detailed environment that has been produced for visual purposes and operational purposes across the Barangaroo South precinct (Fig. 11.10).

Due to the size and scope of the files associated with the Virtual Barangaroo and the OBSI, many of the versions created require a high-powered laptop to run efficiently (Fig. 11.11).

However, due to the utilisation of game engine technology to create the deliverables, one of the benefits that the gaming software provides is to allow

Fig. 11.7 Image capture showing an elevated perspective view of the proposed Lobby fitout in Tower 2 of the International Towers, Barangaroo South. This viewpoint was an incorporated aspect of the one of the fly-through videos produced for Lendlease to communicate design particulars about development. This area of Virtual Barangaroo was also used for interactive view-line testing, aiding in the placement of the concierge and security services

Fig. 11.8 Image capture showing a first-person view of the proposed Level 14 fitout in Tower 2 of the International Towers, Barangaroo South. This viewpoint formed a section of a fly-through video produced for Lendlease to comminute internally to general staff the design features of the Architect designed fitout

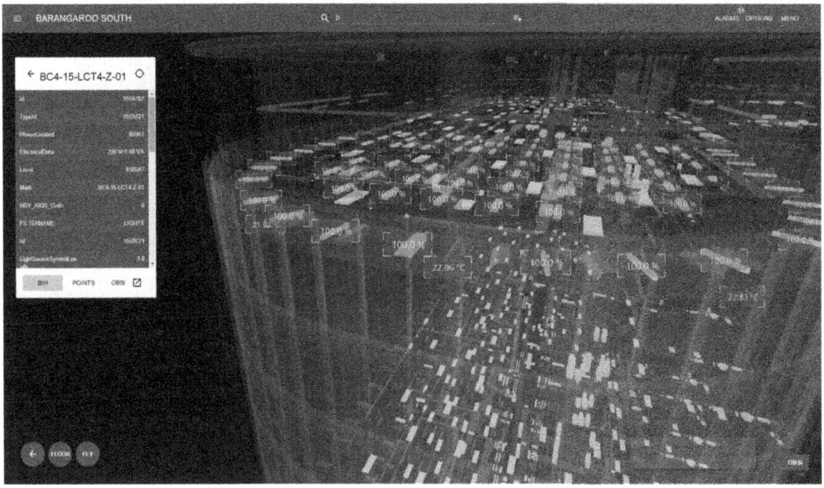

Fig. 11.9 Image capture of an early version of the 3D interface for Lendlease's Open Building Systems Integration (OBSI) platform. In this view of the virtual model, assets (such as lighting fixtures) that are linked to physical real-world equivalents are highlighted in yellow and are showing the live operational states of sensors held within the assets. The values of any data associated to the assets are also contextualised as floating numbers in 3D space. (Colour figure online)

Fig. 11.10 Image capture of an early version of the 3D interface for Lendlease's Open Building Systems Integration (OBSI) platform. This view of the virtual model is showing a summary of alarm statuses across the three International Towers at Barangaroo South

Fig. 11.11 Virtual Barangaroo running on a high-powered gaming laptop showing an interactive digital representation of 4D construction phasing of the Barangaroo International Towers

the output of the environment to various hardware mediums. As can be seen in Fig. 11.12, one of the deliverables was for a lightweight iPad version of Virtual Barangaroo, allowing for easier engagement with stakeholders.

The two main contracted environments—being the Virtual Barangaroo application, used for high-end visual tasks such as leasing and stakeholder engagement, and the OBSI 3D Interface, being for building operations—are both very heavy applications (at least in the initial builds, with the technology stack available at the time of development) and run on high-end gaming computers to assure smooth operation. With the constant improvement of the chosen game engine, this may be ported to mobile versions over time.

The current OBSI 3D Interface application is mainly used for quick visual reference of alarm status across the 3 International Towers and connected basement levels. The current connected systems are fire alarms, door and window open and closed status, and parts of the HVAC system status. There are additional systems integrated into the back-end system that have yet to be integrated into the 3D interface as needs arise.

Fig. 11.12 Approaching photoreal interactive virtual environment running on an iPad of a proposed office floor on Level 48, Tower 1, of the Barangaroo South International Towers

Using colour coding, green for OK, yellow for low-level alarm, orange for mid-level alarm, and red for urgent alarm, an operator can review the 3D model in realtime to ascertain the alarm status of the entire complex at a glance.

One example of a use case is security door open or closed status. As security guards make their regular rounds of the precinct, they open and pass through many doors, some of which have access to sensitive areas. When passing through a sensitive door, an "urgent" alarm will be set off. An operator in the control room can track a security officer's movement through the building, and rather than having to confirm privileges can compare planned security paths with the 3D visual model in realtime and clear urgent alarms as they occur. This allows for more efficiency in operational interactions and assures any critical alarms have a better likelihood of being noticed sooner.

The intention is for Virtual Barangaroo and the OBSI 3D Interface to be available for use at Barangaroo South for the life of the development and is currently being used in the Operational Control Centre for the precinct to allow for efficient identification and interaction of live asset data.

Conclusion

With intuitive interfaces linked to an Open Data framework, we can work towards truly smart cities where inhabitants merge their virtual and real lives seamlessly in an exposed virtual world that is invisible without technology but is just waiting to be accessed with technology.

A rarely addressed aspect of smart cities is that much of the data that is generated in our cities happens within our buildings, which being primarily privately owned, is not part of the city-wide data landscape. Currently the global smart city movement is taking a top down approach to smart cities by trying to make our public services and infrastructure "smart". An argument that can be put forward is that our cities won't be truly smart until we capture our building data and integrate privately owned data sets into our broader urban environments.

As experienced during the development of Virtual Barangaroo, the OBSI, and its associated 3D interface, the massive amounts of data held in multiple systems to operate city-sized developments, and thus smart cites, are overwhelming and require sophisticated interfaces to garner efficiencies.

Extending the value of the BIM into operational activities by utilising the geometry and data to provide context to operational building information provides an avenue to more intuitive interaction with asset data. At the moment this is primarily for private use and better operations of our smart buildings, but as more smart buildings come online, this building data will better inform how our smart cities operate by considering the individuals daily activities and is driving private companies to open their data and make it accessible for public use.

With the volume of data exponentially increasing in our cities, and leading to the smart cities of the future, highly detailed 3D interfaces allow for more people to interact with the built environment in a contextual, intuitive, and meaningful way, and hopefully lead to an increase in health, wellbeing, and liveability of our cities for the betterment of everyone.

Acknowledgement Thank you to the team at ineni Realtime Pty Ltd for providing the information and image captures of Virtual Barangaroo and the OBSI 3D interface.

References

Discover Barangaroo: History. (2011). Retrieved February 5, 2018, from Barangaroo Delivery Authority: http://www.barangaroo.com/discover-barangaroo/history.aspx

Feldman, M. (2016, June 20). *China Races Ahead in TOP500 Supercomputer List, Ending US Supremacy*. Retrieved January 27, 2018, from Top500: https://www.top500.org/news/china-races-ahead-in-top500-supercomputer-list-ending-us-supremacy/

Hyerle, D. (2008). *Thinking Maps®: A visual language for learning*. Retrieved January 15, 2018, from https://link.springer.com/chapter/10.1007/978-1-84800-149-7_4/fulltext.html

Rapley, L. (2014, January 17). *A tale of Barangaroo's boom and bust*. Retrieved January 29, 2018, from Architecture and Design: http://www.architectureanddesign.com.au/features/features-articles/a-tale-of-barangaroo-s-boom-and-bust

Rizzatti, D. (2016, September 14). *Digital data storage is undergoing mind-boggling growth*. Retrieved January 27, 2018, from EE|Times: https://www.eetimes.com/author.asp?section_id=36&doc_id=1330462

Steven, K. L. (2010). *The ultimate history of video games: From Pong to Pokemon and beyond… the story behind the craze that touched our lives and changed the world*. Roseville, CA: Three Rivers Press.

van der Meulen, R. (n.d.). *Gartner says worldwide video game market to total $93 billion in 2013*. Retrieved February 19, 2018, from Gartner: http://www.gartner.com/newsroom/id/2614915

Zakon, R. H. (n.d.). *Hobbes' Internet Timeline 10.1*. Retrieved February 5, 2018, from zakon.org: http://www.zakon.org/robert/internet/timeline/

12

A Dashboard for the Unexpected: Open Data for Real-Time Disaster Response

Ian Tilley and Christopher Pettit

Abbreviations

AJAX	Asynchronous JavaScript and XML
CAL	City Analytics Lab
CRED	Centre for Research on the Epidemiology of Disasters
ETL	Extract, transform and load
FRWG	Fleet Response Working Group
LAMP	Linux, Apache, MySQL and PHP

I. Tilley (✉)
City Futures Research Centre, Faculty of the Built Environment, University of New South Wales, Sydney, NSW, Australia
e-mail: ian@bco.com.au

C. Pettit
Urban Science, Faculty of the Built Environment, University of New South Wales, Sydney, NSW, Australia
e-mail: c.pettit@unsw.edu.au

S. Hawken et al. (eds.), *Open Cities | Open Data*,
https://doi.org/10.1007/978-981-13-6605-5_12

UI	User Interface
UNISDR	United Nations International Strategy for Disaster Risk Reduction
UNSW	University of New South Wales

Highlights

- Real-time tactical city dashboard built and tested in a real disaster even
- Resilience enhanced through the communication of open data on disasters
- Usability of dashboards is critical for effective resilience development
- Responsive design of dashboards important as communicates willing to use such technologies to understand their local area effectively

Introduction

Increasing population density and the effects of climate change present a challenge to pursue real-world solutions that will strengthen resilience in communities and cities. Resilience, as defined by 100 Resilient Cities, is "the capacity of individuals, communities, institutions, businesses, and systems within a city to survive, adapt, and grow, no matter what kinds of chronic stresses and acute shocks they experience" (100 Resilient Cities, 2018).

The United Nations International Strategy for Disaster Risk Reduction (UNISDR) states that there were 7009 climate-related disasters between 1980 and 2011 (Fig. 12.1) (UNISDR, 2018). Annually, the Centre for Research on the Epidemiology of Disasters (CRED) reported an average of 384 natural disasters between 2001 and 2010 (Guha-Sapir, Vos, Below, & Ponserre, 2012). Even though, in 2011, the number of natural disasters was below the average (332), there were 244.7 million victims worldwide and economic damage was the highest ever estimated at US $366.1 billion (Guha-Sapir et al., 2012).

In the geographical context of Australia, where our research focuses, there is a history of cyclonic events and governments are well versed in distributing information about them (Queensland Government, 2017). But many have had devastating impacts on cities and regions. In 1974,

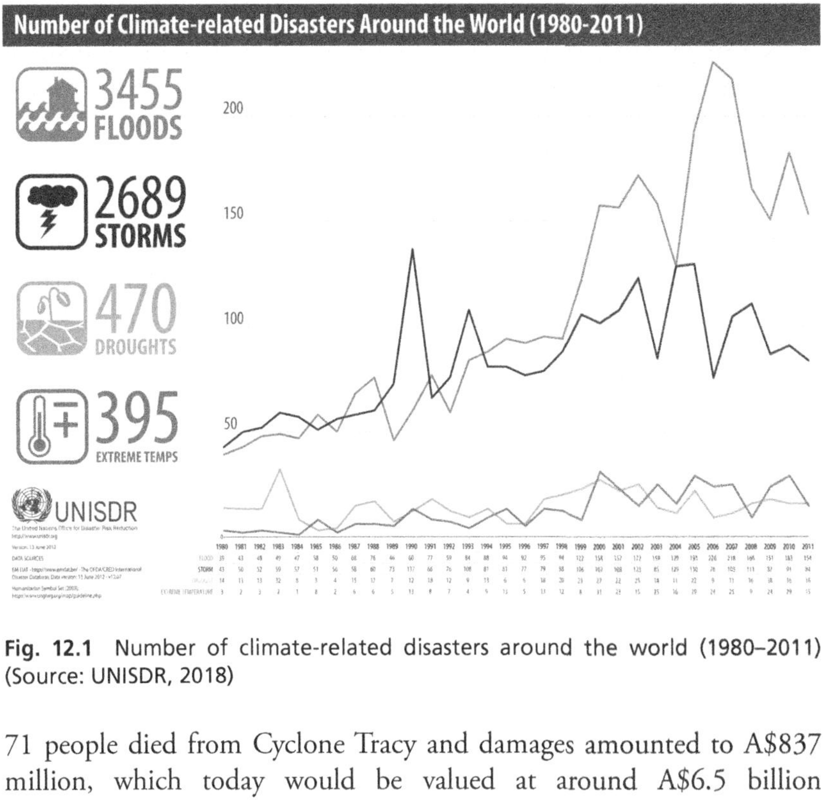

Fig. 12.1 Number of climate-related disasters around the world (1980–2011) (Source: UNISDR, 2018)

71 people died from Cyclone Tracy and damages amounted to A$837 million, which today would be valued at around A$6.5 billion (McDermott & Buckley, 2016). "It destroyed more than 70% of Darwin's buildings, including 80% of houses. Tracy left more than 41,000 out of the 47,000 inhabitants of the city homeless prior to landfall and required the evacuation of over 30,000 people" (McDermott & Buckley, 2016).

In March 2017, severe Tropical Cyclone Debbie crossed the Queensland coast, causing A$ 2.4 billion (US $1.8 billion) in damage (Podlaha, Bowen, Darbinyan, & Lörinc, 2017) and resulting in record-level flooding (Bureau of Meteorology, 2018). Debbie resulted in 14 deaths (Podlaha et al., 2017), making it Australia's deadliest cyclone since Cyclone Tracy (Kamenev, 2011).

Especially amidst changing climates, there is an increasing need for city resilience planning (Godschalk, 2003; Newman, Beatley, & Boyer, 2009). It is important to provide tactical tools to aid resilience in cities, for it is inevitable that communities will face threats and crises that may be inconvenient at best and catastrophic at worst. In this chapter we will investigate the use of tactical citizen-centric dashboards to support citizen engagement during and after such extreme events.

The Social Cost

Australia is not the only country to feel the impacts of severe cyclones and weather events. For example, Bangladesh has suffered many such calamities. Between 1797 and 2009, 35 of the 65 severe cyclones experienced by Bangladesh had storm surges of up to 10 metres (Rana, Gunasekara, Hazarika, Samarakoon, & Siddiquee, 2010). Since 1876, ten different cyclones have had over 5000 fatalities (Bangladesh Meteorological Department, 2007). The deadliest—the 1970 Bhola cyclone—resulted in an estimated 300,000 deaths (Holland, 1993).

The United States has seen several damaging storms over time—more recently Hurricane Sandy (2012) and Hurricane Ike (2008), and less recently, the Great Miami Hurricane (1926) and Galveston Hurricane (1900) (Atlantic Oceanographic and Meteorological Laboratory, 2014). After the 1964 Alaskan Earthquake, in which 124 people died from the subsequent tsunami (National Oceanic and Atmospheric Administration, 2018), the US Federal Government commissioned an assessment on what was known about human occupancy of hazard zones (White & Haas, 1975). The first assessment concluded that the losses and potential losses from natural hazards were increasing due to more people living in unprotected flood plains, seismic zones and coastal regions (Cutter et al., 2008).

In 2005, Hurricane Katrina's impact on New Orleans was truly catastrophic. Deaths of Louisiana residents reached 1570, with the financial impact reaching US $50 billion (Kates, Colten, Laska, & Leatherman, 2006). The reconstruction process has taken a decade and Kates et al. (2006) suggest that rebuilding was "like for like" as opposed to building for long-term sustainability and resilience, to better survive and recover from future major hurricane events. The impact of such a huge loss of life and property, has a long-lasting effect on a nation's economy and psyche.

Post the event, Ursula Martin shows that many communities experienced psychology impacts with depression levels increasing by up to 25% and doctor visits increasing significantly (Martin, 2015). Large-scale disasters have a profound effect on the physical and emotional health of survivors. In regions that repeatedly experience natural disasters, resilience will become ever more important as populations grow (UN News, 2013).

Government Initiatives

But, what government support is there for cities and communities during a crisis? Much of the funding and resources assigned to managing disasters and crises in cities and communities is focused on the very important task of coordinating government and non-government agencies to provide onsite, localised support (Motta, Abelheira, Gomes, Fonseca, & Besen, 2014). The Centro de Operacoes Prefeitura do Rio in Rio de Janeiro, Brazil, is the most widely known system where the Brazilian government uses dashboards, Internet of Things (IoT) and live data to manage developing situations on the ground; it was created in response to the significant death and damage caused by storms in 2010 (Kitchin, Lauriault, & McArdle, 2015). Unfortunately, the general public does not have direct access to the same information utilised by the Centro de Operacoes and are only alerted to emergencies by sirens and short message service (SMS) messages.

Non-government organisations like ReliefWeb (2018) and GISCorps (2009) provide situational data for relief workers in disaster areas around the world but do not directly support getting the same information to the general public being affected by the natural disaster.

In Australia, well-funded government and non-government agencies like the State Emergency Services, Fire Brigade and Police and Ambulance services are working, largely, in a coordinated fashion behind the scenes. However, this important, coordinated effort and the data it generates are, generally, not visible to the public (Bharosa, Janssen, Meijer, & Brave, 2010).

Open Data in Australia

In 2014, the then Federal Government Communications Minister, Malcolm Turnbull, moved to make government data open. He could clearly see the economic benefits of such a move along with budget savings by reducing the duplication of different datasets by different government departments. Later, Prime Minister Turnbull made this government policy, stating that "Publishing, linking and sharing data can create opportunities that neither government nor business can currently envisage" (Turnbull, 2015).

The increased availability of government generated Open Data—for instance, Transport for NSW's Open Data portal (opendata.transport.nsw.gov.au)—has resulted in the arrival of some applications that directly benefit the general public on a day-to-day basis. Examples of these include Fuelcheck, which gives a user the ability to check fuel prices or Tripview that provides access to up-to-date public transport information. In addition, Open Data can be used in ways that can help citizens in real time. Examples include:

- Live Traffic NSW: Live traffic conditions made available through a responsive website. Data made available include traffic accidents, major event notifications and live cameras at significant locations (www.livetraffic.com).
- BoM: Cyclone advice, major storms, floods and warnings (www.bom.gov.au)
- Rural Fire Service: Location of fires and fire status warnings (www.rfs.nsw.gov.au)

In the three examples above, each is provided as a distinct service via a government website or Open Data portal. Yet, there are plenty of scenarios where each of these examples can be used to provide better context to the others. One example is the April 2018 fires in Sydney's South West. Wind direction (BoM) would give an indicator of which way a fire might spread, and updates on the fires' position (Rural Fire Service) would inform residents of the current situation and fires' location. Live Traffic NSW would inform residents of roads that are blocked so they can find the best evacuation route.

With an increasing amount of Open Data being made available, there is the opportunity to access such data feeds programmatically and provide this information in a curated and digestible format. One way to achieve this is through the creation of dashboards, which can provide aggregator-like views into the rich tapestry of data that is continually being created in the era of smart cities, big data and data analytics (Pettit et al., 2018).

Dashboards for City Resilience

Dashboards have their origins in vehicles, where they provide critical real-time information to the driver to ensure they can respond accordingly to changing conditions. Dashboards can be defined as "graphic user interfaces that comprise a combination of information and geographical visualisation methods for creating metrics, benchmarks, and indicators to assist in monitoring and decision-making" (Pettit & Leao, 2017, p. 1).

In the context of cities there are a number of examples where dashboards have been built to provide the real-time pulse of the city. Two such examples include the City of London Dashboard (Gray, O'Brien, & Hügel, 2016) and the City of Sydney Dashboard (Pettit, Lieske, & Jamal, 2017). Both of these dashboards are fuelled through Open Data made publicly available from both government agencies and private sector, as illustrated in Figs. 12.2 and 12.3. These dashboards support the day-to-day functions of city life yet do not specifically focus on crisis events.

In the context of real-time crisis information, there are few that are developed for this specific purpose and that can be accessed by the general public. The Gympie Disaster Dashboard provides weather warnings, power outages and emergency news among other helpful information on the Gympie Regional Council website (Fig. 12.4). The Fleet Response Working Group (FRWG) includes their dashboard as part of the disaster response tools (Fig. 12.5), with the aim of supporting efforts to expedite the movement of private sector repair fleets and resources across multiple state borders in the United States (All Hazards Consortium, 2018).

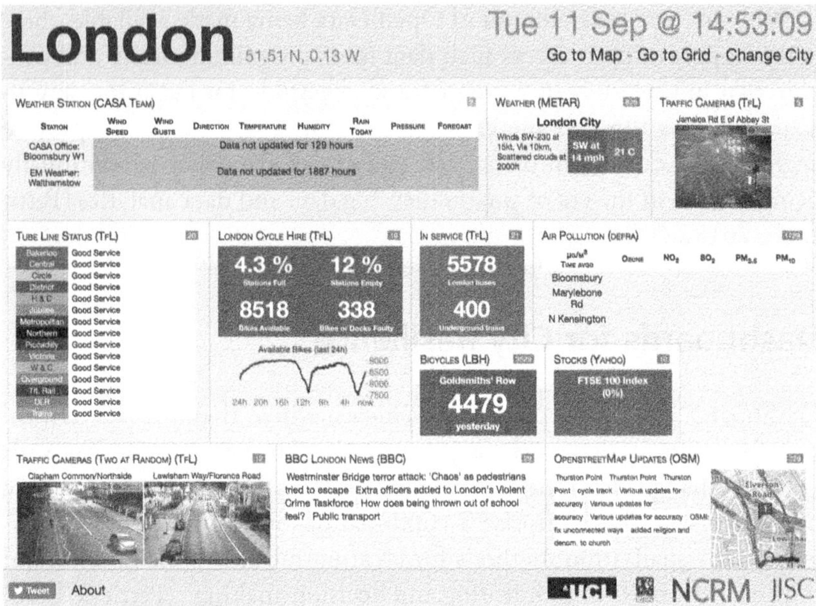

Fig. 12.2 London Dashboard (Source: http://citydashboard.org/)

Fig. 12.3 Sydney Dashboard (Source: http://citydashboard.be.unsw.edu.au/)

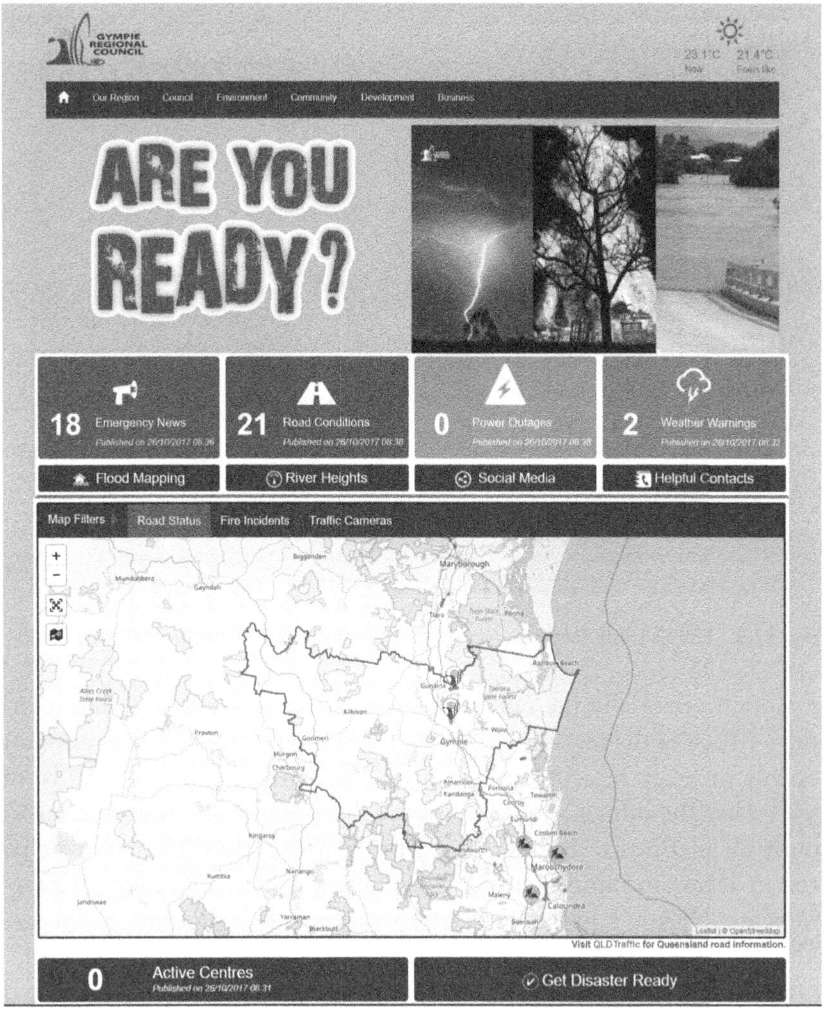

Fig. 12.4 Gympie Disaster Dashboard (Source: Gympie Regional Council, 2017)

Resilience

ResilientCity (2018) describes a resilient city as "one that has developed capacities to help absorb future shocks and stresses to its social, economic, and technical systems and infrastructures so as to still be able to

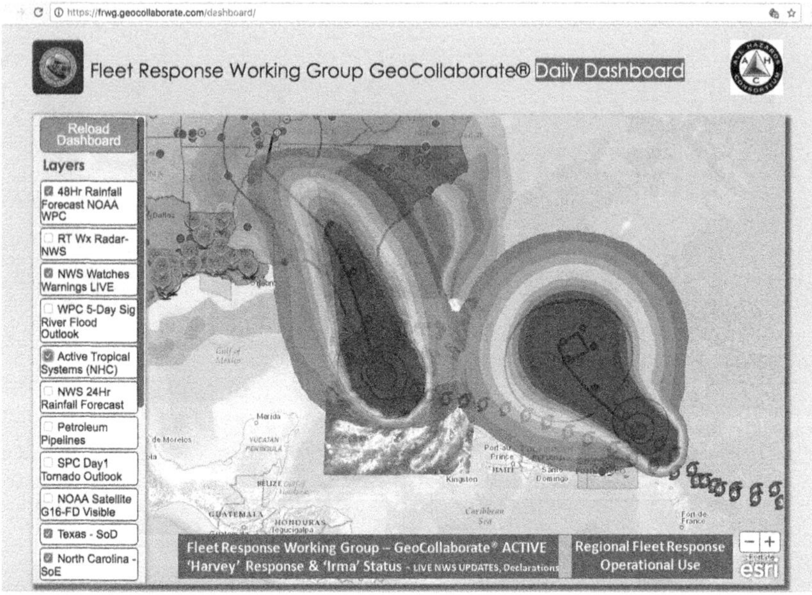

Fig. 12.5 FRWG Daily Dashboard (Source: StormCenter Communications, 2017)

maintain essentially the same functions, structures, systems, and identity". Much of the real-time valuable data about a crisis or disaster is not openly or easily accessible to the people that are caught in the midst of the crisis. The information that is available is generally dispersed across multiple platforms, in different forms of media and difficult for a citizen to access and understand in the context of the crisis they are experiencing. This is a probable barrier to city and community resilience. The dispersal of real-time Open Data limits the ability of a city to anticipate shocks and stresses and communicate the details of unexpected disasters to citizens.

There are many agencies that collect data that can be directly relevant to the types of crises a city may face. A small proportion of this collected data makes its way into a public-facing website that can be utilised by citizens, non–for-profits, aid agencies and others. However, it is often piecemeal and not made available in real time. Furthermore, as noted by Kitchin et al. (2015), the raw data used is also "cooked to some recipe by chefs embedded within institutions that have certain aspirations and goals and operate within wider frameworks" (p. 18). Often there is logic behind the presentation and ready availability of particular datasets that are not obvious to the audience—in other words, the data provided is not necessarily user-centric.

It can be argued that gathering together different data sources that individually promote "limited instrumental rationality" to then focus on crisis events would result in a "powerful realist epistemology" (Kitchin et al., 2015, p. 7). Having this curated focus on key information relating to the crisis being faced by citizens is crucial. It is postulated that such user-centred design dashboards which provide critical real-time information to citizens and the public would provide a useful city resilience tool in responding to crisis events.

Crowdsourced Crisis Data

During extreme weather events, many standard lines of communication fail, as was the case during Cyclone Tracy (Bureau of Meteorology, 2017b). During the 2011 Tōhoku earthquake and tsunami in Japan, citizens used social media to keep in touch during and after the major coastal inundation because the phone networks were completely jammed (Gao, Barbier, & Goolsby, 2011). Some platforms harness social media in the creation of community crisis maps. One such platform is PetaJakarta.org, which is described as a self-organising social-technical system (Perez, Holderness, Turpin, & Clarke, 2015). PetaJakarta.org elicits and facilitates active engagement from the users of the platform by allowing them to post/upload location-tagged content about flooding events in Jakarta. The content, in effect, is created by those who will benefit most from sharing the content.

Another example of a crowdsourced mapping platform is Ushahidi. Ushahidi uses crowdsourcing for social activism and public accountability but its flexibility has facilitated its use by the Australian Broadcasting Company for the 2011 Queensland Floods (Ross & Potts, 2011) and by the Brisbane City Council for a 2013 flood map (Crowdmap, 2013).

Ushahidi has over 120,000 deployments for many different types of crises all around the globe (Gutiérrez, 2018).

The majority of Australian citizens have access to reliable internet services via either fixed-line access or mobile devices (ACMA, 2017). This ubiquitous availability of broadband gives citizens the ability to access real-time crisis information from wherever and whenever needed. It could be argued that up-to-date, curated crisis information along with social media connections can play an important role in helping cities cope and recover from a natural disaster.

Deploying a Tactical Dashboard for Cyclone Events

Tropical Cyclone Debbie

In March 2017, Tropical Cyclone Debbie appeared in the Coral Sea and made its presence felt. Such events are common in Northern Australia and can cause significant damage to housing, city infrastructure, the natural environment and people. Cyclone Debbie was a Category 4 storm, forming on 23 March and dissipating on 7 April—making landfall on 28 March (Bureau of Meteorology, 2017a).

In an era of smart cities, big data and analytics, there has been significant activity in making data more openly available to the public through dashboards (Pettit et al., 2017). To create the Cyclone Debbie Dashboard, several freely available Open Data sources were used (see Table 12.1). Yet, prior to the dashboard, the general public had no easy way to access and visualise these datasets during an extreme event such as a cyclone. Hence on 26 March 2017, a tactical dashboard was made for Cyclone Debbie, incorporating each of the data sources described in Table 12.1. This dashboard is tactical in so far as it was put together "just in time" during the initial impacts of Tropical Cyclone Debbie and in direct response to it. We will proceed to discuss the technical components of the dashboard and undertake a preliminary evaluation based on metrics provided from Google Analytics.

Table 12.1 Cyclone Debbie Dashboard data sources

Data name	Data description (format, API...)	Data custodian
Cyclone advice	.txt direct from BoM	Commonwealth of Australia 2017, Bureau of Meteorology
Cyclone Tracking Maps	Html page Direct from BoM	Commonwealth of Australia 2017, Bureau of Meteorology
Rain Radar	GIF and PNG files FTP every 10 mins	Commonwealth of Australia 2017, Bureau of Meteorology
QLD Traffic webcam	Jpg images refreshed 1 a minute	QLDTraffic.qld.gov.au
QLD Traffic advice	JSON	QLDTraffic.qld.gov.au
Twitter	#TCDebbie	Twitter
Cyclone News	JSON	ABC.net.au
Cyclone preparedness	HTML pages	Regional Council Disaster management
Rainfall projection	Embedded widget	Bureau of Meteorology (via Willy Weather)
Wind projection	Embedded widget	Bureau of Meteorology (via Willy Weather)
Tides projection	Embedded widget	Bureau of Meteorology (via Willy Weather)
Weather projection	Embedded widget	Bureau of Meteorology (via Willy Weather)
Traffic levels	iFramed map	Google

Developing the Dashboard

The hosting platform for the dashboard was a Linux, Apache, MySQL and PHP (LAMP) stack located in an Australia-based data centre. A range of techniques were utilised to extract, transform and load (ETL) the data from the various source providers into the hosting platform in a suitable form for the dashboard to consume. These techniques included:

- Scheduled tasks implemented to acquire radar images and weather data at regular 10-minute and 3-hour intervals
- Extraction of specific webcam locations for custom display widgets
- Utilisation of Asynchronous JavaScript and XML (AJAX) to acquire live event feeds filtered by specific locations
- Purpose-built solutions like Google Maps or OpenStreetMap

The aim of the real-time tactical dashboard was to develop a user-friendly experience through a simple visualisation interface. The performance of the dashboard is also important to ensure a positive user experience. Thus, the beta version of Cyclone Dashboard (cyclonedashboard.com.au) has been designed to be lightweight and quick to download. This reduces the load on the server and networks during times of high use.

The technical components of the Cyclone Dashboard User Interface (UI) were made modular in their composition. The Twitter Bootstrap framework was employed to give control over the positioning of the dashboard elements on the screen. Furthermore, the utilisation of Bootstrap facilitates the ability of the dashboard to be responsive to the device it is being accessed on, thus allowing users to access it on desktops, tablets and smartphones. This is especially important for citizens as it allows them to access the dashboard and view meaningful, relevant data through a range of devices.

Measuring the Impact

Encouraged by the initial server traffic and feedback for the Cairns Cyclone Dashboard (Fig. 12.6), additional dashboards were generated for Townsville, Mackay, Bowen and Brisbane as it was unclear where Debbie would impact. A Facebook page and Twitter account were created and a few posts targeting friends and community groups in the affected areas were made to alert people that the dashboard was available.

Google Analytics was also employed to determine the level of usage the Cyclone Dashboard had received, devices used, specific pages used, along with the locations of the users and the demographics of the users. The Google Analytics results, 48 hours after deployment, indicated that there were 7165 unique users over three days (Fig. 12.7).

A post-event analysis revealed that 97% of the Cyclone Dashboard traffic originated in Australia, and 90% of all traffic was from within Queensland, with the highest amount of page views for the Townsville Dashboard (9422) followed by Bowen (2991), Mackay (1840) and Cairns (1553) from a total of 22,537 page views. The average session

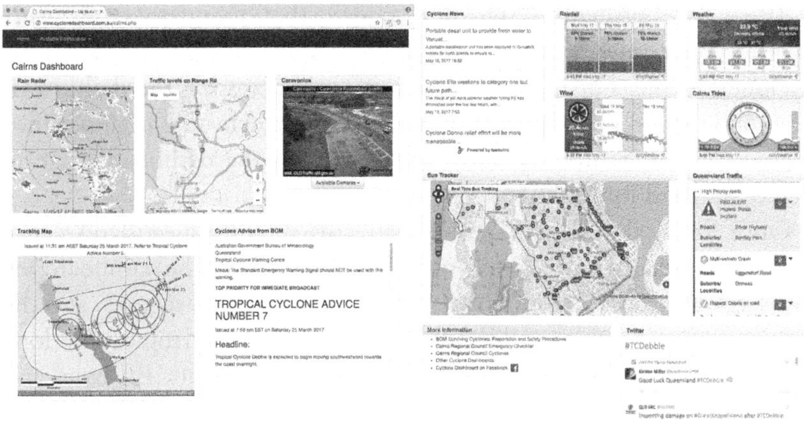

Fig. 12.6 Cyclone Dashboard (Source: http://www.cyclonedashboard.com.au/cairns.php)

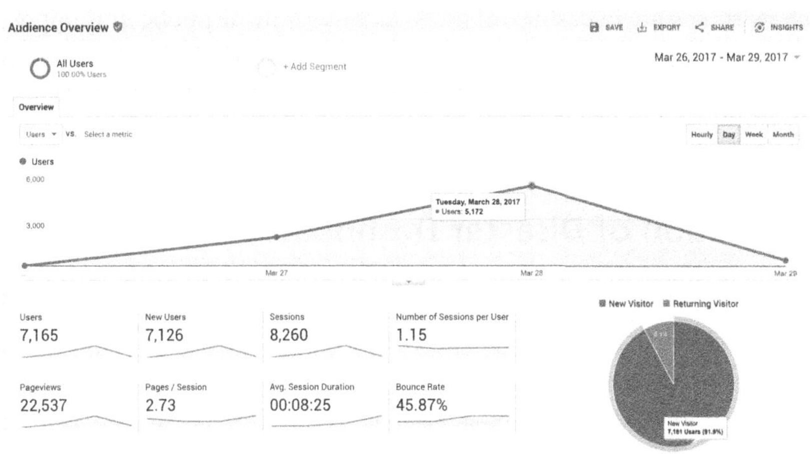

Fig. 12.7 Dashboard usage (Source: https://analytics.google.com/)

length was 8 minutes and 25 seconds. The top two types of browsers used made up 78% of users and were on mobile devices. Only a handful of users (46) viewed the dashboards on a device supporting a 1280 × 800 resolution, which is typically the resolution of a desktop or laptop computer.

These user evaluation statistics suggest that the dashboards were received positively by the communities being directly impacted by Cyclone Debbie. Furthermore, the users were accessing the dashboards on mobile devices while in areas being impacted by the cyclone; they viewed the dashboards regularly and stayed on the dashboard page to receive updated information.

Once Tropical Cyclone Debbie had crossed the Queensland coast, tracked inland and started to dissipate, traffic levels for the dashboard fell away, as was expected. The threat passed for the coastal regions, and the dashboards, at that stage, did not cover inland areas. Tropical Cyclone Debbie's impact continued through inland Queensland, Brisbane and across to New Zealand.

In utilising freely available Open Data and putting this data into a free public dashboard, the value of the data, in this holistic context, was more than the perceived value of the sum of its parts. This was confirmed by the swift increase in dashboard users once it was made public. It is important to note that indicators, as presented in crisis dashboards, show trends in conditions but they do not tell us what to do. "They are indicators, not answers" (Innes & Booher, 2000).

Evaluation of Disaster Dashboards

During the Tropical Cyclone Debbie event (23 March 2017–7 April 2017), the author could not source any real-time Australia-based dashboards that covered this type of crisis in the Australian context. During the Hurricane Harvey (17 August–3 September) and Hurricane Irma (30 August–13 September) events, there were opportunities to assess real-time crisis dashboards in the United States.

All of these dashboards (Table 12.2) were created to help the affected communities and the sponsors of these dashboards could be broadly grouped into three categories.

1. Government agencies and community-conscious corporations
2. Broadcasters
3. Committed individuals

Table 12.2 Evaluated dashboards

Sponsor	Name	Open Data
Google	Google Crisis Response	Map based. Showing shelters, hurricane path, flooding
NBC 2	First Alert Hurricane Tracker	Tracking map
CNN	Interactive Strom Tracker	Tracking map
Weather.com	Hurricane Central	Tracking map, storm maps, bulletins and alerts
NBC 2	Tropical Tracker	Tracking map, live-streamed TV
NBC 2	First Alert live radar	Tracking map, Doppler radar, watch and warnings
National Oceanic and Atmospheric Administration	Hurricane Harvey	Tracking map, impact zones, warnings, forecasts
Pacific Disaster Center	Situational Awareness Products—Florida	Projected flooding maps, peak wind gusts, hospital damage, evacuation zones, estimated impacts, disaster aware near real-time updates
NOAA/National Weather Service	Enhanced Radar Mosaic	Doppler radar, warnings
Fleet Response Working Group	Daily Dashboard	Rainfall, watches and warnings, active tropical systems, tracking map
News 24/7	Live Updates	Streamed CNN Live
Crowd tangle	WeatherEventFacebook	Dashboard with Facebook and Twitter streams from government agencies and the media
Jeff Piotrowski	Hurricane Irma Live Naples FL	Live video stream and chat channel
NBC Miami	Interactive Radar	Radar and satellite
News Flood Huston	Live Hurricane Irma Threatens Florida	Live-streamed CNN

The vast majority of data utilised in these dashboards was generated by government agencies though it is unclear if the utilised data was open or acquired under licence.

Conclusions and Future Research

In this research, a tactical dashboard has been built and used in the context of a real-world extreme event, Tropical Cyclone Debbie. The tactical dashboard successfully aggregated and made available important real-time information for citizens being impacted by the event. The usage metrics, which report 7165 users over three days for the dashboard, clearly indicate the usefulness of this type of dashboard to support communities during a crisis event. While the beta version of Cyclone Dashboard was event-specific, further research should be pursued to consider the usefulness of similar dashboards in supporting communities as they deal with significant weather, social or terrorism events that can impact the function and resilience of cities.

In the context of Australia, it is clear that Federal, State and Local government policy settings should support the development of resilient communities by making more data open and accessible. Some State and Local governments across Australia have provisioned multi-agency emergency response platforms, but they provide limited information to the general public. Through the provision of real-time, tactical extreme event dashboards, which aggregate data from across multiple sources, there is an opportunity to support the community in responding to extreme events.

Future research will refine the tactical dashboard, including the incorporation of mapping data through OpenStreetMap to provide location intelligence information to communities to assist with disaster response. The next phase of the research will evaluate the beta version of Cyclone Dashboard's utility and usability from the perspective of two distinct user groups: (i) decision-makers and (ii) citizens and communities. User testing will be conducted through the University of New South Wales (UNSW) City Analytics Lab (CAL). Finally, with more Open Data becoming increasingly available, it is anticipated that further tactical dashboards will become richer in content and will hence have great utility in assisting society in responding to and therefore recovering from an increasing number of extreme events.

References

100 Resilient Cities. (2018). Frequently Asked Questions (FAQ) about 100 resilient cities. Retrieved May 29, 2018, from http://www.100resilientcities.org/100RC-FAQ/#/-_/

All Hazards Consortium. (2018). Multi-state fleet response working group. Retrieved May 31, 2018, from http://www.ahcusa.org/fleet-response.html

Atlantic Oceanographic and Meteorological Laboratory. (2014). Frequently Asked Questions. Retrieved May 31, 2018, from http://www.aoml.noaa.gov/hrd/tcfaq/E9.html

Australian Communications and Media Authority. (2017). Communications Report 2016/17.

Bangladesh Meteorological Department. (2007). Numerical simulations of storm surges in the Bay of Bengal. In *First JCOMMM Scientific & Technical Symposium*: Seoul, Korea, 2–6 October 2007. Retrieved May 31, 2018, from http://tidepool.nos.noaa.gov/pub/outgoing/hurricane/NRAP/Wallace/New_Product_Papers/Numerical%20Simulations%20in%20Bay%20of%20Bengal.pdf

Bureau of Meteorology. (2017a). Tropical cyclone Debbie. Retrieved June 21, 2017, from http://www.bom.gov.au/cyclone/history/debbie17.shtml

Bureau of Meteorology. (2017b). Tropical cyclone Tracy. Retrieved June 21, 2017, from http://www.bom.gov.au/cyclone/history/tracy.shtml

Bureau of Meteorology. (2018). Tropical cyclone Debbie Technical Report. Retrieved May 30, 2018, from http://www.bom.gov.au/cyclone/history/database/Tropical-Cyclone-Debbie-Technical-Report-Final.pdf

Bharosa, N., Janssen, M., Meijer, S., & Brave, F. (2010). Designing and evaluating dashboards for multi-agency crisis preparation: A living lab. In M. A. Wimmer, J. L. Chappelet, M. Janssen, & H. J. Scholl (Eds.), *Electronic Government. EGOV 2010* (Lecture Notes in Computer Science, 6228). Berlin, Heidelberg: Springer.

Cutter, S. L., Barnes, L., Berry, M., Burton, C., Evans, E., Tate, E., & Webb, J. (2008). A place-based model for understanding community resilience to natural disasters. *Global Environmental Change, 18*(4), 598–606.

Crowdmap. (2013). Brisbane city council used Crowdmap (Ushahidi) for the floods in 2013. Retrieved May 17, 2018, from http://crowdmap.com/map/brisbaneweathereventoct2013/

GISCorps. (2009). What we do. Retrieved May 28, 2018, from http://www.giscorps.org/what-we-do/.

Godschalk, D. R. (2003). Urban hazard mitigation: Creating resilient cities. *Natural Hazards Review, 4*(3), 136–143.

Gray, S., O'Brien, O., & Hügel, S. (2016). Collecting and visualizing real-time urban data through city dashboards. *Built Environment, 42*(3), 498–509.

Guha-Sapir, D., Vos, F., Below, R., & Ponserre, S. (2012). *Annual disaster statistical review 2011: The numbers and trends.* Centre for Research on the Epidemiology of Disasters (CRED).

Gutiérrez, M. (2018). *Data activism and social change.* London: Palgrave Pivot.

Gympie Regional Council. (2017). Gympie Disaster Dashboard. Retrieved October 26, 2017, from http://disaster.gympie.qld.gov.au/

Gao, H., Barbier, G., & Goolsby, R. (2011). Harnessing the crowdsourcing power of social media for disaster relief. *IEEE Intelligent Systems, 26*(3), 10–14.

Holland, G. J. (1993). Ready Reckoner—Chapter 11, Global guide to tropical cyclone forecasting. In *WMO-No. 1194, World Meteorological Organization, Geneva, Switzerland.* Retrieved May 31, 2019, from https://www.wmo.int/cycloneguide/pdf/Global-Guide-to-Tropical-Cyclone-Forecasting.pdf

Innes, J. E., & Booher, D. E. (2000). Indicators for sustainable communities: A strategy building on complexity theory and distributed intelligence. *Planning Theory & Practice, 1*(2), 173–186.

Kamenev, M. (2011). Australia's worst cyclones: Timeline. *Australian Geographic.* Retrieved May 30, 2018, from http://www.australiangeographic.com.au/topics/science-environment/2011/02/australias-worst-cyclones-timeline/

Kates, R. W., Colten, C. E., Laska, S., & Leatherman, S. P. (2006). Reconstruction of New Orleans after Hurricane Katrina: A research perspective. *Proceedings of the National Academy of Sciences of the United States of America, 103*(40), 14653–14660.

Kitchin, R., Lauriault, T. P., & McArdle, G. (2015). Knowing and governing cities through urban indicators, city benchmarking and real-time dashboards. *Regional Studies, Regional Science, 2*(1), 6–28. https://doi.org/10.1080/21681376.2014.983149

Martin, U. (2015). Health after disaster: A perspective of psychological/health reactions to disaster. *Cogent Psychology, 2*(1), 1–6.

McDermott, M., & Buckley, R. (2016). Cyclone Tracy disaster—Darwin, Northern Territory, Australia December 24–25, 1974: The disaster effect and lessons learnt. *Presented at FIG Working Week 2016, 2–6 May 2016, Christchurch, NZ.* Retrieved May 31, 2018, from http://www.fig.net/resources/proceedings/fig_proceedings/fig2016/techprog.htm

Motta, M., Abelheira, M., Gomes, O., Fonseca, W., & Besen, D. (2014). Rio de Janeiro Community Protection Program. *Procedia Economics & Finance, 18*, 128–135.

National Oceanic and Atmospheric Administration. (2018). Tsunami events search. Retrieved May 31, 2018, from https://www.ngdc.noaa.gov/nndc/struts/results?bt_0=1964&st_0=1964&type_7=Like&query_7=prince&d=7&t=101650&s=7

Newman, P., Beatley, T., & Boyer, H. (2009). *Resilient cities: Responding to peak oil and climate change*. Washington, DC: Island Press.

Perez, P., Holderness, T., Turpin, E. R., & Clarke, R. (2015). Citizen-Driven Flood Mapping in Jakarta: A Self-organising Socio-technical System. In *Proceedings from 2015 IEEE International Conference on Self-Adaptive and Self-Organizing Systems Workshops*, pp. 174–178.

Pettit, C., Bakelmun, A., Lieske, S. N., Glackin, S., Thomson, G., Shearer, H., ... Newman, P. (2018). Planning support systems for smart cities. *City, Culture and Society, 12*, 13–24.

Pettit, C. J., & Leao, S. Z. (2017). Dashboard. In L. A. Schintler & C. L. McNeely (Eds.), *Encyclopedia of big data* (pp. 1–7). Cham: Springer International Publishing.

Pettit, C. J., Lieske, S., & Jamal, M. (2017). CityDash: Visualising a changing city using open data. In S. Geertman, J. Stillwell, A. Andrew, & C. J. Pettit (Eds.), *Planning support systems and smart cities, lecture notes in geoinformation and cartography* (pp. 337–353). Basel, Switzerland: Springer International Publishing.

Podlaha, A., Bowen, S., Darbinyan, C., & Lörinc, M. (2017). Global Catastrophe Recap—April 2017. *Aon Benfield Analytics*. Retrieved July 30, 2017, from http://thoughtleadership.aonbenfield.com/Documents/20170509-ab-analytics-if-april-global-recap.pdf

Queensland Government. (2017). Cyclones and severe storms. Retrieved July 1, 2017, from http://www.qld.gov.au/emergency/dealing-disasters/cyclone-severe-storm

Rana, S., Gunasekara, K., Hazarika, M., Samarakoon, L., & Siddiquee, M. (2010). Application of remote sensing and GIS for cyclone disaster management in coastal area: A case study at Barguna district, Bangladesh. *International Archives of the Photogrammetry, Remote Sensing and Spatial Information Sciences—ISPRS Archives, 38*, 122–126.

ReliefWeb. (2018). What is ReliefWeb?. Retrieved May 28, 2018, from http://reliefweb.int/about

ResilientCity. (2018). Resilience. Retrieved May 17, 2018, from http://www.resilientcity.org/index.cfm?id=11449

Ross, N., & Potts, M. (2011). ABC's crowd sourced flood-mapping initiative. Retrieved October 26, 2017, from http://www.abc.net.au/technology/articles/2011/01/13/3112261.htm

StormCenter Communications. (2017). Fleet Response Working Group GeoCollaborate® Daily Dashboard. Retrieved October 26, 2017, from http://frwg.geocollaborate.com/dashboard/

Turnbull, M. (2015) Australian Government Public Data Policy Statement. https://www.pmc.gov.au/sites/default/files/publications/aust_govt_public_data_policy_statement_1.pdf.

UNISDR. (2018). Disaster statistics. Retrieved May 29, 2018, from https://www.unisdr.org/we/inform/disaster-statistics

UN News. (2013). World population projected to reach 9.6 billion by 2050—UN report. Retrieved May 29, 2018, from http://www.un.org/apps/news/story.asp?NewsID=45165

White, G. F., & Haas, J. E. (1975). *Assessment of Research on Natural Hazards*. Cambridge, MA: MIT Press.

Part III

Civic Innovation and Transparency

13

An Information Management Strategy for City Data Hubs: Open Data Strategies for Large Organisations

Pascal Perez, Christopher Pettit, Sarah Barns, Jonathan Doig, and Carmela Ticzon

P. Perez (✉)
SMART Infrastructure Facility, University of Wollongong, Wollongong, NSW, Australia
e-mail: pascal_perez@uow.edu.au

C. Pettit
Urban Science, Faculty of the Built Environment, University of New South Wales, Sydney, NSW, Australia
e-mail: c.pettit@unsw.edu.au

S. Barns
Institute for Culture and Society, Western Sydney University, Sydney, NSW, Australia
e-mail: s.barns@westernsydney.edu.au

J. Doig • C. Ticzon
City Futures Research Centre, Faculty of the Built Environment, University of New South Wales, Sydney, NSW, Australia
e-mail: j.doig@unsw.edu.au; c.ticzon@unsw.edu.au

S. Hawken et al. (eds.), *Open Cities | Open Data*,
https://doi.org/10.1007/978-981-13-6605-5_13

Highlights

- Our case study shows that the creation of a useful city data store can be hampered by the confusion between data and information, the lack of data standards and systems interoperability, or poor internal communication.
- The implementation of a robust information management system is necessary to provide a sound foundation for Open City–Open Data initiatives.
- A strategic alignment of administrative processes, technological capability, financial commitment, and adequate human resources is necessary to support such an implementation.
- The proposed information management framework can be used to benchmark various city data stores and hubs.

Introduction

The proliferation of digital services in recent years has seen a rapid upswing of interest in the uses of urban data to improve urban management. As Goldsmith and Crawford write in The Responsive City (2014, p. 3), our growing capacity to collect, analyse, and share information today has great potential to transform and even reinvigorate the governance of cities. The authors argue that abundant sources of data—whether government data released in open, machine-readable formats, data created and contributed by citizens, or data contributed by private data providers—are allowing governments to move beyond what they call the 'compliance model' of local or municipal government towards more active, problem-solving capabilities 'that truly value the intelligence and dedication of its employees and the imagination and spirit of its citizens' (6). Goldsmith and Crawford advocate the adoption of collaborative, data-driven models of governance that 'open up the machinery of government to its people, letting them collaborate to create solutions coproduced by public servants and their constituents' (6). Data, they argue, can 'deliver government whenever and however citizens need it', replacing the bureaucratic and centralised structures that have frustrated citizens and officials alike for decades (9).

The rise of the smart city movement worldwide is testament to the growing recognition that urban data plays an increasingly strategic role in the governance of contemporary cities. However, challenges remain for government agencies in designing and delivering data management services that both support their own internal knowledge management requirements and foster collaborative governance approaches between governments, citizen users, commercial delivery partners, and other data custodians.

In the context of many government organisations, the current urban data landscape is characterised by avoidable duplications, large data gaps, and spatial or temporal mismatch. Moreover, many data assets remain siloed within and between organisations, so their value is far from being fully realised (Zuiderwijk, Janssen, Choenni, Meijer, & Alibanks, 2012).

Challenges for governments are compounded by the rate of disruption in digital delivery services, which increases the risks associated with digital obsolescence. In addition, with the advent of big data analytics, there is now greater risk of ill-informed manipulation of various datasets by end-users, resulting in misguided analyses and potential decisions. Conversely, end-users are rarely included in the early stage design of urban data stores and dashboards, leading to their relatively low return on investment. Henceforth, effective information management strategies are needed to support better urban governance in the digital age (Pettit, Lieske, & Jamal, 2017; Shaoyun et al., 2010). The implementation of a proper information management system, in the context of Open City–Open Data initiatives, requires a strong alignment of the administrative processes with technological, financial, and staff capacity (Dawes & Helbig, 2010). Furthermore, as Barns (2018) has argued, the design and management of urban data interfaces by city governments must be closely aligned to broader governance agendas for investments to be properly realised.

This chapter therefore investigates best-practice (BP) and end-user requirements in city data infrastructure, with a focus on data stores or hubs. First, we present a recent study undertaken with Landcom NSW—a state-owned agency in charge of managing large urban transformation programmes in New South Wales (NSW, Australia)—aimed at developing a roadmap towards an urban knowledge-sharing platform called the

Connected City Data Hub. Interviews with key stakeholders confirm a series of impediments to the design and implementation of a purposeful and useful city data hub. Then, we draw upon several sources to propose an information management framework that is then used to review several international city data stores or hubs. We conclude with key considerations that should guide a best-practice approach to city data stores.

Case Study: The *Connected City Data Hub* Initiative

The Organisation: Landcom NSW

Landcom NSW is a state-owned land and property development organisation in charge of managing large urban transformation programmes in NSW (Australia). Its role is to develop, enable, and demonstrate new homes and neighbourhoods across NSW. Its overall vision is to deliver city-shaping projects that create more affordable, vibrant, connected, and sustainable places to live (Urban Growth NSW, 2017).

The *Connected City Data Hub Scoping Study and Roadmap* project, undertaken by the authors, aimed at contributing to Landcom NSW's *City Transformation Lifecycle* agenda by establishing a clear understanding of the digital ecosystem and user requirements needed to inform future-forward investment in data delivery services.

As part of the scoping exercise, the project team conducted interviews with eight representatives of Landcom NSW. Interviewees spanned diverse operational and management roles across the organisation, with responsibilities including IT services and records management, government relations, stakeholder engagement, sustainability, and project management.

Interview Outcomes

The interviews were undertaken using a semi-structured interview format and addressed current experiences of data management within the organisation. In particular, the discussions addressed the following items:

- Types of data generated by the organisation, including the collection, management, and use of data across different stages of the urban transformation lifecycle
- Existing data management platforms and infrastructure in operation across the organisation
- Provisions in place to support Open Data; inter-agency data-sharing arrangements; relevant provisions supporting the confidentiality and privacy of data
- Levels of self-identified data literacy across the organisation
- Areas for improvement in the practices of data management and use across the organisation; practices supporting value-add to data being generated through the development lifecycle

Overall, interviewees identified different types of data relevant to specific phases or aspects of an urban development project: (1) planning and design phase, (2) stakeholder consultation and community engagement, (3) monitoring and evaluation, and (4) sales monitor. They also commented on data literacy across the organisation, data management platforms and data governance procedures.

Most interviewees described their data literacy levels as 'low' or 'low to medium' and suggested relatively low levels of data literacy across the organisation, leading to a poor understanding of the benefits of improved data management systems for business processes. One interviewee declared: '*We generate all sorts of data that we don't realise we generate, because we don't see it as data necessarily. [...] I don't think this organisation appreciates how data rich it is potentially and therefore what the other uses of this data could potentially be put to, by itself and others.*'

Concerning the planning and design phase, several interviewees noted that the majority of the data being collected is accessed from external agencies and that Landcom NSW itself is rarely the data custodian. This is particularly the case for larger urban transformation projects that require multi-agency collaborations. As simply put by one participant: '*most of the data [I use] comes from elsewhere*'. This has clear implications for the amount of data the organisation is able to release via a future *Connected City Data Hub*.

Landcom NSW uses *Objective Connect* as its official record management service platform. It supports the organisation's records management requirements but also the levels of external access to company reports and other documentation associated with a project. Landcom is required by legislation to be able to document key decision-making and compliance processes. Henceforth, *Objective Connect* is used to ensure documents are maintained securely and appropriately for each project. According to one interviewee: '*It's a compliance thing. It stops users from being able to say I'll put that on the H-drive and destroy it whenever I feel like it. Working for a Government agency, [records] need to be captured in a record keeping system and we use Objective as a management tool for that.*'

Stakeholder consultation and community engagement were perceived by interviewees as key areas where Landcom NSW collates its own data and information. A Communications and Engagement team uses a variety of communication platforms, ranging from traditional customer relationship management tools (*Consultation Manager*) to community consultation platforms like *Bang the Table* or *Social Pinpoint*. These platforms collate a mix of qualitative and quantitative information, including the location of the respondent, basic demographic information, and feedback from the respondent. As described by one interviewee: '*[This is] the digital equivalent of standing in a room with a bunch of people with some post-it notes and a map*'. Information is then exported into Excel spreadsheets and used for internal reporting exercises. Sentiment analyses associated with social media platforms (*Facebook* or *Twitter*) are also regularly performed and forwarded to project or site managers.

As Landcom NSW is required by legislation to report on key environmental sustainability indicators, the organisation is also collecting a large amount of data to support its sustainability reporting as part of its monitoring and evaluation process. The sustainability strategy that informs data collection around key environmental indicators has been in place since 2013. The Sustainability team uses digital tools like *CCAP Precinct* or *Ecological* to predict the environmental, social, and economic impacts of residential, commercial, and mixed-use developments. While sustainability reporting is a data-rich arena for Landcom NSW, most interviewees considered that current approaches to data management across the different reporting and benchmarking platforms remained inconsistent and needed improvement.

Sales and market research data collated by Landcom NSW was broadly identified as a potentially under-used resource by several interviewees. A participant declared: '*Through sales data you get the additional data of where people are coming from. Through this you can ultimately identify whether people are using that place [they've bought] as their billing address, and from there you can determine what level of purchasers are investors. That in itself is an interesting piece of data, how many people ultimately move to the place that they buy versus how many just buy it and rent it.*'

Overall, participants reported heavy reliance on third-party software services, reflecting upon Landcom's corporate IT strategy, which focuses the resources of IT management away from internal service provision to the governance and procurement of external platforms. An interviewee summarised this strategic change: '*Our business is not an IT business; it's built on [urban] transformation and land development. Before I arrived the problem was the old IT team had turned into this monolithic centre-piece of the business and nothing could happen unless it came through their change committee and business processes were dictated by the software system that IT built themselves bespoke—very 1990's. [The idea was] let's get away from that get into the 21st century.*'

It was also reported that Landcom NSW does have a cloud-based enterprise data warehouse capability in place. The original intention in establishing this capability was to ensure that the underlying data generated across business units could be exported and integrated into a range of business management processes. Most interviewees admitted that it was largely under-used due to poor data literacy across the organisation. One interviewee shared the following story: '*[Initially] the general audience of mid-level managers in my experience hadn't come across [online] dashboards, real-time analytics tools, and those types of things. [For example] someone that we presented to said, "Cool, how do I select all parameters, for all time, for all criteria and dump it into a spreadsheet?" I said, "Why?" They replied, "so that I can work with the data?" I said, "No, no this is the tool that will allow you to much more effectively work with the data", and they were like, "No, no I want a spreadsheet! To prove the point we did it, the spreadsheet crashed", and the poor person was saying, "Oh no how am I going to survive, my spreadsheet doesn't work". I explained, "that's why these enterprise tools exist, they can deal with volumes more data than your spreadsheet can".*'

Finally, participants were asked to describe whether they believed Landcom NSW should be making data available to the general public or to research organisations. The diversity of responses revolved around the appropriate level of data access (type and granularity) for different categories of users (general public, researchers, other public agencies, or the private sector). The following quotes illustrate the diversity of perspectives.

Interviewee: '*We don't do that. The closest we get to shared data is the [GIS] business unit, the spatial data analysis stuff, which integrates data layers from [a variety of sources] and then they can share that as well, but other than that, no.*'

Interviewee: '*I guess there's a bit of reluctance here to do that. I just can't see, I just don't think we've got to that level of sophistication and as we were saying before, we're not even aware of half of the data we have. We wouldn't even have the means to do it particularly well I would have thought. There is no consensus around what data is safe to release.*'

Interviewee: '*I don't think anyone should own data, I think it should be open source. But people, individuals, should not be able to be identified through that open source. [...] It just needs to be in a place that is secure, that its integrity is verified, and where individuals whose zeros and ones, individuals whose profiles are in those zeros and ones that they're accessing, individuals profiles cannot be identified. But beyond that, I think government should own the data and should own it only to make it freely available to others.*'

A majority of participants were mostly aware of Open Data policies across the NSW Government. While some were highly supportive of Landcom NSW adopting a very transparent approach to the release of its data, others expressed concerns around how this data could be mismanaged or misused.

Recommendations from Interviewees

Interviewees also made recommendations to improve data management practices across the organisation and facilitate the adoption of an Open Data paradigm:

- Expand the enterprise-wide data warehousing capability of the organisation
- Invest in a Chief Information Officer

- Ensure business practices are aligned with technology design
- Update the existing Client Relation Management software
- Adopt more sophisticated Geographic Information System (GIS) software to improve spatial analyses and share spatial information more easily with other agencies and stakeholders
- Conduct longitudinal monitoring of sentiment data in order to better understand people's acceptance of urban transformation
- Invest in more internal data analytics capability
- Improve internal standards for data collection, storage, and retrieval.

Our case study confirmed most of the issues listed by Zuiderwijk et al. (2012) and associated with a global move by government agencies towards Open Data paradigms in general and city data stores in particular. A mix of poor data literacy within the organisation, the poor interoperability of legacy systems, and inadequate data management procedures might hinder Landcom NSW's efforts to develop its *Connected City Data Hub*. In the first instance, it seems that Landcom NSW needs to develop a proper information management strategy and framework (Dawes & Helbig, 2010).

Information Management Framework

Integrating information and communication technologies (ICT) into an organisation's business and operations is broadly covered by the establishment of an information management strategy (IMS) (Chatfield & Reddick, 2016; Gold, Malhotra, & Segars, 2001; Kalampokis, Tambouris, & Tarabanis, 2011). The various dimensions affecting the drafting and implementation of the strategy can be grouped into three categories: (1) people and culture, (2) process, and (3) technology (Nam & Pardo, 2011). The following sections summarise the nature and content of these three dimensions.

People and Culture

This dimension addresses the 'human capital' part of IMS; it includes the technical skills and competencies required, the fostering of good data management habits and sense of responsibility for data, and the provision of incentives for such behaviour, as well as the presence of a collaborative culture at work to promote knowledge transfer. Komninos (2011) describes this aspect of IMS as the dimension where *amplification* can take place; where mechanisms are put in place to ensure that members of the enterprise or community that run, support, and use the knowledge infrastructure have the skills or can upskill to be able to contribute and co-create information—thereby amplifying the capabilities of the information and data infrastructure. Appropriate organisational structure and culture are fundamental to making the most of data.

Process

This dimension is congruent with Komninos' description of the capacity of an organisation for *orchestration* (Komninos, 2011). Orchestration is characterised by the coordination of institutions and human capital via political mechanisms and organisational protocol to effect collaboration and knowledge transfer and produce innovative solutions. This dimension deals with putting proper protocols in place for maintaining data asset security, using standards for procuring and processing data across the enterprise, and establishing efficient practices for providing and controlling access to data assets.

Technology

This dimension addresses the requirement of ICT instrumentation to be able to collect spatial data, as well as the integration of data and metadata formats and collaborative platforms/software. The technology used to store, catalogue, and make data accessible must allow for the data asset's interoperability between different platforms, across different agencies, and perhaps even across different time periods. The system architecture must be able to provide contingencies for making existing data accessible and usable by other existing systems and other potential uses of the data in the future, as well as have mechanisms to allow the integration of data from various sources and accommodate new types of data. At the heart of a data hub is data and the tools to manage, access, and use it. An open standards-based approach facilitates collaboration and efficiency.

Best-Practice Framework

Given that the urban data landscape is subject to disruption in digital delivery services and that enterprises face risks associated with digital obsolescence, there is a practical need for an open-ended framework that perpetuates continuous self-review and optimisation. Frameworks such as the *Capability Maturity Model* can be adopted as a roadmap for the organisation's IMS (Paulk, Weber, Curtis, & Chrissis, 1995). This kind of framework has been used by businesses in the ICT industry to assess their organisational performance, as well as in other project management contexts (Gold et al., 2001). The authors propose to adapt this open-ended framework for organisations that are seeking to incrementally implement and optimise their IMS strategy. The proposed framework combines elements from project management models (Yazici, 2009), emergent collaborative organisation models (Morgan, 2012), and ICT enterprise capability maturity models (Carcary, 2013; Ekuobase & Olutayo, 2015). Figure 13.1 synthesises the capability maturity framework whereby the *People & Culture*, *Process*, and *Technology* dimensions are assessed against successive maturity stages (from ad hoc to optimised stages).

As various aspects of the IMS are improved—as threads of capabilities are interwoven—the robustness and integration of the IMS mature. The progress of the IMS towards maturity may be viewed as an iterative exercise whereby standards for *People and Culture*, *Process* and *Technology* are planned, implemented, and thereafter monitored, evaluated, and reviewed until the system reaches its optimal state. It is within this cycle that data governance is formed and improved and where city indicators and benchmark measures may be further informed.

Figure 13.2 summarises best-practice principles that should guide the formation and implementation of a purposeful and useful information management system (IMS). Although the framework and these guiding principles are applicable to any IMS, they are particularly relevant to city data stores and hubs as our case study showed that a set of tools and a new paradigm (Open Data—open government) cannot deliver a purposeful product as long as people, culture, and processes align.

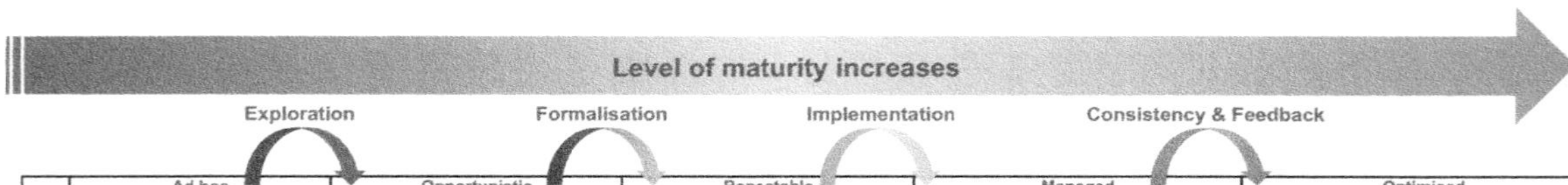

	Ad hoc	Opportunistic	Repeatable	Managed	Optimised
People/Culture	• Culture with regards to data openness is unaddressed • No established leadership • Intra- and inter-organisational departments' capacities with respect to supporting a KM enterprise are unknown • Intra- and inter-organisational departments' potential roles have not been discussed.	• Incorporation of openness in the organisation's objectives and change management strategy is discussed. • Leadership within the organisation is identified • Capacities and potential roles of departments within and without the organisation are identified • Potential collaborations (including the public) for data acquisition and analysis are identified. • Executive sponsorship is explored	• Openness and the organisational and operational qualities that are entailed in such policy is communicated to every level of the organisation. • Leaders are engaged • Intra- and Inter-organisational departments' roles and responsibilities are defined • Agreements for collaboration are made. • Measures of success are defined • Executive sponsorship is secured	• Openness culture is modelled and championed • Consistency in fulfilling roles and responsibilities is observed • Training is provided to staff to ensure that protocols and standards are executed • Inter-organisational collaboration is implemented • Performance metrics are observed and reported • Executive sponsorship is continuous	• Openness policy is reassessed. Wider stakeholder engagement may be involved in this process. • Leaders' roles have grown/evolved • Strategies are in place to ensure the continuity of KM management and inter-organisational collaboration in case of leadership changes • Training of staff may be enhanced further • Collaborations with other organisations is evaluated and other potential collaborations are explored • Executive sponsorship is continuous, but other long-term funding options are also explored • Overall performance is evaluated • Performance metrics are reviewed in light of wider institutional strategy (i.e. city indicators)
Process	• There is no consistent or repeatable workflow in collecting and managing data. • Protocols for collecting, storing, managing, and sharing information are not discussed	• Work flows are being reviewed • Protocols used by other exemplar organisations are explored	• Work flows and protocols are defined and communicated to every level of the organisation. • Where there is intra- and inter-organisational collaboration, protocols are also communicated. • Performance measures are defined • Processes are documented.	• Work flows and protocols are consistently implemented • Intra- and inter-organisational collaboration is integrated into the work flow. • Teams within the organisation may be established to lead and report on certain parts of the work flow. • Any changes to protocol are documented, implemented throughout the organisation, and communicated to external stakeholders	• Areas of improvement are assessed • If found necessary, new functions integrated into work flows in light of changing information needs, or organisational changes • Currency of protocols with international standards and practices is assessed • Adequacy of performance metrics is reviewed
Technology	• The organisation is unaware of technical standards that need to be adhered to for maximising data assets. • The organisation is unaware of tools and software solutions for managing information.	• Technical standards employed by local and international organisations are explored. • Available tools and software solutions are explored.	• Appropriate technical standards, tools and software solutions are selected. • Targets are defined • Selected standards and technology are documented.	• Technical standards are implemented and adhered to throughout the organisation. • Staff routinely use selected tools and software • Upgrades and integrations are planned for and documented • Technical issues are reported, resolved, and documented • Performance against targets is monitored • Upgrades/integrations planned	• Implementation results are evaluated • Upgrades and integration are continuous • Technology adoption strategy is reviewed considering emergent technology and standards, and organisational capacity • Adequacy of targets is reassessed

Fig. 13.1 Framework for evaluating the maturity of an information management system (IMS), with respect to knowledge management dimensions

People/culture
1. Establish an organizational strategy to implement a knowledge management system
2. Provide support for training and upskilling
3. Engage with internal staff and wider community to establish trust
4. Address leadership: establish information governance

Process

5. Conduct and communicate a data inventory
6. Establish data agreements
7. Publish standards and tools
8. Document and publish project information
9. Establish protocols addressing security and privacy concerns
10. Seek client feedback on the IM system and its implementation

Technology

11. Adopt Open Data formats
12. Adopt standards for metadata
13. Make datasets discoverable via catalogue
14. Adopt 5-Star Open Data Scheme
15. Publish web services for data visualization and analysis

Fig. 13.2 Fifteen best-practice principles to serve as general guidelines for forming and implementing an IMS strategy

Finally, the best-practice principles are incorporated into the Capability Maturity Model in order to develop a coherent IMS strategy for the organisation (Fig. 13.3). Most of the recommendations made by interviewees from Landcom NSW can easily be mapped onto this model: (1) expand the data warehousing capability (BP11–13), (2) invest in a Chief Information Officer (CIO) (BP4), (3) align business practices and technology (BP5–9), (4) perform advanced analysis and sharing of spatial data (BP14–15), (5) invest in internal analytics capability (BP2), and (6) improve internal standards for data collection, storage, and retrieval (BP1).

Benchmarking Existing City Data Stores

While the primary intent of the proposed IMS framework is to provide guidelines for the design and implementation of a purposeful and useful urban data store like Landcom NSW's *Connected City Data Hub*, it can also be used as a benchmarking tool to assess existing city data stores.

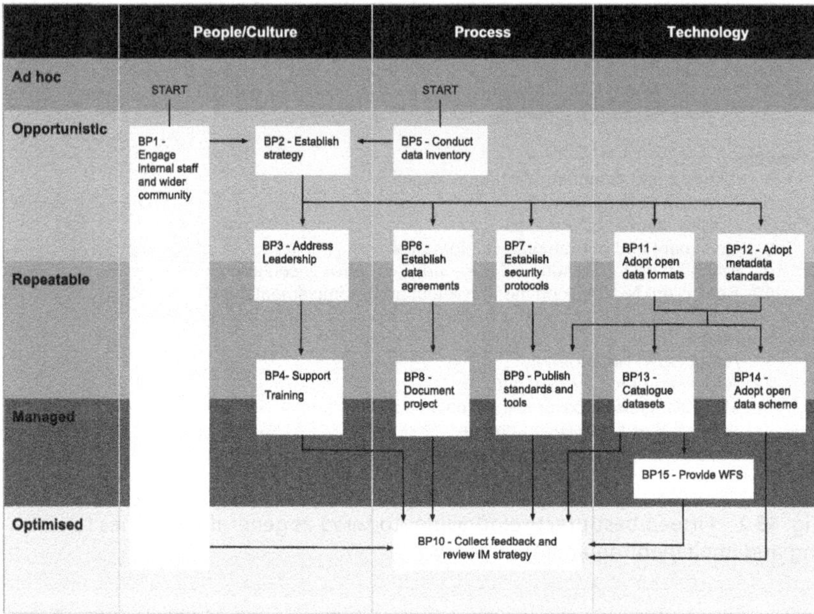

Fig. 13.3 The 15 best-practice principles (BP) incorporated into the capability maturity model

There are an increasing number of (partly) urban data hubs around the world. Some of them have a national or continental focus, like Geoplatform, developed by the American Federal Geographic Data Committee (FGDC; see: https://www.geoplatform.gov/), or the Australian Urban Research Infrastructure Network (AURIN), developed by a consortium of Australian universities (see: https://aurin.org.au/), or the INSPIRE geoportal, developed by the European Commission (see: inspire-geoportal.ec.europa.eu). However, city-specific Open Data portals are more closely aligned with our Landcom NSW case study as they share the same objectives: (1) to support an open government strategy, (2) to contribute to a smart city approach, and (3) to enhance local innovation and social engagement. Four well-documented examples of city data stores were selected to be assessed against our best-practice criteria: (1) London Datastore, (2) Singapore Datastore, (3) Geneva SITG, and (4) Paris Data.

London Datastore

https://data.london.gov.uk/

The London Datastore is a free and Open Data sharing portal where anyone can access data relating to the capital city. It was created in 2013, using the *Witan* platform developed by Mastodon C. Potential barriers to the public release of data such as government structures, privacy protection, and legal liabilities were identified early on. In 2014, the Greater London Authority (GLA) released the second-generation London Datastore with the following principles: (1) the intellectual property and technical specification of London Datastore are non-proprietary, (2) the London Datastore is a place where organisations can publish their data in open format, and (3) where publishers may seek support and peer advice in doing so. The London Datastore today serves as a focal point for the city's data strategy and smart city initiative, which is still unfolding (Greater London Authority, 2016). So far, the London Datastore contains 700 city-related data sets and their associated metadata records, contributed by various government agencies like Transport for London (TfL) that has already shared more than 30 data sets—including real-time travel feeds—with third-party developers, powered around 200 travel information products that benefit public transport customers. Data is made available through CKAN's (Comprehensive Knowledge Archive Network) Action API (Application Programming Interface), a powerful RPC (Remote Procedure Call) style API that exposes core features to API clients, including the website's core functionality. All spatial data sets are WFS (Web Feature Service) 2.0 compliant. In 2015, the London Datastore won the ODI Open Data Publisher Award. In August 2017, the Mayor of London appointed the capital's first ever Chief Digital Officer (CDO) whose role includes overseeing and encouraging the development of the London Datastore.

Singapore Datastore

https://data.gov.sg/

The Singapore Datastore was launched in 2011 as the government's one-stop portal to its publicly available datasets from 70 public agencies. It was built and is maintained by GovTech, a government agency tasked with harnessing info-communications technology and related engineering for public sector transformation. To date, the portal contains 1271 data sets and their associated metadata records, as well as more than 100 apps have been created using the government's Open Data. A new version of the city data store (beta version only) aims to make government data relevant and understandable to the public through the active use of charts and articles, seeking active feedback from members of the public to improve the user experience for all. The portal aims to: (1) provide one-stop access to the government's publicly available data, (2) communicate government data and analysis through visualisations and articles, (3) create value by catalysing application development, and (4) facilitate analysis and research by third-party developers. Like the London Datastore, the portal uses CKAN's Action API, a powerful RPC-style API that exposes core features to API clients. All spatial data sets are WSF 2.0 compliant. The Singapore Datastore is an initiative from the Ministry of Finance and is managed by the Government Technology Agency of Singapore.

Geneva SITG

http://ge.ch/sitg/

The *Système d'Information du Territoire à Genève* (SITG), Geneva SITG in short, is both a data store and a multi-agency organisation aimed at coordinating and promoting an Open Data–open government strategy within Geneva's territory. Although the initial decision to create a SITG was acted in 1991, it only came into existence in 2013, developed by the *Service de Géomatique et de l'Organisation de l'Information* of Geneva's Territorial Authority. The data store includes all the data and associated analytical tools related to land use planning and management. The Open Data sets are meant to stimulate innovation and foster the creation of digital services that will benefit the private sector and local communities. So far, Geneva SITG includes 604 data sets and their associated metadata records. Geneva SITG is intended to (1) become the digital memory of Geneva's territory, (2) support spatial data analyses and dissemination, (3) enhance communication and coordination amongst territorial partners, and (4) stimulate regional innovation. Unlike the London and Singapore Datastores, Geneva SITG uses ESRI proprietary technology, exposing Open Data and digital services through the ArcGIS Representational State Transfer-based Application Programming Interface (REST-API). As a consequence, spatial data sets use the Warehouse Management System (WMS) format instead of WFS 2.0. However, the JavaScript Open Layers library provides a core of functionalities that allows for web mapping client applications. The SITG organisation itself coordinates a network of 13 public sector agencies and institutions. Its role is to collate and disseminate territorial data to a broader audience. The SITG is governed by a steering committee and includes a technical commission, a public forum, and a secretariat. Its mandate includes the organisation of events and training workshops. Since 2014, Geneva SITG benefits from a legal framework recognising its Open Data principles.

Paris Data

https://opendata.paris.fr/page/home/

The Paris Data portal was launched in 2011 by the Paris City Council to collate and disseminate data provided by various Council agencies and services. It is powered by the OpenDataSoft software, a pay-as-you-go and turnkey solution, designed for non-technical users to share, publish, and reuse data. It is available as a commercial-off-the-shelf, cloud-hosted product used by other cities and local governments (Paris, Durham, Stadt Mannheim). OpenDataSoft uses REST-API to exchange data and services with third-party client applications and its own data query language (ODSQL) for complex filtering and data aggregation. The platform only supports the WFS 1.1 format. Paris Data includes 232 data sets available for external use through an Open Database License (ODbL). A team of three technicians maintains and develops the platform; they rely on a network of correspondents in each relevant agency or service. In 2010, the Mayor of Paris and his Council ratified the Open Data-open government strategy. In 2014, his successor imposed an Open Data clause for public markets and procurements. Paris Data is closely related to the DataCity Paris initiative, an open-innovation forum for local start-ups and the Open Government Partnership, a consultation channel aimed at seeking social feedback about data releases.

Assessing the City Data Stores

Figure 13.4 summarises the findings of the assessment against 15 best-practice (BP) criteria of the proposed IMS framework. The assessment relied on a thorough review of the content of the data stores, associated documentation, and links to other sources. While the process was identical across the four examples, it cannot be ascertained that the search was exhaustive. Henceforth, these results have to be interpreted with caution and should only be used in the context of the original aim of the study: testing the use of the proposed framework to benchmark existing city data stores and hubs.

Only Geneva SITG makes an explicit and comprehensive reference to training (BP2), providing direct access to online tutorials. The fact that the platform was designed and is maintained by a local team from the *Service de Géomatique et de l'Organisation de l'Information* could explain this singularity, although Singapore Datastore and Data Paris also receive significant contributions from local teams. Data Paris is the only example that doesn't clearly identify a technological leadership (BP4) while associated material insists on the political drive that led to its creation.

It is relatively surprising that three out of four examples don't explicitly and clearly address data security issues (BP9) when Open Data raises so many concerns amongst public services. The link established from Data Paris portal to OpenDataSoft's website provides access to several references about data security. This singularity might be driven by the Cloud-based and pay-as-you-go nature of the platform, a strong incentive for the company to reassure its clients. Geneva SITG is the only example that fully relies on the ESRI technological solution for spatial data management and visualisation, preferring the WMS standard to WFS 2.0. Data Paris offers a mixed approach but limits its service to WFS 1.1.

Data store	BP1	BP2	BP3	BP4	BP5	BP6	BP7	BP8	BP9	BP10	BP11	BP12	BP13	BP14	BP15
London	✓	✗	✓	✓	✓	✓	✓	✓	✗	✓	✓	✓	✓	✓	✓
Singapore	✓	✗	✓	✓	✓	✓	✓	✓	✗	✓	✓	✓	✓	✓	✓
Geneva	✓	✓	✓	✓	✓	✓	✓	✓	✗	✓	✗	✓	✓	✓	✗
Paris	✓	✗	✓	✗	✓	✓	✓	✓	✓	✓	✓	✓	✓	✓	✗

Fig. 13.4 Assessing four city data stores against the 15 best-practice criteria (BP) of the proposed IMS framework (tick: BP addressed in publicly available material; cross: no explicit reference to BP in publicly available material)

Overall, the proposed IMS framework allows for a relatively in-depth assessment of a given city data store or hub, based on publicly available information. It allows to shed light on the relationships between conceptual intentions, technological choices, or organisational characteristics, as well as some of their consequences.

Conclusion

Governments play a critical role in realising the potentials of data-driven urbanism and smart cities, and yet challenges remain for many agencies in defining clear data management pathways capable of meeting a range of internal and external stakeholder needs. In this chapter we have discussed the challenges identified within Landcom as they relate to the management and use of data and information services.

Through a series of interviews with identified representatives, our research identified a range of challenges, including levels of data literacy in an organisation not specifically established with a data management or smart city remit and lack of clarity around the relative 'openness' of data used by the agency in meeting its remit to facilitate land development on behalf of the NSW Government. With a view to supporting Landcom in playing a more active role in supporting the data ecosystem within which it operates, we have set out an information management strategy (IMS) that draws on best-practice approaches to data store and dashboard design. In doing so, we emphasised the need to address the three dimensions of people and culture, process, and technology as each being critical to success. The list of 15 best-practice criteria proposed in this chapter is by no means a comprehensive list as IMS implementation is a dynamic process, disrupted and evolved over time by technological, cultural, and organisational changes. IMS also varies according to the needs and purpose of the implementing organisation. It is advisable for organisations like Landcom NSW to treat this list as a guideline that can and should evolve as their experience in IMS implementation matures.

Ultimately, while the promises of responsive, data-driven services are many, governments require effective internal and external strategies for enhancing and enabling their roles as urban data custodians and facilitators, if we are to see these potentials realised.

Acknowledgements The *Connected City Data Hub Scoping Study and Roadmap* is a collaborative project between Landcom NSW, the University of New South Wales, Western Sydney University, and the University of Wollongong. The authors thank all participants—interviewees, workshop attendees, and researchers—for their time and effort.

References

Barns, S. (2018, March). Smart cities and urban data platforms: Designing interfaces for smart governance. In *City Culture and Society Special Issue: Innovation and identity in next-generation smart cities* (Vol. 12, pp. 5–12). Elsevier.

Carcary, M. (2013). IT risk management: A capability maturity model. *The Electronic Journal Information Systems Evaluation, 16*(1), 3–13.

Chatfield, A. T., & Reddick, C. G. (2016). Open data policy innovation diffusion: An analysis of Australian Federal and State Governments. In *Proceedings of the 17th International Government Research Conference on Digital Government Research*, 8–10 June 2016. Available from ACM Digital Library. [16 February 2017].

Dawes, S., & Helbig, N. (2010). Information strategies for open government: Challenges and prospects for deriving government transparency. In: M. A. Wimmer, J. L. Chappelet, M. Janssen, H. J. Scholl (eds.), *Electronic Government. EGOV 2010.* Lecture Notes in Computer Science, vol 6228. Springer, Berlin, Heidelberg, pp. 50–60. Available from: Springer. [1 May 2017].

Ekuobase, G. O., & Olutayo, V. A. (2015). Study of Information and Communication Technology (ICT) maturity and value: The relationship. *Egyptian Informatics Journal, 17*(3), 239–249.

Gold, A. H., Malhotra, A., & Segars, A. H. (2001). Knowledge Management: An organizational Capabilities Perspective. *Journal of Management Information Systems, 18*(1), 185–214. Available from: Taylor and Francis Online.

Goldsmith, S., & Crawford, S. (2014). *The responsive city: Engaging communities through data-smart governance* (208p.). San Francisco, CA: Wiley.

Greater London Authority. (2016). *Data for London: A city data strategy* (35 p.). London: GLA. Retrieved from https://files.datapress.com/london/dataset/data-for-london-a-city-data-strategy/

Kalampokis, E., Tambouris, E., & Tarabanis, K. (2011). Open government data: A stage model. In M. Janssen, H. J. Scholl, M. A. Wimmer, & Y. Tan (Eds.), *Electronic Government. EGOV 2011* (Lecture Notes in Computer Science, vol. 6846) (pp. 235–246). Berlin, Heidelberg: Springer.

Komninos, N. (2011). Intelligent cities: Variable geometries of spatial intelligence. *Intelligent Buildings International, 3*(3), 172–188.

Morgan, J. (2012). *The collaborative organization* (p. 286). McGraw-Hill.

Nam, T., & Pardo, T. A. (2011). Conceptualizing smart city with dimensions of technology, people, and institutions. In *The Proceedings of the 12th Annual International Conference on Digital Government Research*, 12–15 June 2011. Available from ACM Digital Library.

Paulk, M. C., Weber, C. V., Curtis, B., & Chrissis, M. B. (1995). *The capability maturity model: Guidelines for improving the software process*. SEI Series in Software engineering. Reading, MA: Addison-Wesley. ISBN 0-201-54664-7.

Pettit, C., Lieske, S. N., & Jamal, M. (2017). City dash: Visualizing a changing city using Open Data. In *Planning support science for smarter urban futures* (pp. 337–353). Springer International Publishing.

Shaoyun, G., Xuena, H., Junling, L., Yongxin, Q., Jun, Y., & Qngyan, K. (2010). The Data Management of E-government System in the Deepening Application Phase. In *2010 International Conference on Intelligent Computation Technology and Automation*, May 2010, vol. 3, pp. 747–750. Available from: IEEEXplore.

UrbanGrowth NSW. (2017). *Annual Report 2017* (p. 105). Sydney, Australia: UrbanGrowth NSW.

Yazici, H. J. (2009). The role of project management maturity and organizational culture in perceived performance. *Project Management Journal, 40*(3), 14–33.

Zuiderwijk, A., Janssen, M., Choenni, S., Meijer, R., & Alibanks, R. S. (2012). Socio-technical impediments of Open Data. *Electronic Journal of e-Government, 10*(2), 156–172. Available from: http://www.ejeg.com

14

Tell Me How My Open Data Is Re-used: Increasing Transparency Through the Open City Toolkit

Auriol Degbelo, Carlos Granell, Sergio Trilles, Devanjan Bhattacharya, and Jonas Wissing

Abbreviations

APIs	Application Programming Interfaces
ICT	Information and Communication Technology
OCT	Open City Toolkit
SUS	System Usability Scale
URL	Uniform Resource Locator

A. Degbelo (✉) • J. Wissing
Institute for Geoinformatics, University of Münster, Münster, Germany
e-mail: degbelo@uni-muenster.de; jonas.wissing@uni-muenster.de

C. Granell • S. Trilles
Institute of New Imaging Technologies, Universitat Jaume I, Castellón de la Plana, Spain
e-mail: carlos.granell@uji.es; strilles@uji.es

D. Bhattacharya
NOVA Information Management School (NOVA IMS), Universidade Nova de Lisboa, Lisbon, Portugal
e-mail: dbhattacharya@novaims.unl.pt

S. Hawken et al. (eds.), *Open Cities | Open Data*,
https://doi.org/10.1007/978-981-13-6605-5_14

Highlights

- Dashboard to visualize apps and dataset usage in a city context
- Visualization of spatial locations from which apps and datasets are accessed
- Interactive guidelines as problem-solution patterns to communicate Open Data-based innovations
- Interactive guidelines to empower citizens to participate in and shape the future of their cities

Introduction

The Open Data movement has been gaining momentum in recent years, with increasingly many public institutions making their data freely accessible. Open Data is improving governments around the world, empowering citizens, creating new economic opportunities, and solving big public problems (see Young & Verhulst, 2016). As of May 2018, the Open Data Inception (opendatainception.io)[1] lists no less than 2600 Open Data portals all around the world; the US Open Data Portal[2] lists about 190,000 available datasets; the European Union Open Data Portal[3] offers about 12,000 datasets; the data portal of the Australian Government[4] contains about 57,000 datasets; and the UK's Open Data portal provides about 45,000 datasets to browse through. These figures are indicative of the amplitude of the Open Data movement. The term "open" may have different interpretations (for a recent review, see Pomerantz & Peek 2016) but is used in this chapter to denote data "that anyone can freely access, use, modify, and share for any purpose".[5]

Open Data has also attracted a significant amount of scholarly attention in recent years. A detailed presentation of Open Data ecosystems in Europe was done by Schade, Granell, and Perego (2015). Attard, Orlandi,

[1] https://opendatainception.io/ (last accessed: May 19, 2018).

[2] https://www.data.gov/ (last accessed: May 15, 2018).

[3] http://data.europa.eu/euodp/en/home (last accessed: May 15, 2018).

[4] http://data.gov.au/ (last accessed: May 15, 2018).

[5] http://opendefinition.org/ (last accessed: May 15, 2018).

Scerri, and Auer (2015) provided a systematic survey of open (government) data initiatives with a detailed description of processes within the open government data lifecycle. Taking Chile as a case study, Gonzalez-Zapata and Heeks (2015) identified two main types of stakeholders of open government data: primary stakeholders (i.e., politicians, public officials, public sector practitioners, and international organizations) and secondary stakeholders (i.e., civil society activists, funding donors, Information and Communication Technology (ICT) providers, and academics). Susha, Zuiderwijk, Charalabidis, Parycek, and Janssen (2015) organized workshops with experts from the field of open government and Open Data to identify factors influencing the success or failure of Open Data initiatives. They provided a list of 47 success factors for Open Data publication and 18 success factors for Open Data use. Hartog, Mulder, Spée, Visser, and Gribnau (2014) interviewed different types of stakeholders (e.g., civil servants, data source holders, and policymakers) to uncover the "readiness" for Open Data of two governmental bodies: the municipality of The Hague and the province of South Holland. Citizens' motivations to participate was the subject of Wijnhoven, Ehrenhard, and Kuhn (2015), where the authors found that strong belief that their suggestions will be applied correctly, perception of fun, and ideology (i.e., the person's attitude towards civic duties) are key factors of citizen engagement in open government projects. Additional work in the context of open government data has looked into open government portals' support for transparency and political accountability (Lourenço, 2015), openness and maturity indices for e-government (Veljković, Bogdanović-Dinić, & Stoimenov, 2014), a measurement framework to quantitatively assess the quality open government data (Vetrò et al., 2016), and visualization tools for open government data (Graves & Hendler, 2013), to name but a few.

Despite much attention of the scholarly community, many datasets being already open and more to come, finding information about the *actual usage* of these open datasets is still a challenge. Platforms such as CKAN (Comprehensive Knowledge Archive Network) offer a plugin (i.e., the *stats* extension[6]) to retrieve summary statistics about the most viewed datasets. This is valuable information, but there is still a need for

[6] http://docs.ckan.org/en/ckan-2.7.3/maintaining/tracking.html (last accessed: May 15, 2018).

techniques, which enable re-use tracking beyond dataset views. The rate of re-use was mentioned in Attard, Orlandi, and Auer (2016) as one of the aspects of Open Data value creation not sufficiently addressed at the moment. Benitez-Paez, Degbelo, Trilles, and Huerta (2018) found the lack of re-use examples to be one of the issues encountered by users while navigating Open Data portals. Having more information about the re-use of open datasets is critical to unveil their true value as: "[o]pen data on its own has little intrinsic value; the value is created by its use" (Janssen, Charalabidis, & Zuiderwijk, 2012). Open Data re-use information is also necessary for effective planning in the city context. For instance, it provides public institutions with a better idea of the types of datasets that are highly demanded (and by whom) and helps them prioritize the types of datasets to curate or regularly update.

This chapter introduces two software tools intended to advance the state of the art on open (government) data re-use: a tool to increase transparency and interactive guidelines. The tools tackle the re-use problem at two levels: automatic re-use tracking (the former) and re-use documentation (the latter). Both tools are part of the Open City Toolkit (OCT), a collection of datasets, tools, services, specifications, and guidelines to deliver services based on Open Data that are useful for citizens, businesses, and governing bodies (Degbelo, Granell, et al., 2016). The OCT combines technology-driven and citizen-centric strategies. It purports, as indicated in Degbelo, Bhattacharya, Granell, and Trilles (2016), to address the lack of integrated and open collections of software components to realize smart cities.

OCT Transparency Tool

The OCT transparency tool is useful to answer the questions: what are datasets available in my city? How often are these datasets used? And which apps use these datasets? An essential technical means of realizing this is the use of semantic Application Programming Interfaces (APIs). The design of semantic APIs and their different layers were discussed in detail in Degbelo, Trilles, et al. (2016). The main features of the OCT transparency module are:

- App registration: each developer (individual or organization) can register its app by getting an API key. This API key is used later to identify apps that access some datasets.
- Dataset registration: through this functionality, developers can register their own dataset to the OCT transparency module, so as to make it visible to other users (e.g., citizens, city councils, companies, developers).
- Logging: this functionality involves recording all activities related to an app (i.e., topics of datasets accessed, frequency of access, and spatial locations from which the datasets are accessed).

As a proof of concept for the idea, eight web applications were created based on existing open government data (e.g., population, migration, and referendum data). The applications and the process of their creation were presented in Degbelo and Kauppinen (2018). The datasets used are available on Zenodo (https://doi.org/10.5281/zenodo.293201). Figure 14.1 presents a dashboard visualization illustrating information about dataset usage provided by the OCT transparency tool. The tool also informs about the places from which an app has accessed datasets and places from which datasets were called (see Fig. 14.2). It is a dashboard in the sense of Matheus, Janssen, and Maheshwari (2018), who define dashboards as "the visualization of a consolidated set data for a certain purpose, which enables to see what is happening and to initiate actions". The next subsections report on some tests about the usability, usefulness, and scalability of the tool.

Usability

Two rounds of usability tests were conducted in February 2017 and October 2017. Each of the rounds involved seven people, leading to a total of 14 usability test participants. The usability tests were summative (see Lewis 2014 for a definition of summative usability), focusing on efficiency and effectiveness. In the first round, students were asked to register an app and a dataset and provide informal feedback about their experience doing so. Their feedback was integrated into the development of the

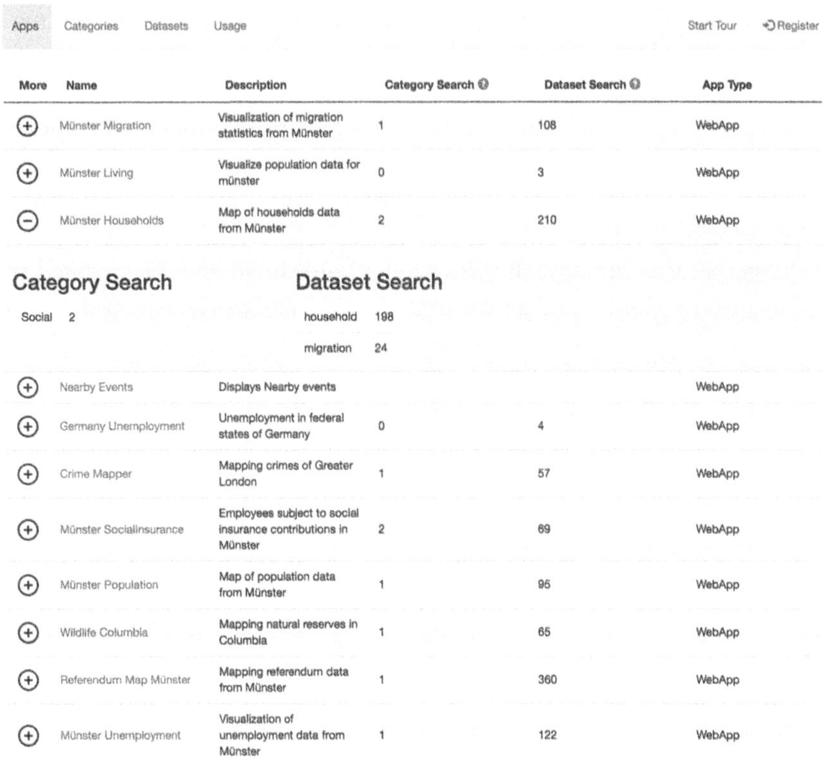

Fig. 14.1 Dashboard visualization about datasets usage provided by the OCT transparency tool

second version of the transparency tool, which was used during the second round of tests. In this second round, participants were asked to register their app, register a dataset, and build their first OCT app. The three tasks were completed successfully by all seven participants in less than 30 minutes (see Fig. 14.3). The System Usability Scale (SUS) score for the participants was 67.14, which means (following the scale introduced in Bangor, Kortum, & Miller, 2008, 2009) that the participants rated the usability of the OCT transparency module as "good". Using SUS as a usability questionnaire is suitable in this case because previous work (Brooke, 2013; Sauro, 2013; Tullis & Stetson, 2004) pointed out that it produces acceptable results even with a small number of participants.

Fig. 14.2 Example visualization of spatial locations from which one specific app (i.e., Referendum Map Münster) is accessed

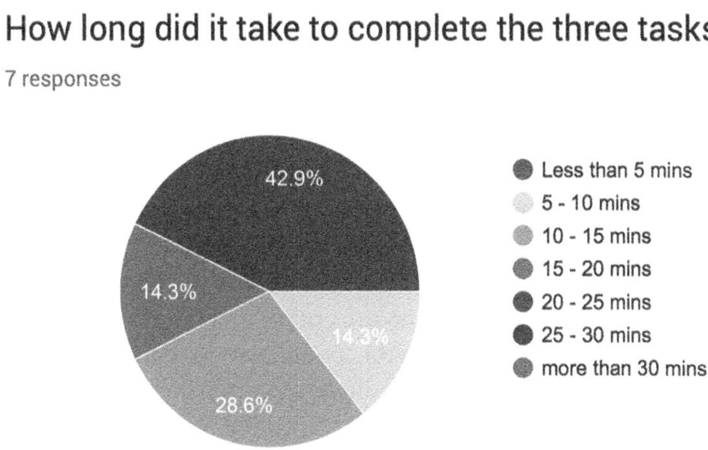

Fig. 14.3 Registering an app, a dataset, and building one's first OCT app can be done within 30 minutes

Utility

Eight semi-structured interviews of employees from two different city councils (Lisbon and Münster) were conducted in October 2017[7]. The purpose of the interviews was to (1) gather insights from people working on (or having worked with) Open Data about the importance of Open Data re-use for city councils, and (2) collect some feedback from domain experts on the OCT transparency tool. Participants were recruited through snowball sampling (for a description of the sampling method, see Lazar, Feng, & Hochheiser, 2010). The OCT transparency tool was used as a probe during the interviews.

The interview protocol was adapted from Roth (2009) and included an introductory question, five key questions, and an ending question. The first three key questions were: (1) How important is information about Open Data re-use for your institution?; (2) What are you currently doing to collect information about Open Data re-use?; (3) What issues do you face while collecting information about Open Data re-use?; (4) In your opinion, what could be the benefits of the module for Open Data re-use?; and (5) In your opinion, what could be the limitations of the module for Open Data re-use? Questions (4) and (5) were asked after showing an introductory video of 90 seconds about the tool.

The interviews lasted an average of about 30 minutes. Table 14.1 reports on the results of questions (4) and (5), which are directly related to the transparency tool. The table illustrates that the participants, overall, saw more pros than cons. The pros often mentioned included feedback to the city council about the popularity of datasets and easier discoverability of datasets. Cons often reported included meta(data) maintenance (existing apps and datasets must be registered again on the tool to be made visible), as well as the current lack of quality checks by the tool. There is no guarantee that data saturation[8] has been reached

[7] The ideal number of participants for interviews is purpose-dependent (see, e.g., Guest, Bunce, & Johnson, 2006), but a common range is between 8 and 15 participants (Lopez & Whitehead, 2013). When doing qualitative research, "the 'richness' of data collected is far more important than the number of participants" (Lopez & Whitehead, 2013).

[8] See, Fusch and Ness (2015) for a brief introduction to data saturation.

Table 14.1 Interviewees' feedback about the OCT transparency tool

Participant ID	Current role	Advantages mentioned	Limitations mentioned
#1	Head of department	(a) Data publishers can get some feedback about the most popular datasets and categories (b) Knowledge about most popular categories can inform about the types of datasets to make open	Data and metadata maintenance
#2	Project manager	(a) Knowledge about datasets, which the city council does not need to publish (b) Knowledge about the most popular categories can inform about the types of datasets to make open (c) Asking new questions (e.g., why someone accesses datasets from a place?)	Maintenance
#3	Team leader	(a) Facilitate the discoverability of datasets (b) Show politicians that Open Data is the way to go	None
#4	Manager Open Data portal	See the datasets and apps that are used	(a) Module currently lacks information quality checks (b) Module currently lacks verification of data
#5	Technical lead	Helps understand data use	(a) No verification (b) Coherency of the data
#6	Head of division	Easier discoverability of datasets	Module currently lacks notifications to users about crashes and data additions
#7	Head of library	Easy to gather statistics about the data that is being used	No answer
#8	Geologist	Information about data usage	No idea

with the sample of eight participants, that is, that the eight interviewees have listed all possible pros and cons pertaining to the tool. This notwithstanding, their feedback is useful: the pros mentioned validate the utility of the tool while the cons point at areas where work is still needed in order to facilitate its adoption in the city context.

Scalability

Several tests were conducted to assess the performance of the platform under a growing number of requests. The tests measured the response time of simultaneous database accesses on the system. Each test involved a group of queries to the endpoint of the API. Each group of queries was executed five times, and the response time was averaged over the five executions. The tests simulated 1, 5, 10, 25, 50, 100, 250, 500, and 1000 concurrent uses respectively. The data packets retrieved were kept constant (7 KB) during the whole test sessions. As Fig. 14.4 suggests, the scalability of the platform is better than linear. The code source of the application is available on GitHub (https://github.com/geo-c/OCT-Core).

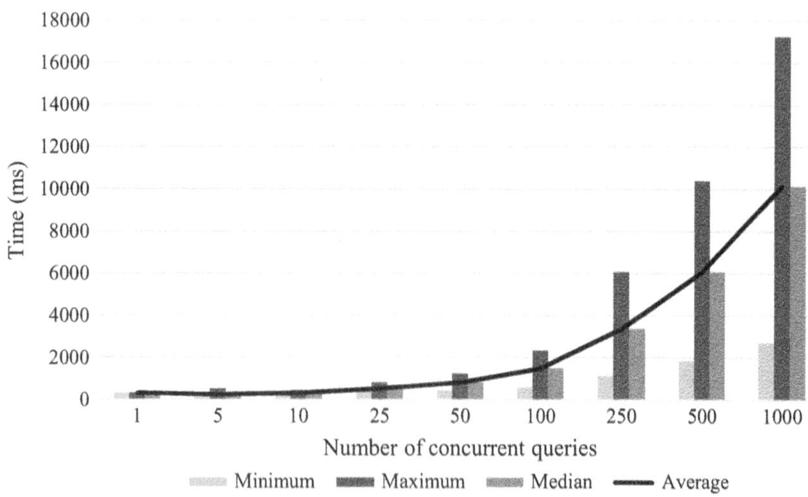

Fig. 14.4 The OCT transparency tool in reaction to growing instances of concurrent requests

OCT Interactive Guidelines

While the OCT transparency tool enables the monitoring of cities' open datasets usage, the OCT interactive guidelines tool deals with the following question: what can I do with all these datasets? The main audiences of both tools are different: decision-makers, data publishers, and managers are typically the focus of the OCT transparency tool. Of course, citizens can also seek for the impact of open datasets. Yet, they often measure the usefulness and impact of open datasets through a different lens, namely "how can open datasets enhance their daily activities?" Both tools are complementary, addressing Open Data usage and Open Data-based innovations (though the interpretation of these two terms may vary depending on the target stakeholders).

The Need for Guidelines

We can explain the role and purpose of the OCT interactive guidelines by borrowing an analogy from MIT (Massachusetts Institute of Technology) researcher Cesar Hidalgo, who compared Open Data sites and supermarkets[9]: "Imagine shopping in a supermarket where every item is stored in boxes that look exactly the same. Some are filled with cereal, others with apples, and others with shampoo. Shopping would be an absolute nightmare!" Hidalgo argued that most, if not all, Open Data sites are organized like arrays of "brown boxes" in supermarkets, that is, arrays of links to public datasets that quite often are published as they were collected. This way, most of these sites look like they are only addressing a small portion of the whole population: those with technical skills (programmers, researchers, etc.) or professionals (e.g., data-driven journalists, civic agents, etc.), that is, those few specialists who are able to handle and transform datasets to tell stories to the rest of people. The OCT has a CKAN-based module (not introduced in this chapter, see Degbelo, Bhattacharya, et al., 2016), which is not that far off this strategy; research resources are registered and made publicly available

[9] What's Wrong with Open-Data Sites—and How We Can Fix Them, by Cesar Hidalgo. https://blogs.scientificamerican.com/guest-blog/what-s-wrong-with-open-data-sites-and-how-we-can-fix-them/ (last accessed: Oct 4, 2017).

as endpoints that can be queried via well-documented data access and retrieval APIs. The expected stakeholders of the CKAN-based OCT module are other scientists, researchers, and civic hackers/programmers who feel comfortable (programmatically) handling Open Data and coding.

If we do not consider the tech elite, which is the remaining 95% of the population (OECD, 2016), Open Data sites become difficult to understand (see e.g., Benitez-Paez et al., 2018; Beno, Figl, Umbrich, & Polleres, 2017). Returning to Hidalgo's analogy of the supermarket, imagine you (citizen) are asking for "cannelloni" in the food section and the clerk delivers you a bag with all the raw ingredients to cook them yourself. Like most of the Open Data sites, Open Data is delivered in the way in which it was collected. Next, you look again at the clerk and order cannelloni "ready to be eaten" because you do not have time or do not know to cook them. Like most Open Data sites, Open Data is not delivered in the way it can best be used and/or understood. Rather, open datasets are often delivered with no clue on how to process them, manage them, or, even worse, whether they can be useful for citizens at all. In sum, citizens demand "ready-to-consume, easy-to-understand products" rather than raw ingredients like open datasets. Sometimes these products take the form of apps or can be expressed as interactive guidelines. The OCT interactive guidelines tool seeks to make city problems and subsequent actions understandable to citizens.

Most Open Data sites do not deliver elaborated stories that emerge from the combination of their contained open datasets. However, most people are looking for stories ("cannelloni") that can be easily comprehended ("eaten"). In case people want to know the details (e.g., raw ingredients to cook cannelloni themselves), they can directly download or access data sets through the corresponding data access API. What we pursue here is the design and creation of "stories" that bring together, behind the scenes various datasets and other types of resources and transform them into interactive city guidelines, which help a large portion of the society to understand their benefits and impacts regardless of the complexity of the details.

The OCT interactive guidelines tool is intended to help city stakeholders walk through a story. On the one hand, the term "guidelines" is seen as narratives that refer to problem-solution patterns by presenting challenges, benefits, and impacts in an understandable manner, that is, everyone may share and refer to them when talking to others. Problems may be of diverse nature, such as social, mobility, environment, and cultural; solutions may involve a combination of datasets, code, apps, services, and any other relevant resource that helps to sort out the current problem. On the other hand, the qualifier "interactive" underlines the ability of users to dynamically explore (to a certain degree) the guideline through a set of blocks for different purposes such as graphs, plots creation, maps visualization, custom JavaScript code, p5 code (a sort of JavaScript wrapper for processing), and the inclusion of text and markdown formats. We intentionally avoid static guidelines, as in the form of tutorials or paper-based posters, to let stakeholders engage dynamically with the content of the visual narratives.

Conceptual Architecture

Figure 14.5 shows the conceptual architecture to materialize the OCT interactive guidelines tool. Designers of stories are one type of users. These could be for instance researchers, data journalists, or data publishers; they use "storytelling" formats for creating visual and interactive narratives of

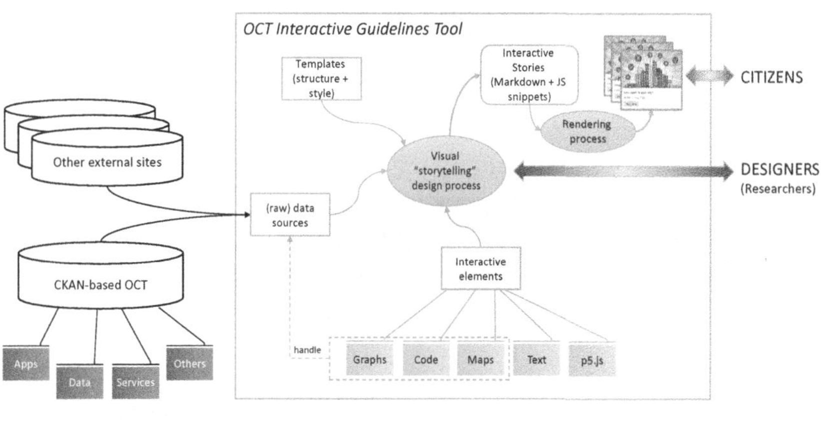

Fig. 14.5 Architecture of the OCT interactive guidelines tool

how smart city solutions are being installed and deployed in cities. That is, they design and tell stories based on external or own datasets and other research resources. The tool provides an edit mode to create and easily update each story and publish it into the catalogue of guidelines (Fig. 14.6). Citizens are the second type of users. They can pick a guideline from the catalogue and explore it through interactive elements at their disposal. For example, via interactive plots, charts, and maps and through on-the-fly annotations as a way to provide feedback about the story being visualized (this feature will be released shortly). The source code of the OCT interactive guidelines tool is also available on GitHub (https://github.com/geo-c/OCT-Guidelines).

Technically, each guideline is stored as a markdown file (like a regular text file). Markdown tags specify the sections of a guideline and keep information about the author, last update, title, and a list of the data sources used. To ease the process of creating guidelines, the tool provides

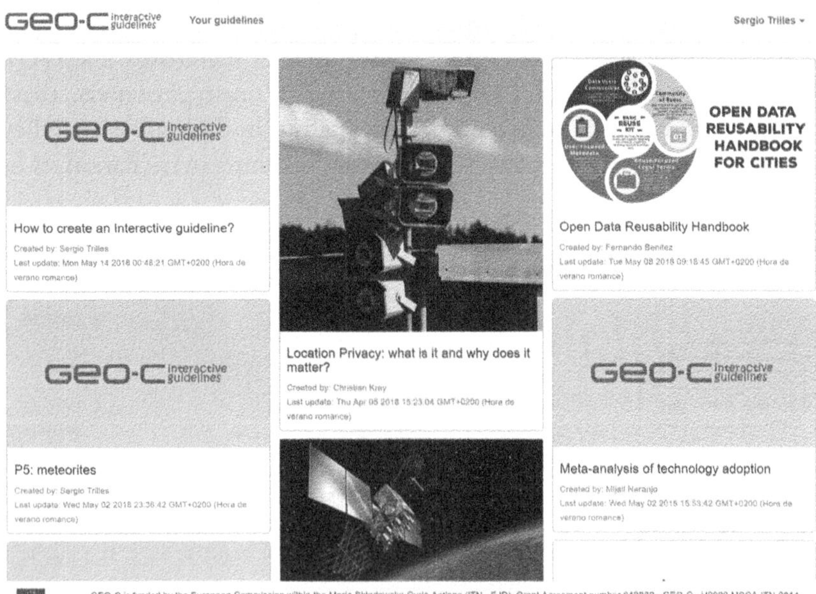

Fig. 14.6 Catalogue of interactive guidelines showing examples of OCT interactive guidelines

a range of templates with predefined sections and style. Each guideline is exportable as a regular markdown file for the offline edition, which may be uploaded again later on. Besides, a guideline contains a collection of interactive elements (codified as JavaScript snippets). Currently, the supported elements are graphs and plots, custom JavaScript code, maps, and text; adding annotations is part of ongoing work.

These interactive elements are able to handle data sources, both deployed in a CKAN-based instance platform and published elsewhere. For example, a graph can take as input a data source available in the OCT catalogue, permitting users to interact with the graph and thereby with the associated data source. Furthermore, any resource registered in the OCT catalogue (http://giv-oct.uni-muenster.de:5000/) is potentially an input source for interactive guidelines by only specifying its access point (e.g., Uniform Resource Locator [URL]). Moreover, interactive guidelines can be registered in the OCT catalogue as any other public and open resource. This way, the OCT interactive guidelines tool augments the capabilities of the OCT catalogue to deliver not only datasets but stories to a wide range of stakeholders. On the downside, designing compelling, understandable, and thought-provoking guidelines requires authors with proven communication and design skills so that the intended messages are effectively transmitted to the public.

Examples of these interactive guidelines are available at http://elcano.init.uji.es/guidelines. At the moment of this writing, there are ten guidelines; some of them are like "tutorials" to guide user creators to use and combine the different available blocks. Others intend to help stakeholders to solve particular problems or show a list of suggestions to follow. For instance, the guideline "Location Privacy: what is it and why does it matter?" attempts to communicate to citizens the importance and implications of sharing the location using smartphones. The "Checklist and tools when publishing open data" guideline tries to explain what an Open Data needs to earn some stars of Tim Berners-Lee's five-star model (Berners-Lee, 2006). To achieve that, some blocks such as text and p5.js blocks are used. The latter blocks are particularly useful to provide interactivity to the guidelines (e.g., they help to generate buttons, where users can click and see what each star category means).

In sum, the interactive guidelines do not specifically target technologically savvy people such as Open Data advocates and programmers. These guidelines aim to inform people about problems that matter in their cities, making them understandable and presenting potential solutions. Interactive guidelines, when designed as effective narratives, can raise awareness about certain problems that matter to citizens, even to the point to persuade and reframe thinking. In this sense, interactive guidelines could have an educational footprint in the long run.

Conclusion

There is an increasing amount of open datasets available through open government portals, but still much work to be done to inform about the actual usage of these open datasets. This chapter has introduced two software components to enable progress on this issue: the Open City Toolkit transparency module, as well as interactive guidelines. The former aims at informing about the rate of Open Data re-use and the latter purports to communicate innovations ensuant on Open Data. The chapter has presented the key ideas behind the two components and (evolving) prototypical implementations illustrating them. The work introduced is relevant to Open Data publishers and citizens at large.

Immediate directions for future work, based on feedback from the participants, include further improving the usability of the tool and devising means to automatically check the quality of the datasets. Developing metrics for (subjective) aspects of data quality such as "fitness for purpose", "trustworthiness", or "understandability" is a challenge, but other aspects of quality such as "dataset availability" or "dataset currency" are easier to assess and can be implemented. Automatically recommending (possibly) relevant datasets to new apps registering on the OCT transparency tool seems also a promising direction for future work.

Acknowledgements Comments from three anonymous reviewers have helped improve the clarity of the article. The authors gratefully acknowledge funding from the European Union through the GEO-C project (H2020-MSCA-

ITN-2014, Grant Agreement Number 642332, http://www.geo-c.eu/). Carlos Granell is funded by the Ramón y Cajal Programme (grant number RYC-2014-16913). Sergio Trilles is funded by the postdoctoral programme Vali+d (GVA) (grant number APOSTD/2016/058). We thank participants of the course "Geoinformation in Society" who volunteered to do the usability tests and employees from the city councils of Münster and Lisbon for their feedback. Finally, we thank Christian Kray and Marco Painho for the support provided during the course of this work.

References

Attard, J., Orlandi, F., & Auer, S. (2016). Value creation on open government data. In *2016 49th Hawaii International Conference on System Sciences (HICSS)* (pp. 2605–2614). Koloa, Hawaii, USA: IEEE. https://doi.org/10.1109/HICSS.2016.326.

Attard, J., Orlandi, F., Scerri, S., & Auer, S. (2015). A systematic review of open government data initiatives. *Government Information Quarterly, 32*(4), 399–418. https://doi.org/10.1016/j.giq.2015.07.006

Bangor, A., Kortum, P., & Miller, J. (2009). Determining what individual SUS Scores mean: Adding an adjective rating scale. *The Journal of Usability Studies, 4*(3), 114–123.

Bangor, A., Kortum, P. T., & Miller, J. T. (2008). An empirical evaluation of the system usability scale. *International Journal of Human-Computer Interaction, 24*(6), 574–594. https://doi.org/10.1080/10447310802205776

Benitez-Paez, F., Degbelo, A., Trilles, S., & Huerta, J. (2018). Roadblocks hindering the reuse of open geodata in Colombia and Spain: A data user's perspective. *ISPRS International Journal of Geo-Information, 7*(1), 6. https://doi.org/10.3390/ijgi7010006

Beno, M., Figl, K., Umbrich, J., & Polleres, A. (2017). Open data hopes and fears: Determining the barriers of open data. In P. Parycek & N. Edelmann (Eds.), *2017 Conference for E-Democracy and Open Government (CeDEM)* (pp. 69–81). Krems, Austria: IEEE. https://doi.org/10.1109/CeDEM.2017.22

Berners-Lee, T. (2006). Linked data—Design issues. Retrieved May 18, 2018, from http://www.w3.org/DesignIssues/LinkedData.html

Brooke, J. (2013). SUS: A retrospective. *Journal of Usability Studies, 8*(2), 29–40.

Degbelo, A., Bhattacharya, D., Granell, C., & Trilles, S. (2016). Toolkits for smarter cities: A brief assessment. In R. García, P. Caballero-Gil, M. Burmester, & A. Quesada-Arencibia (Eds.), *UCAmI 2016—10th International Conference on Ubiquitous Computing & Ambient Intelligence* (pp. 431–436). Las Palmas, Gran Canaria, Spain: Springer International Publishing. https://doi.org/10.1007/978-3-319-48799-1_47

Degbelo, A., Granell, C., Trilles, S., Bhattacharya, D., Casteleyn, S., & Kray, C. (2016). Opening up smart cities: Citizen-centric challenges and opportunities from GIScience. *ISPRS International Journal of Geo-Information, 5*(2), 16. https://doi.org/10.3390/ijgi5020016

Degbelo, A., & Kauppinen, T. (2018). Increasing transparency through web maps. In P.-A. Champin, F. L. Gandon, M. Lalmas, & P. G. Ipeirotis (Eds.), *Companion of Proceedings of the Web Conference 2018—WWW '18* (pp. 899–904). Lyon, France: ACM Press. https://doi.org/10.1145/3184558.3191515

Degbelo, A., Trilles, S., Kray, C., Bhattacharya, D., Schiestel, N., Wissing, J., & Granell, C. (2016). Designing semantic application programming interfaces for open government data. *JeDEM—EJournal of EDemocracy and Open Government, 8*(2), 21–58.

Fusch, P. I., & Ness, L. R. (2015). Are we there yet? Data saturation in qualitative research. *The Qualitative Report, 20*(9), 1408–1416.

Gonzalez-Zapata, F., & Heeks, R. (2015). The multiple meanings of open government data: Understanding different stakeholders and their perspectives. *Government Information Quarterly, 32*(4), 441–452. https://doi.org/10.1016/j.giq.2015.09.001

Graves, A., & Hendler, J. (2013). Visualization tools for open government data. In S. Mellouli, L. F. Luna-Reyes, & J. Zhang (Eds.), *Proceedings of the 14th Annual International Conference on Digital Government Research—dg.o '13* (p. 136). Quebec, Canada: ACM Press. https://doi.org/10.1145/2479724.2479746

Guest, G., Bunce, A., & Johnson, L. (2006). How many interviews are enough? An experiment with data saturation and variability. *Field Methods, 18*(1), 59–82. https://doi.org/10.1177/1525822X05279903

Hartog, M., Mulder, B., Spée, B., Visser, E., & Gribnau, A. (2014). Open data within governmental organisations. *EJournal of EDemocracy and Open Government, 6*(1), 49–61.

Janssen, M., Charalabidis, Y., & Zuiderwijk, A. (2012). Benefits, adoption barriers and myths of open data and open government. *Information Systems Management, 29*(4), 258–268. https://doi.org/10.1080/10580530.2012.716740

Lazar, J., Feng, J. H. J., & Hochheiser, H. (2010). *Research methods in human-computer interaction*. John Wiley & Sons.

Lewis, J. R. (2014). Usability: Lessons learned ... and yet to be learned. *International Journal of Human-Computer Interaction, 30*(9), 663–684. https://doi.org/10.1080/10447318.2014.930311

Lopez, V., & Whitehead, D. (2013). Sampling data and data collection in qualitative research. In Z. Schneider, D. Whitehead, G. LoBiondo-Wood, & J. Haber (Eds.), *Nursing and Midwifery Research: Methods and critical appraisal for evidence-based practice* (pp. 124–140). London: Elsevier Health Sciences.

Lourenço, R. P. (2015). An analysis of open government portals: A perspective of transparency for accountability. *Government Information Quarterly, 32*(3), 323–332. https://doi.org/10.1016/j.giq.2015.05.006

Matheus, R., Janssen, M., & Maheshwari, D. (2018). Data science empowering the public: Data-driven dashboards for transparent and accountable decision-making in smart cities. *Government Information Quarterly*. https://doi.org/10.1016/j.giq.2018.01.006

OECD. (2016). *Skills matter: Further results from the survey of adult skills*. OECD Skills Studies. Paris, France: OECD Publishing. Retrieved October 9, 2017, from https://doi.org/10.1787/9789264258051-en.

Pomerantz, J., & Peek, R. (2016). Fifty shades of open. *First Monday, 21*(5). https://doi.org/10.5210/fm.v21i5.6360

Roth, R. E. (2009). A qualitative approach to understanding the role of geographic information uncertainty during decision making. *Cartography and Geographic Information Science, 36*(4), 315–330. https://doi.org/10.1559/152304009789786326

Sauro, J. (2013). 10 things to know about the system usability scale (SUS). Retrieved May 17, 2018, from https://measuringu.com/10-things-sus/.

Schade, S., Granell, C., & Perego, A. (2015). Coupling public sector information and public-funded research data in Europe: A vision of an open data ecosystem. In C. Reddick & L. Anthopoulos (Eds.), *Information and Communication Technologies in public administration: Innovations from developed countries* (pp. 275–298). Routledge.

Susha, I., Zuiderwijk, A., Charalabidis, Y., Parycek, P., & Janssen, M. (2015). Critical factors for open data publication and use: A comparison of city-level, regional, and transnational cases. *EJournal of EDemocracy and Open Government, 7*(2), 94–115.

Tullis, T., & Stetson, J. (2004). A comparison of questionnaires for assessing website usability. In *Proceedings of the Usability Professionals Association (UPA) 2004 Conference* (pp. 7–11). Minneapolis, Minnesota, USA.

Veljković, N., Bogdanović-Dinić, S., & Stoimenov, L. (2014). Benchmarking open government: An open data perspective. *Government Information Quarterly, 31*(2), 278–290. https://doi.org/10.1016/j.giq.2013.10.011

Vetrò, A., Canova, L., Torchiano, M., Minotas, C. O., Iemma, R., & Morando, F. (2016). Open data quality measurement framework: Definition and application to Open Government Data. *Government Information Quarterly, 33*(2), 325–337. https://doi.org/10.1016/j.giq.2016.02.001

Wijnhoven, F., Ehrenhard, M., & Kuhn, J. (2015). Open government objectives and participation motivations. *Government Information Quarterly, 32*(1), 30–42. https://doi.org/10.1016/j.giq.2014.10.002

Young, A., & Verhulst, S. (2016). *The global impact of open data—Key findings from detailed case studies around the world* (1st ed.). O'Reilly Media, Inc.

15

From Repositories to Switchboards: Local Governments as Open Data Facilitators

Irina Anastasiu, Marcus Foth, Ronald Schroeter, and Markus Rittenbruch

Highlights

- This chapter presents a novel approach to Open Data that is influenced by the concept of a switchboard and moves away from Open Data repositories towards the interactive and bidirectional exchange of information between local governments, civil society, academia and business.
- This chapter proposes an iterative, exploratory and participatory approach to the identification of Open Data use-cases as part of tackling internal local government challenges. By engaging with the citizenry as early and continuously as possible, we argue that benefits for both parties can be achieved.
- This chapter positions the role of local government as a custodian, facilitator and synthesiser for both Open Data itself and the co-creative practices across four sectors of society required for successful

I. Anastasiu (✉) • M. Foth • R. Schroeter • M. Rittenbruch
QUT Design Lab, Queensland University of Technology, Brisbane, QLD, Australia
e-mail: irina.anastasiu@qut.edu.au; m.foth@qut.edu.au; r.schroeter@qut.edu.au; m.rittenbruch@qut.edu.au

S. Hawken et al. (eds.), *Open Cities | Open Data*,
https://doi.org/10.1007/978-981-13-6605-5_15

use-case-driven Open Data delivery: local government itself, civil society, academia and business.
- This chapter critiques neoliberal smart city policies and emphasises the need to explore alternative policies, practices and regulations for Open Data interventions from other perspectives in order to realise their full potential and to avert the risk of exacerbating existing inequalities.

Introduction

In 2009, the systematic provision of public sector information gained traction after a decade-long incubation phase. Following a series of legislative interventions, free information infrastructure conferences and campaigns by newspapers to free government data, the first "data.govs" sprouted (Heimstädt, Saunderson, & Heath, 2014; Lathrop & Ruma, 2010). The envisioned economic, social and political hopes of "abandoning models of false scarcity that restrict access, interpretation, and re-use" were high (Bates, 2012).

Although merely a subset of Open Data, open government data is the major point of focus in academic literature (Davies & Perini, 2016; Heimstädt et al., 2014). The Open Knowledge Foundation (OKFn) has been pivotal in defining the term Open Data and shaping its trajectory (Heimstädt et al., 2014). It promotes government as the most significant Open Data provider to contribute to the realisation of the promises of Open Data. The foundation's Open Data Handbook (Open Knowledge Foundation, 2017b) predominantly draws on open government data to support its main arguments and examples. In what some refer to as the "Open Data value chain" (Carrara, San Chan, Fischer, & van Steenbergen, 2015), governments are placed in the role of supplier on the one side while on the other side businesses, non-governmental organisations (NGOs), community groups and individuals are seen as consumers, sometimes with an intermediary in between. This chapter will explore and critique the linear sequencing of this data value chain, and explore a novel role for local government within the Open Data ecosystem.

Open Data is often seen as a central technical component to achieve various smart city goals and is therefore inextricably linked to the vision of a smart city. It is seen as critical to spur the economy, innovation, public participation; improve the efficiency and effectiveness of government services; and lead to self-empowerment, new knowledge and new private products and services. (Open Knowledge Foundation, 2017b). Open Data also finds a place within a people-centric smart city vision through smart citizens (Cardullo & Kitchin, 2017). In this vision, citizens are empowered by available data. It allows them to gain insights into their surroundings and helps them to make more informed, healthier, more sustainable decisions in the context of their complex urban environments (Foth, Brynskov, & Ojala, 2015; Kitchin, 2016; Robinson, Rittenbruch, Foth, Filonik, & Viller, 2012). Civic hacking allows for the creation of transparency tools to hold governments accountable by the citizenry (Schrock, 2016). Open Data fits equally well into a technocratic smart city vision, with data as an end in itself. A fundamental building block of the system that enables optimisation and efficiency in urban management and service delivery through urban science, data analytics and machine learning (Batty, 2013; Thrift, 2014; Vickery, 2011). This chapter will challenge some of the conventional thought processes and rationales behind the Open Data and smart cities movement, and explore how the novel role of local government proposed in this chapter can contribute to the delivery of their promises.

This chapter will challenge some of the conventional thought processes and rationales behind the Open Data and smart cities movement, and explore how the novel role of local government proposed in this chapter can contribute to the delivery of their promises.

We propose a progressive shift that sees local governments transitioning from their current role of being a one-way service provider to being an enabler and facilitator of a two-way exchange. This two-way exchange comprises deliberate collaborations between different sectors of society, including the public and private sector, academia and, in particular, civil society. In the specific case of Open Data, local governments do not necessarily need to own all the data that is being made accessible. Instead, they could embrace the role of custodians and synthesisers, and facilitate processing datasets provided by other sectors into new, richer and more targeted open datasets. This steers local government away from the current trend of creating and hosting Open Data repositories. Instead, the new role can be conceptualised as a data switchboard provider, taking leadership in establishing and maintaining open standards, rather than outsourcing this duty to commercial operators and thus risking vendor lock-in on proprietary software platforms. In this scenario, smart strategies for Open Data actively contribute towards the effective use of Open Data (Gurstein, 2003, 2011) by the public and the self-empowerment of citizens and community organisations.

The active engagement of local governments with end-users and potential beneficiaries of Open Data can not only identify Open Data use-cases at an early stage but create opportunities for an ongoing engagement around which further initiatives required for the effective use of Open Data can flourish, such as the collaborative implementation of upskilling and educational activities, or co-development of thoughtful, just and thorough Open Data policy, practice and legislation. Such engagement strategies then address not only the technicalities of data provision but also the social, educational, institutional, policy, practice and regulatory changes required for Open Data to be effectively used by individuals and organisations that tackle some of the most urgent social, urban, economic or environmental challenges.

Re-examining the Promises of Open Data and the Smart City Agenda

The Organisation for Economic Co-operation and Development (OECD) estimates the market for public sector information in the European Union to have a mean value between 27 billion (Ubaldi, 2013) and 325 billion Euro (Carrara et al., 2015). Many scholars praise the potential of the Open Data movement to reshape democracy through transparency, empowerment and participation (Baack, 2015). While this strong potential impact invites further exploration, there are emerging drawbacks, risks and challenges. These range from accessibility issues in terms of technical and social factors (Gurstein, 2003, 2011) to privacy concerns (Floridi, 2014; Kitchin, 2014b) and questioning whether well-intentioned Open Data initiatives might result in the aggravation of the very issues they sought to solve, such as exacerbating inequalities or negatively affecting current governance structures (Bates, 2012; Gurstein, 2011; Lessig, 2009). What if Open Data perpetuates instead of decentralising existing power distribution (Benjamin, Bhuvaneswari, & Rajan, 2007; Gurstein, 2011; Wright, Prakash, Abraham, & Shaw, 2010)?

Davies and Perini (2016) critique the assumption that governments are the ones holding the data capable of delivering on the socio-economic promises of Open Data. This assumption is prevalent especially in developed countries. In developing countries this pool of Open Data actors includes NGOs, international agencies and private actors. Davies and Perini's (2016) framework for differentiating Open Data (Davies & Perini, 2016) comes in response to Fung and Weil's (Fung & Weil, 2010) call for an "open society," encouraging private actors to share data alongside governments. For an open society, the role of the government needs to be elaborated upon. Specifically, it is key to explore how the network of institutions currently tasked with representing public interest should evolve and operate. As a practice, policy and legislation around open government data are grounded in smart city agendas, they can lead to the realisation either of a just and sustainable materialisation of these models or of their dystopian counterparts.

Data collection by private corporations such as Google and Facebook is increasing rapidly. In addition to data collection and mining, large commercial smart city technology providers put cities at risk of suffering from vendor lock-in, operating behind proprietary machine learning algorithms, where their logic and the specific data they are trained on are protected due to commercial interests (Kitchin, 2016) and risk obsoleteness (Sassen, 2015). Rethinking how data and algorithms can be put back into the hands of organisations accountable to the public, as well as of the public itself, is of utmost priority. This rings true especially now that algorithms and artificial intelligence are increasingly used to influence life-altering decisions with potentially dangerous consequences (Markou, 2017), such as prison sentences (Liptak, 2017) or measuring and using people's trustworthiness scores as part of a "social credit system" (Hatton, 2015). Beyond their lack of transparency, the values, context and biases introduced into proprietary algorithms, the training data sampling and the interpretation of results (Kitchin, 2014a) raise serious concerns with regard to ethics (Kitchin, 2016) and social justice.

The Open Data and open-source movement could play an active role in averting a smart city dystopia, where power is concentrated in the hands of the few behind private corporations that are not accountable to the public by any democratic mechanism (Shaw & Graham, 2017). It has the potential to decentralise this power by allowing citizens partaking in the Open Data and open-source movement to access, use and co-produce not only the data but also the software systems and algorithms that provide and make use of it. This kind of civic hacking could be understood as an appropriation of Lefebvre's revolutionary principle of the right to the city (1996) as the right to the smart city. Similar to the right to urban space and to shape the process of urbanisation itself (Harvey, 2008), citizens not only have access to smart city data, services and technologies, but co-produce these while at the same time shaping the process of "technologisation" itself. This includes active influence with regards to the principles, values and politics embedded in this process. Following Lefebvre's concept, this implies challenging the capitalist and profit-oriented mode of production of such infrastructure.

At a first glance, the values and practices of the Open Data and open-source movement seem beneficial towards the goal of free technical, financial and legal access to data, as well as altruistic, meritocratic principles in the creation of digital software towards "world improvement" (Levy, 2010). In some ways, the open-source movement is at odds with current smart city ideology by defying capitalist principles. In other ways, it follows suit. As Brenner and Schmid (2015) observe, and Melgaço and Willis (2017) eloquently summarise, the smart city agenda has been shaped by a mindset grounded in neoliberalism, favouring free-market capitalism as an approach to apply a "technical 'fix' to intractable governance problems" (Brenner & Schmid, 2015). Sadowski (2016) provides an elaborate deconstruction of the discourses of smart city technology purveyors and the "reflected and reinforced" ideologies: technocracy and neoliberalism. As much as the smart city is framed as "pragmatic, neutral and non-ideological," the kinds of urban transformations that can result from Open Data ultimately are political decisions and the smart city a deeply political and ideological project (Sadowski, 2016; Vanolo, 2014).

Deconstructing the smart city discourse of the EU, Vanolo (2014) evidentiates how it represents a disciplinary strategy and may act as "a powerful tool for the production of docile subjects and mechanisms of political legitimisation."

In contradiction to this political nature of the smart city and of Open Data stands the disdain of free and open-source hackers to be involved in "politics" in a formalised manner, out of fear action may be corrupted by ideology (Coleman, 2004). Schrock (2016) offers further thoughts on hacker politics, concluding civic hackers are utopian realists "transgressing established boundaries of political participation." De-politicised civic hacking interventions can raise concerns as they take us deeper into technocratic and techno-utopian smart city territory.

On a superficial level, the Open Knowledge Foundation's (Open Knowledge Foundation, 2017a) open definition seems a sensible starting point for an Open Data definition and is widely used by practitioners and academics (Bates, 2012; Davies & Perini, 2016). Requiring legal and technical access as well as rights of usage and reproduction at free cost for everyone, it does not seem to consider the complex and often imbalanced power relations present in contemporary society. By simplistically advocating for unrestricted access irrespective of the organisation using it and for what purpose, the definition presumes equality of opportunity where it does not exist. Both this definition and the open-source movement's emphasis on meritocracy embrace the kind of laissez-faire and individual responsibility principles of neoliberalism. Yet, Open Data that cannot be effectively used is closed data, even when it is free of cost, comes without legal obligations and is in a machine-readable format. We should critically reflect on whether making data genuinely open should be an individual responsibility or rather a collective one. Looking further into the future, we must ask whether an "open society" (Fung & Weil, 2010) should be built on these values or rather different ones.

The Open Knowledge Foundation (Open Knowledge Foundation, 2017b) presents how Google is using official, open and free EU documentation to train its translation algorithms. Such large technology corporations are the ones best-equipped with the resources and expertise to exploit Open Data towards their own profit (Cole, 2012). Not only do these corporations have sufficient economic power to hire data specialists and cover the costs of expensive software and infrastructure (Johnson, 2014), but they also accrue additional power aided by insights gained through Open Data usage. In this context, Open Data further empowers the empowered. A legitimate question then becomes "What return do those funding Open Data—primarily governments and taxpayers—get when corporations use Open Data to advance their business?"

The most obvious mechanism of value capture is tax. Yet, the EU, for instance, is reactively looking into tax reforms after identifying that Google, Facebook and Amazon pay less than half the amount of tax compared to traditional companies (Rankin, 2017). In 2014 Apple has paid just 50 euros in tax for every million in profit obtained in the European Union and was ordered to repay ten years' worth of tax, about 13 billion euros (Kanter & Scott, 2016). We may get better products, but in turn we allow monopolies to develop which become gatekeepers for the ways we understand, make sense of and interact with the city and each other (Shaw & Graham, 2017).

In parallel, NGOs, non-profit organisations, community groups and small businesses with strong local ties, often at the heart of tackling social, economic, ecological and humanitarian issues, struggle to develop the data literacy, technical skills and digital competences (Cole, 2012) that are required for the access and "effective use" (M. B. Gurstein, 2011) of Open Data. Considering their attempt to address wicked problems where there is little financial profit to be made, we need a more nuanced Open Data movement that differentiates on the basis of the stakeholders' starting context, who will benefit from their use of Open Data, and in which way. The movement should seek to level the playing field for all actors—elevating the less empowered, equipping them with the prerequisite for effective Open Data use and ensuring Open Data benefits them equally. The concepts and delivery models presented in this chapter must be considered and realised through this lens of power inequality to develop policies, practices and regulations that, in addition, ensure the power holders return benefit to the community from their use of Open Data relative to their own benefit from it and that effective use of Open Data across a diverse set of actors is employed as a key measure of success.

This requires transcending the triple-helix model, calling for collaboration between government, business and academia for the "reinvention of cities," that, among others, Leydesdorff and Deakin (2011) advocate for. This can be achieved through the addition of civil society as a high-priority element of the helix (Fig. 15.1) towards a quadruple helix. Depending on the city's ecosystem, the interaction between these sectors in a real urban context and the extent of participation of citizens may vary.

A four-stranded helix must go beyond the reductionism of considering the citizenry as mere users of smart city innovation led by businesses. Instead, civil society should be acknowledged as an equal, if not leading, partner, comprising citizens who hold valuable tacit knowledge and insight regarding the problems and opportunities present within their immediate surroundings (Foth & Brynskov, 2016). Community groups, activists and individuals are already coming together on a voluntary basis to envision and shape the kind of city they want to live in through methods such as organising events, protests, writing letters to representatives or undertaking do-it-yourself and do-it-with-others urban design. In spite of this willingness to act, and a sense of responsibility and ownership for their urban environment, they are often perceived as a threat by members of traditional public-private partnerships and even in academic

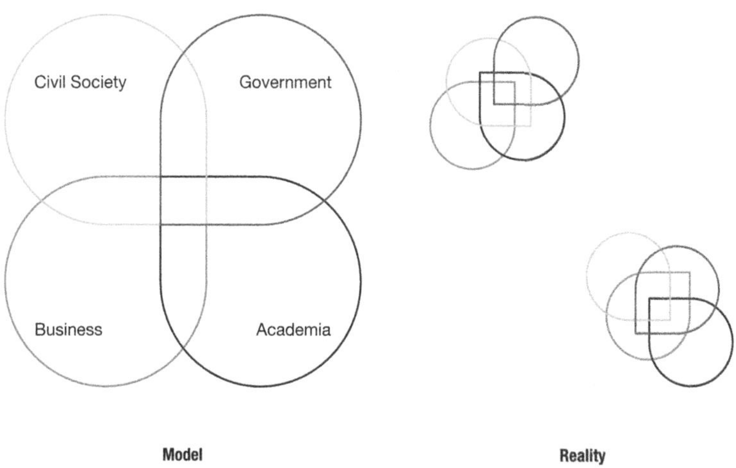

Fig. 15.1 Quadruple helix as a model and in reality. Depending on priorities and ecosystem, cities achieve various levels of overlap between the four sectors. Nonetheless, civil society must be recognised as an equal partner, rather than as a mere consultant in shaping the city

circles (Douglas, 2014). A model of co-creative practices builds on the energy and effort of such groups and individuals, connects them with other stakeholders already at the table and openly embraces friction and disagreement as a catalyst for outcomes that provide benefit to a variety of interest groups (Özdemir & Tasan-Kok, 2017). By employing dialectical processes, there is a capacity to build superior and more sustainable outcomes and expand the area of consensus between stakeholders (Dick, 1990).

Realising this mindset is a fundamental stepping stone towards more comprehensive models of urban governance, such as the quintuple helix (Carayannis, Barth, & Campbell, 2012) which calls for socio-ecological transition and includes the environment, and a new urban paradigm of the post-anthropocentric city (Forlano, 2016; Foth, 2017).

Integrating Design Thinking and Action Research Towards Co-creation

To adapt current citizen participation practices and policies towards co-creative approaches requires an understanding of existing opportunities and limitations through first-hand experiences. The results presented in this chapter emerged from insights gained from a Code for Australia fellowship conducted over a three-month period in 2016 at Moreton Bay Regional Council in South-East Queensland, Australia. This allowed the main researcher to embed herself as a fellow into the organisation to explore opportunities for local government as a facilitator of collaboration, as well as to comprehend the limitations and barriers within current practices.

The initial brief called for the design and implementation of a technical solution to improve the Council's database of local businesses, which only covered 0.06% of the over 25,000 businesses estimated to be operating in the region (identified through comparison with the Australian business register data obtained from the Australian Bureau of Statistics) and contained outdated information. Particularly, outdated contact details posed a challenge for the economic development department seeking to connect these businesses with growth opportunities.

Therefore, the project was initially approached from a design thinking perspective using the double diamond process. This consists of a mix of divergent and convergent patterns of thinking in order to deliver solutions for given problems (British Design Council, 2007).

The double diamond (Fig. 15.2) is primarily focused on tangible outcomes such as products or services. An essential aspect of this approach is to challenge the initial problem definition, approaching it from a point of openness to possible alternatives. It also encourages engagement with a wide variety of relevant stakeholders; in this case, those internal to the local council but also local business owners. This flexibility and openness act as a catalyst for the identification and addressing of root causes rather than the treatment of symptoms of problems, yielding better-tailored solutions. The process consists of four steps alternating between exploratory and synthesising phases: Discover, Define (first diamond), Develop and Deliver (second diamond) (British Design Council, 2007).

At the same time, the fellowship functioned as an intensive, day-to-day immersion into and study of local government practices and challenges. This was achieved through participant observation (DeWalt, DeWalt, & Wayland, 1998; Emerson, Fretz, & Shaw, 2001)—a range of ethnographic methods to better understand group dynamics, practices and values. These included informal interviews, participation in the life of the

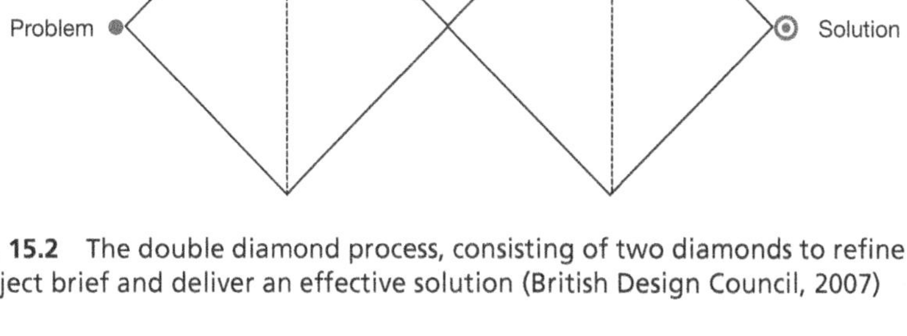

Fig. 15.2 The double diamond process, consisting of two diamonds to refine the project brief and deliver an effective solution (British Design Council, 2007)

group, collective discussions and results from group activities (Atkinson, Coffey, Delamont, Lofland, & Lofland, 2001). This ethnography-inspired practice was combined with action research (Bradbury, 2015).

Action research (Fig. 15.3) is a cyclical, reflective and adaptive method of inquiry geared towards provoking social change through purposeful and well-informed action. It consists of four stages: plan, act, observe and reflect (Hearn, Tacchi, Foth, & Lennie, 2009). The learnings from each reflection flow back into the planning stage of the next cycle. This allows for immediate and continuous theorising, data collection and reflection, which contributes towards the progressive emergence of structure. Beyond being a research method, action research requires an attitude of welcoming change, which we embraced in this study and which was shared by our project partners and stakeholders. Action research has an ethnographic branch, termed Ethnographic Action Research (Hearn et al., 2009). It was influential in allowing us to understand the organisational, social and cultural requirements for the successful adoption of a technical solution.

The Code for Australia fellowship balanced the double diamond approach and the action research approach as problem definitions and project goals adapted to emerging findings. The resulting process combining the two is shown in Fig. 15.4, consisting of two phases: exploration and co-creation.

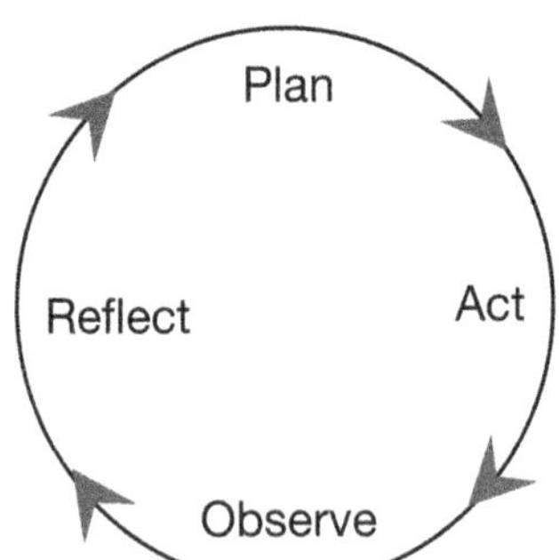

Fig. 15.3 The action research process, consisting of four phases: plan, act, observe and reflect. The findings of each cycle feed back into the planning of the next cycle

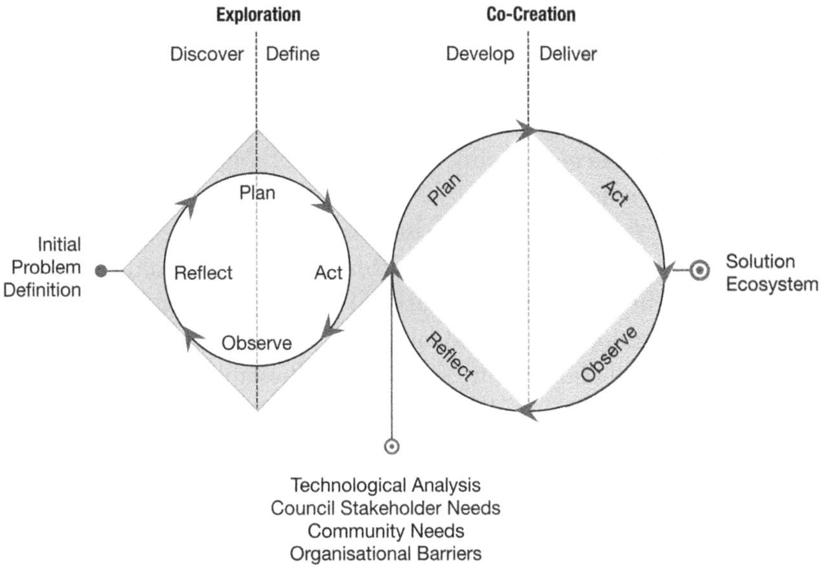

Fig. 15.4 The research approach integrating the double diamond and action research into two project phases. The double diamond approach dominated the first phase, whereas action research took centre stage in the second phase

The exploration phase integrated ethnographic action research within the first diamond concerned with gaining a more comprehensive picture of the initial problem identified with the business register, but also an understanding of the social and organisational practices informing the maintenance and usage of the database. Informal one-on-one conversations between researcher and Council stakeholders revealed insights into their role within the organisation, how and why they interacted with the database, their interactions with other colleagues and, where applicable, their interactions with the local business community. An online survey with open-ended questions was distributed via the Council's mailing lists for local businesses, seeking to (1) understand the challenges they faced in their day-to-day business; (2) gauge whether they were aware of the online business directory and the business resources on the Council's website; and (3) ask for ideas and recommendations for the business register on the Council's website. It yielded 39 valid responses. The researcher investigated various technical approaches for quality assurance and management of the data through desktop research. In order to regularly synthesise the project findings, weekly meetings were held with the main contact person for the project, the chief digital officer. As part of these meetings, any new insights from the endeavours above were discussed, enhanced with strategic insights from the Chief Digital Officer (CDO) perspective and reflected upon.

By the time the first project phase was completed, the newly gained insights indicated that providing a technical solution would result in a moderate benefit, as long as organisational barriers persisted. Overcoming such barriers required organisational change and emphasised the value of the action research component, which is a well-established methodology in the facilitation of this kind of transformation (Reason & Bradbury, 2001).

As a result, action research and double diamond swapped priorities in the co-creation phase of the project, seeking to implement the second diamond within an action research framework to ideate and deliver potential solutions through collaboration and engagement between and with both Council and business community stakeholders. This shift was possible due to the project team's openness to alternative outcomes and pathways from the onset of the project. Combined with findings from the researcher's engagement with local businesses, this further allowed the project team to explore the business register database as Open Data of benefit to both local businesses and the economic development department.

In order to advance the organisational transformation, the researcher ran two workshops in response to the lack of content ownership, collaboration deadlock and the unexploited potential that lay between the IT and economic development departments. They aimed to promote mutual exploration, and understanding of the work conducted within the two departments involved in the project and the collaborative development of a process to deliver a visualisation and exploration tool for the business register dataset.

To progress the community-oriented aim of the project, the researcher revisited the 39 responses and analysis of the online survey, and how the results could inform an Open Data use-case. Six members of the business community had additionally agreed in the survey to a 30-minute interview. The semi-structured interviews focused on gaining deeper insights on the challenges faced by individual businesses, internet practices and business register improvements mentioned in the survey response. Insights from this formed the basis for ideas towards a draft concept for the use of the business register as Open Data within an online solution to benefit both the local business community and the Council.

Under this paradigm of co-creation, rather than being the main action-taker, implementer or designer, the role of the researcher centred around mediation, facilitation and enabling: brokering between the stakeholders; facilitating the emergence, exchange and distillation of existing knowledge from and between stakeholders; enabling collective action and reflection; and encouraging the incorporation of emerging knowledge into co-created outcomes with shared ownership.

Key Insights and Outcomes Towards the Effective Use of Open Data

For those on the sidelines, the culture, processes, struggles and dependencies within local governments often exist neatly simplified and tucked away behind the all-encompassing term "the council." In stark contrast to this perspective stands the complexity the researcher was exposed to during this project. From the inside, especially when positioned across departments and across hierarchy levels, the daily challenges became increasingly obvious. People across teams aimed to deliver the best possible results within the given circumstances. Their eagerness to participate, contribute, try out and learn was striking. The researcher was in a unique position to take an experimental approach as an external academic embedded within the organisation. These experiments were met with openness and curiosity by members of the growing project team. The status as an outsider made the researcher both less threatening and less subjected to organisational norms, procedures and responsibilities. This situation was conducive for producing actionable knowledge and facilitating change.

This project has shown that approaching an internal local government challenge through an exploratory and moment-to-moment process can act as a source of use-cases for Open Data that create benefits for both local government and citizens. Technically, Open Data is delivered by an IT team, but the expertise around the particular dataset—the contents it describes, the context of these contents—resides with domain experts such as the affiliated business unit or citizens, local businesses and community groups. The dataset is the element of shared interest, albeit for different reasons. If business units manage to engage with questions of how community and residents can contribute to a data challenge internal to the Council and to actively collaborate with them to explore mutually beneficial solutions, this can lead to Open Data use-cases. Collected feedback from the community on data format, fields, mode of delivery and so on can then be consulted on with the technical team and subsequently delivered. Open Data—as a technical building block—needs to be delivered across community, business units and technology departments, not in technical isolation. Through cooperation across local governments, learnings can be shared and new standards built from the bottom up.

This capacity of unobtrusive, equidistant and adaptive facilitation could be a key to the successful advancement of use-case-driven Open Data standards and implementations in local governments, as advocated for by international alliances such as the Open and Agile Smart Cities (OASC) initiative (Kitchin, Coletta, Evans, Heaphy, & MacDonncha, 2017). This kind of facilitation then is also a key element for the organisational transformations required for effective Open Data delivery, breaking down as many barriers of engagement between business units of the smart city and community, as well as between those business units and IT teams as possible. Further, this kind of facilitation can also be the key to co-creating and progressively embracing agile processes in complex organisations, as long-term, linear and sequential projects carried out under waterfall models (Royce, 1987) come to completion and capacity is freed up as a result.

The particular context of the fellowship, situated between the economic development team and the IT team, and the business register database around which they coalesce, could serve as a pilot project to do things differently. This represents a shift away from the business register database as the central point of focus and towards the identification, conceptualisation and implementation of novel ideas, concepts and approaches for the use and delivery of Open Data through collaborative experimentation and learning. The ultimate goal is that orchestration becomes mostly obsolete as more supple and adaptive processes are adopted and their ownership is increasingly shared. The researcher identified that business analysts can work alongside external facilitators in setting up the basis for such orchestration, as they are excellent observers and inquirers into practices and challenges of staff and business units, are familiar with the use of qualitative inquiry methods resembling ethnography and their professional circle within the Council spans far and wide across departments.

Open Data use-cases emerging through this kind of approach can be beneficial for both the community and business units. In the case of Moreton Bay, as the economic profile of the region suggested (Australian Bureau of Statistics, 2015), almost all survey respondents and interviewees owned what would be categorised as a small or medium enterprise, with a strong tendency towards sole trader businesses, micro-businesses (1–4 employees) and small businesses (5–19 employees). A small number were initiators or leaders of community organisations and not-for-profits. Their main challenges were centred around customer acquisition and business partnerships, as well as finding sponsors and donors. At the same time, a significant emphasis was set on regional solidarity and pride, mutual respect and retention of business in the local area.

While almost all participants did not relate to an abstract concept of (Open) Data beyond contact details, or goods and services provided or did not articulate how specifically Open Data could contribute to enhancing regional solidarity and local business retention, their input provided a clear use-case. Open Data on local businesses should be geared towards regional self-empowerment and mutual discovery towards strengthening of business partnerships, collaborations and acquisition of local goods and services, rather than from the nearby Brisbane region. This could also create opportunities for greater mutual support between local businesses and local community organisations.

Such self-empowerment and regional solidarity require a bidirectional flow of data both from the local government to residents and vice-versa.

This is a critical element of the switchboard model for Open Data we will introduce shortly. While we do not argue that a bidirectional flow is always desirable, further explorations are required to identify contexts in which it is advantageous and in which it is not.

As a first step to level the playing field, a range of pre-conditions for effective access to opportunities resulting from Open Data by the wider public is offered by Gurstein (2011). Insights presented in this chapter contribute particularly towards two of these pre-conditions: first, the provision of "content services" by encouraging Open Data contents adapted to particular use-cases; and second, "social facilitation" by advocating for continuous engagement, exchange and reciprocal learning between partners, as well as local government taking a leadership role towards more co-creative approaches and more sustainable business models as a result of Open Data use (Bonina, 2013).

A Switchboard for Open Data

Over the course of the three months, by approaching the fellowship from an exploratory as well as an agile perspective, the project developed the ability to not only create internal benefits and tackle particular internal challenges but also generate value for the local small business community through a novel Open Data concept.

Figure 15.5 shows a simplified diagram of this novel Open Data concept, specifically around the business register data use-case. It does not aim to illustrate all the various technical complexities or be an accurate and detailed depiction of technical components. Rather, it seeks to convey the role of local government as a type of switchboard for Open Data in a smart city based on the business register database use-case. In this concept, Open Data is not only released by the government but also opened up for contributions from external parties. A given dataset is provided through an application program interface (API) that both provides access to the dataset and allows data to be written back into the system.

This way, the original dataset is enhanced through applications that are making use of the API, potentially adding new entries or enriching existing ones. For example, an application that matches local suppliers with

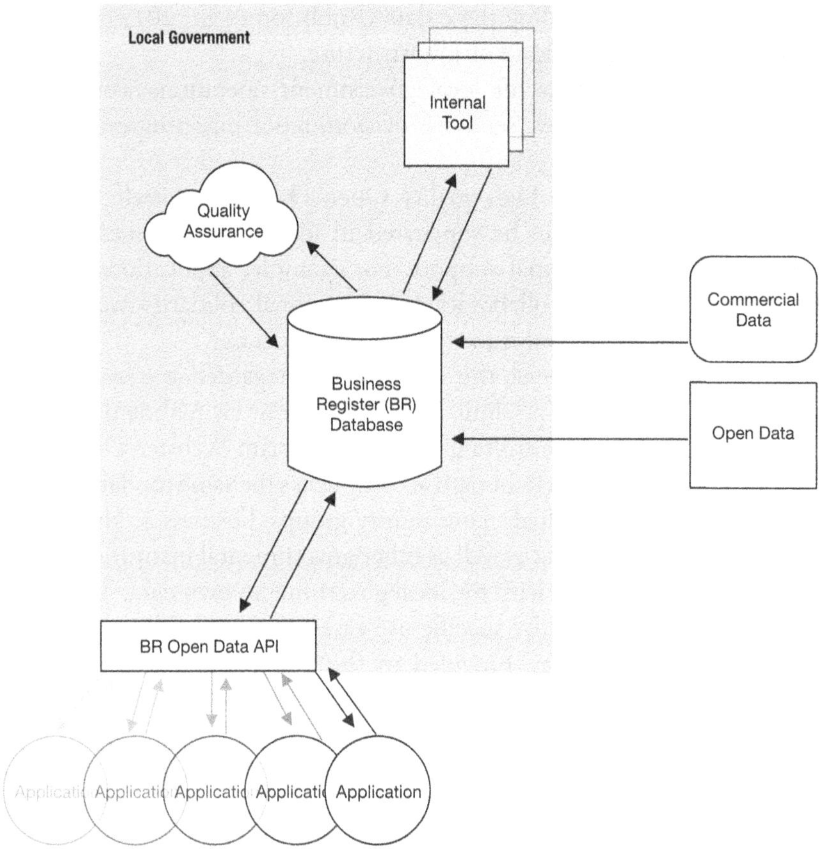

Fig. 15.5 Simplified diagram of opening up the business register dataset to the public via a bidirectional API and enhancing the data through additional external datasets while ensuring quality standards

local buyers could add information on which partnerships are actualised, as well as incentivise business to keep their information up to date. Further datasets from external parties, both commercial and open, can be added into the mix and fully or partially incorporated into the overall register of Open Data that this approach can make available to users, depending on privacy settings and regulations.

Periodic quality assurance routines would need to be instated to ensure data consistency, accuracy and integrity. Exploring ways of composing, harmonising and providing these data (Robinson et al., 2012) is critical in the provision of this kind of infrastructure.

Finally, internal tools for local government operations would draw upon the data to improve service provision, reporting, progress tracking and decision-making.

By gaining access to high-quality Open Data about itself, the local business community can be supported in identifying avenues for self-empowerment and mutual support. For example, applications or tools for mutual discovery, collaboration and regional solidarity would lend themselves as a pilot project for this particular dataset.

On a more abstract level, this model can be regarded as a switchboard for Open Data. Operating similarly to a telephone switchboard connecting, combining and controlling circuits (Merriam-Webster Dictionary, 2017), an Open Data switchboard accomplishes the same for data streams coming in from individuals, community groups, businesses, NGOs and not-for-profits, academia as well as other governmental institutions, mixing and matching them with the local government's own data. These new combinations informed by specific use-cases, both internal and external to the Council, are then provided to the public. Similar to electrical switchboards with protective fuses, an Open Data switchboard possesses quality and privacy assurance mechanisms for reliability and safety. Open Data is no longer a static data copy in a repository, but rather a dynamic and ever-changing asset and tool for various sectors of society. The open-source movement, as we will discuss in more depth shortly, can provide principles and processes to actualise such a switchboard—both in terms of data exchange and system implementation.

A Model for Delivering Open Data Infrastructure Across IT Departments and Smart City Business Units in Collaboration with the Public

The delivery of an Open Data switchboard requires ongoing collaboration and adaptation across three main stakeholders: a particular local government business unit, the IT department and the public. The latter ranges from individuals, to groups and initiatives, businesses, NGOs and not-for-profits, as well as academia and other research institutions. These relationships need to be gradually built, nurtured and expanded towards a form of ongoing co-creative practices. Traditional collaboration between governments and businesses need to be expanded to go beyond working with influential technology businesses and hackers to include more diverse actors currently on the periphery.

Figure 15.6 illustrates this collaboration as three ongoing cycles of iterations in its essential elements and shows the progressive overlaps of collaboration as part of two phases. In the first stage, they intersect one-on-one. Local government business units and the public are in ongoing contact with regards to matters concerning use-cases for Open Data—this includes identifying use-case and clarifying requirements for the data and its exchange mechanisms to materialise its potential and various ways of upskilling to allow for effective use of the Open Data. The identified requirements are further discussed between the business unit and the IT department. Finally, the IT department both provides access to the data and receives data from the public. As relationships and connections are fostered, the collaboration transitions to a second stage, where ownership

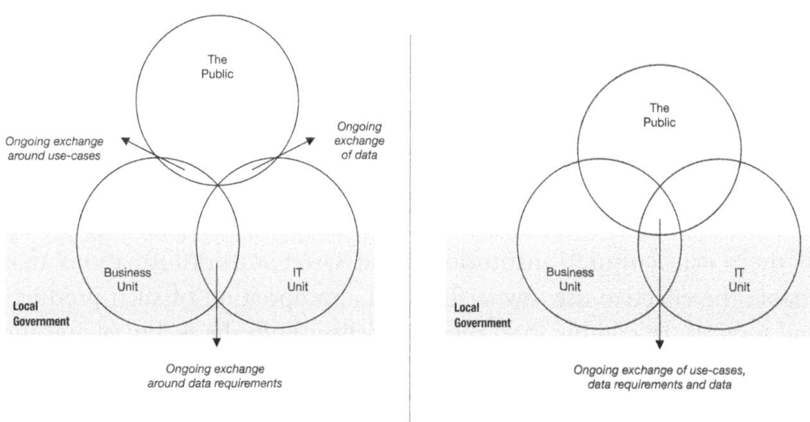

Fig. 15.6 Open Data collaboration as three ongoing cycles of iterations across its basic stakeholders: the local government's business unit, its IT unit and the public. Relationships need to be gradually built to achieve stronger interactions across various Open Data aspects, such as use-cases, data requirements and the data exchange itself

and responsibilities around the above aspects of Open Data are shared across all three partners and co-created. As they evolve, this partnership will not only need to address the logistics of Open Data delivery but, by opening to the public, can invite their expertise and collaboratively provide the kind of information, education and resources required to bridge the Open Data access gap.

The realisation of such a model must be accompanied by a widespread shift in mindset: from local government as service provider towards collaborator, and from residents as passive service recipients towards active, knowledgeable and diverse contributors (Foth & Brynskov, 2016). As contributors, they could not only create benefit for themselves but potentially also for local government. This shift requires effective policies and interaction tools at each of the first three, then four, interfaces. In this chapter, we provided inceptive input towards each of them.

Conclusion

As we learned from the methodological approach of this project, we see a fusion between the double diamond and action research approaches as a viable way to deliver a more participatory and interactive Open Data strategy. The double diamond and action research share many characteristics, such as an iterative, insight-based, critical, observant and open-minded nature. They are in many ways complementary, as the double diamond focuses on the design and delivery of products and services, while action research can contribute to the understanding and realisation of the social, cultural, institutional and systemic transformations that enable the effective use, ownership and appropriation of such products and services. In combination they can deliver both the technical and the social component of more sustainable, legitimate and equitable smart city transformations. While the double diamond contributes to iterations on the specificities of the Open Data switchboard system, action research can assist in the implementation of the change necessary to bring smart city units and the public closer together.

This project demonstrates that, as much as Open Data provided by a smart city can act as the basis for a co-created set of digital services that empower local businesses to develop a stronger, tightly linked local economy, this Open Data can be substantially enriched by local businesses feeding data back into the system and subsequently generate benefits for the city to operate on a richer, more comprehensive and accurate dataset. At first, this dataset may be messy as stakeholders gradually ease into formalised and well-defined co-creative practices. By repeating this cycle indefinitely, both operational data for the city, as well as Open Data increase in quality and value for their respective stakeholders. With the emergence of new use-cases, the API, system and model behind it can be adapted collaboratively to address an increasing number of issues and mesh up an increasing number of datasets towards wider benefit.

The system is thus put in service of the city and its citizenry to implement solutions to complex challenges, rather than a solution in itself, as current smart city technology is often framed.

Acknowledgements We, in particular the main author, would like to express our gratitude to Code for Australia and Moreton Bay Regional Council for the opportunity and collaboration in conducting this research. We would also like to thank the anonymous reviewers for their highly valuable comments on a previous version of this chapter.

References

Atkinson, P., Coffey, A., Delamont, S., Lofland, J., & Lofland, L. (2001). Handbook of *ethnography*. 1 Oliver's Yard, 55 City Road, London England EC1Y 1SP, UK: SAGE Publications Ltd.

Australian Bureau of Statistics. (2015). Moreton Bay region summary. Retrieved October 30, 2017, from http://stat.abs.gov.au/itt/r.jsp?RegionSummary&region=35010&dataset=ABS_REGIONAL_LGA&geoconcept=REGION&datasetASGS=ABS_REGIONAL_ASGS&datasetLGA=ABS_REGIONAL_LGA&regionLGA=REGION&regionASGS=REGION

Baack, S. (2015). Datafication and empowerment: How the Open Data movement re-articulates notions of democracy, participation, and journalism. *Big Data & Society, 2*(2). https://doi.org/10.1177/2053951715594634

Bates, J. (2012). "This is what modern deregulation looks like": Co-optation and contestation in the shaping of the UK's Open Government Data

Initiative. *The Journal of Community Informatics, 8*(2). Retrieved from http://ci-journal.net/index.php/ciej/article/%20view/845/916

Batty, M. (2013). *The new science of cities*. MIT Press.

Benjamin, S., Bhuvaneswari, R., & Rajan, P. (2007). Bhoomi: "E-governance", or, an anti-politics machine necessary to globalize Bangalore? *CASUM—m Working Paper*.

Bonina, C. M. (2013). New business models and the value of Open Data: definitions, challenges and opportunities. *RCUK Digital Economy Theme, www.nemode.ac.uk/wpcontent/uploads/2013/11/Bonina-Opendata-Report-FINAL.pdf*. Available at: https://ofti.org/wp-content/uploads/2014/08/Bonina-Opendata-Report-FINAL.pdf

Bradbury, H. (2015). *The SAGE handbook of action research*. SAGE.

Brenner, N., & Schmid, C. (2015). Towards a new epistemology of the urban? *Cityscape, 19*(2–3), 151–182.

British Design Council. (2007). *Eleven lessons: Managing design in eleven global brands. A study of the design process*. British Design Council.

Carayannis, E. G., Barth, T. D., & Campbell, D. F. J. (2012). The Quintuple Helix innovation model: Global warming as a challenge and driver for innovation. *Journal of Innovation and Entrepreneurship, 1*(1), 2.

Cardullo, P., & Kitchin, R. (2017). Being a "citizen" in the smart city: Up and down the scaffold of smart citizen participation. Retrieved from https://osf.io/preprints/socarxiv/v24jn

Carrara, W., Chan, W. S., Fischer, S., & van Steenbergen, E. (2015). *Creating value through Open Data. European Commission*. https://doi.org/10.2759/328101

Cole, R. J. (2012). Some observations on the practice of "Open Data" as opposed to its promise. *The Journal of Community Informatics, 8*(2). Retrieved from http://ci-journal.net/index.php/ciej/article/view/920

Coleman, G. (2004). The political agnosticism of free and open source software and the inadvertent politics of contrast. *Anthropological Quarterly, 77*(3), 507–519.

Davies, T., & Perini, F. (2016). Researching the emerging impacts of open data: Revisiting the ODDC conceptual framework. *The Journal of Community Informatics, 12*(2). Retrieved October 24, 2017, from http://ci-journal.net/index.php/ciej/article/view/1281

DeWalt, K. M., DeWalt, B. R., & Wayland, C. B. (1998). Participant observation. In H. R. Bernard (Ed.), *Handbook of methods in cultural anthropology* (pp. 259–299). Walnut Creek, CA: Altamira Press.

Dick, B. (1990). *Building agreement from disagreement: The anatomy of dialectical processes*. Paper presented at International Congress of Action Research and Process Management, Brisbane, Australia.
Douglas, G. C. C. (2014). Do-it-yourself urban design: The social practice of informal "improvement" through unauthorized alteration. *City & Community, 13*(1), 5–25.
Emerson, R. M., Fretz, R. I., & Shaw, L. L. (2001). Participant observation and fieldnotes. In *Handbook of ethnography* (pp. 352–368). SAGE.
Floridi, L. (2014). Open data, data protection, and group privacy. *Philosophy & Technology, 27*(1), 1.
Forlano, L. (2016). Decentering the human in the design of collaborative cities. *Design Issues, 32*(3), 42–54.
Foth, M. (2017). The next urban paradigm: Cohabitation in the smart city. In *Building a new economy for Australia conference*. Retrieved November 13, 2017, from https://eprints.qut.edu.au/109238/
Foth, M., & Brynskov, M. (2016). Participatory action research for civic engagement. In E. Gordon & P. Mihailidis (Eds.), *Civic media: Technology, design, practice*. Cambridge, MA: MIT Press.
Foth, M., Brynskov, M., & Ojala, T. (2015). *Citizen's right to the digital city: Urban interfaces, activism, and placemaking*. Springer.
Fung, A., & Weil, D. (2010). Open government and open society. *Open Government: Collaboration, Transparency, and Participation in Practice, 105*, 106–108.
Gurstein, M. (2003). Effective use: A community informatics strategy beyond the digital divide. *First Monday, 8*(12). Retrieved from http://firstmonday.org/ojs/index.php/fm/article/viewArticle/1107/1027
Gurstein, M. B. (2011). Open data: Empowering the empowered or effective data use for everyone? *First Monday, 16*(2). Retrieved from http://firstmonday.org/htbin/cgiwrap/bin/ojs/index.php/fm/article/viewArticle/3316/2764
Harvey, D. (2008). The right to the city. *New Left Review, 53*, 23–40.
Hatton, C. (2015). China sets up huge "social credit" system. *BBC News*. Retrieved November 1, 2017, from http://www.bbc.com/news/world-asia-china-34592186
Hearn, G. N., Tacchi, J. A., Foth, M., & Lennie, J. (2009). *Action research and new media: Concepts, methods and cases*. Cresskill, NJ: Hampton Press.
Heimstädt, M., Saunderson, F., & Heath, T. (2014). Conceptualizing Open Data ecosystems: A timeline analysis of Open Data development in the

UK. In *CeDEM14: Proceedings of the International Conference for E-Democracy and Open Government* (pp. 245–255).

Johnson, J. A. (2014). From open data to information justice. *Ethics and Information Technology, 16*(4), 263–274.

Kanter, J., & Scott, M. (2016). Apple owes $14.5 billion in back taxes to Ireland, E.U. says. *The New York Times*. Retrieved May 20, 2018, from https://www.nytimes.com/2016/08/31/technology/apple-tax-eu-ireland.html

Kitchin, R. (2014a). Big data, new epistemologies and paradigm shifts. *Big Data & Society, 1*(1). https://doi.org/10.1177/2053951714528481

Kitchin, R. (2014b). The real-time city? Big data and smart urbanism. *GeoJournal, 79*(1), 1–14.

Kitchin, R. (2016). The ethics of smart cities and urban science. *Philosophical Transactions. Series A, Mathematical, Physical, and Engineering Sciences, 374*(2083). https://doi.org/10.1098/rsta.2016.0115

Kitchin, R., Coletta, C., Evans, L., Heaphy, L., & MacDonncha, D. (2017). Smart cities, urban technocrats, epistemic communities and advocacy coalitions. Retrieved from https://osf.io/preprints/socarxiv/rxk4r

Lathrop, D., & Ruma, L. (2010). *Open government: Collaboration, transparency, and participation in practice*. O'Reilly Media, Inc.

Lefebvre, H. (1996). The right to the city. In E. Kofman & E. Lebas (Eds.), *Writings on cities*. Cambridge, MA: Wiley-Blackwell, 147i59.

Lessig, L. (2009). Against transparency. *New Republic*. Retrieved November 1, 2017, from https://newrepublic.com/article/70097/against-transparency

Levy, S. (2010). *Hackers: Heroes of the computer revolution—25th Anniversary Edition*. O'Reilly Media, Inc.

Leydesdorff, L., & Deakin, M. (2011). The triple-helix model of smart cities: A neo-evolutionary perspective. *Journal of Urban Technology, 18*(2), 53–63.

Liptak, A. (2017). Sent to prison by a Software Program's Secret Algorithms. *The New York Times*. Retrieved November 1, 2017, from https://www.nytimes.com/2017/05/01/us/politics/sent-to-prison-by-a-software-programs-secret-algorithms.html

Markou, C. (2017). Why using AI to sentence criminals is a dangerous idea. *The Conversation*. Retrieved November 1, 2017, from http://theconversation.com/why-using-ai-to-sentence-criminals-is-a-dangerous-idea-77734

Melgaço, L., & Willis, K. S. (2017). Editorial: Social smart cities: Reflecting on the implications of ICTs in urban space. *plaNext—Next Generation Planning, 4*, 5–7.

Merriam-Webster Dictionary. (2017). Definition of Switchboard. Retrieved November 13, 2017, from https://www.merriam-webster.com/dictionary/switchboard

Open Knowledge Foundation. (2017a). The Open Definition. Retrieved October 24, 2017, from http://opendefinition.org/

Open Knowledge Foundation. (2017b). Why Open Data? Handbook of Open Data. Retrieved October 24, 2017, from http://opendatahandbook.org/guide/en/why-open-data/

Özdemir, E., & Tasan-Kok, T. (2017). Planners' role in accommodating citizen disagreement: The case of Dutch urban planning. *Urban Studies*. https://doi.org/10.1177/0042098017726738

Rankin, J. (2017). EU to find ways to make Google, Facebook and Amazon pay more tax. *The Guardian*. Retrieved November 1, 2017, from http://www.theguardian.com/business/2017/sep/21/tech-firms-tax-eu-turnover-google-amazon-apple

Reason, P., & Bradbury, H. (2001). *Handbook of action research: Participative inquiry and practice*. SAGE.

Robinson, R., Rittenbruch, M., Foth, M., Filonik, D., & Viller, S. (2012). Street computing: Towards an integrated Open Data Application Programming Interface (API) for Cities. *Journal of Urban Technology, 19*(2), 1–23.

Royce, W. W. (1987). Managing the development of large software systems: Concepts and techniques. In *Proceedings of the 9th international conference on Software Engineering* (pp. 328–338). IEEE Computer Society Press.

Sadowski, J. (2016). Selling *smartness*: Visions and *politics of the smart city*. Doctor of Philosophy, Arizona State University.

Sassen, S. (2015). Urbanizing technology. In M. Foth, M. Brynskov, & T. Ojala (Eds.), *Citizen's right to the digital city: Urban interfaces, activism, and placemaking* (p. 253). Springer.

Schrock, A. R. (2016). Civic hacking as data activism and advocacy: A history from publicity to open government data. *New Media & Society, 18*(4), 581–599.

Shaw, J., & Graham, M. (2017). An informational right to the city? Code, content, control, and the urbanization of information. *Antipode*. https://doi.org/10.1111/anti.12312

Thrift, N. (2014). The promise of urban informatics: Some speculations. *Environment & Planning A, 46*(6), 1263–1266.

Ubaldi, B. (2013). Open government data: Towards empirical analysis of open government data initiatives. *OECD Working Papers on Public Governance No. 22*, 0_1.

Vanolo, A. (2014). Smartmentality: The smart city as disciplinary strategy. *Urban Studies, 51*(5), 883–898.
Vickery, G. (2011). Review of recent studies on PSI re-use and related market developments. Retrieved from http://www.umic.pt/images/stories/publicacoes6/psi_final_version_formatted-1.pdf
Wright, G., Prakash, P., Abraham, S., & Shah, N. (2010). *Report on open government data in India*. Centre for Internet and Society, Bangalore. Retrieved from http://cis-india.org/openness/publications/ogd-report/.

16

Understanding the Open Data Challenge for Building Smart Cities in India

Sarbeswar Praharaj and Saswat Bandyopadhyay

Highlights

- Indian national initiatives for city data infrastructure development are unable to create a reliable and periodic data assemblage system at the local level.
- Open city datastores developed by fewer municipalities in India provide limited access to data on critical sectors, for example, health, environment, transport and so on.
- This chapter recommends the development of City Data Hubs (CDHs) at the regional level where cities show shared governance and political culture.

S. Praharaj (✉)
Faculty of the Built Environment, University of New South Wales, Sydney, NSW, Australia
e-mail: s.praharaj@unsw.edu.au

S. Bandyopadhyay
Faculty of Planning and Public Policy, CEPT University, Ahmedabad, India
e-mail: saswatb@cept.ac.in

S. Hawken et al. (eds.), *Open Cities | Open Data*,
https://doi.org/10.1007/978-981-13-6605-5_16

- A series of recommendation guides aspiring smart cities to develop an Open Data roadmap to address urbanisation challenges.

Introduction

The urban fabric of our cities is increasingly being pervaded by networks, sensors and systems that support greater monitoring and efficient management of urban spaces and services. These new developments are shaping what Mitchell (1995) calls "City of Bits". The traces of the digital bits produced through millions of impressions over the internet remain in time and have spatial location dimensions, and therefore provide a unique opportunity to reveal the manifold and complex facets of the city in real time (Kloeckl, Senn, & Ratti, 2012). Although organisations and cities globally produce nearly 1.7 million billion bytes of digital datasets per minute, it is estimated that only about 4 per cent of the data is accessible in the public domain (Rial, 2013). While the private corporations are seeking to protect and harness massive commercial opportunities offered by the current information revolution, many of the public agencies are pushing for an alternative mechanism called "Open Data".

Open Data is on the one hand a philosophy and on the other hand a set of policies. The philosophy of Open Data underpins the fact that providing public access to data promotes transparency and accountability in the functioning of civic authorities (O'Hara, 2012; Peled, 2011). Open Data as a policy advocates the publishing of data in a machine-readable format to ensure that computer applications can retrieve the data in a structured way and through open licences to allow both commercial and personal use without restriction (Barns, 2016; Mellouli, Luna-Reyes, & Zhang, 2014). Beyond the transparency agenda, researchers use Open Data as a strategy to support the foundations of smart cities. Batty (2012), for example, believes that opening up of city information provides the opportunity for start-ups, entrepreneurs and urban enthusiasts to use the data and engage in frugal innovation and the development of smart solutions. Indeed, as Kitchin (2014) says, the various data streams produced in the everyday city, matched with dynamic analytic capabilities, make a city knowable and controllable in ways that improve the performance and delivery of public services.

While the value of Open Data in providing a practical way of planning informed and innovative cities is undisputed, the availability and organisational culture of data across the world are uneven. The availability of and access to data at the scale of cities and their sub-systems are especially limited in emerging economies due to a lack of an organised data capturing and dissemination strategy at the local level (Chakraborty, Wilson, Sarraf, & Jana, 2015; Rambaldi, Kyem, McCall, & Weiner, 2006). Siloed urban institutional systems (Praharaj, Han, & Hawken, 2017) further limit the scope of data aggregation and the consistency of data formats published. Much of the city data in developing countries is made available through a decennial census, which quickly becomes outdated and loses its reliability as a fact base for urban management. The lack of granularity in these macro census data leads to the mushrooming of ad-hoc datasets developed by local agencies for planning that has limited perceived legitimacy. Often the local agencies lack the confidence to open up such data for common use that can face public backlash. The issues of availability of urban data are acute in the case of slums and informal settlements, which exist in a legally contested space. Also, as Elwood (2008) outlines "data are a powerful source of influence" and are generally viewed by the political community as a threat to existing power structures, which could also help explain the reluctance of many public agencies and non-governmental profit organisations to share the data they collect.

In this chapter, we critically examine the policies and initiatives around Open Data in India, both at the national and at the city level. We analyse the urban Open Data platforms, their usability, nature of datasets published and their limitations. We evaluate how the various initiatives and upcoming data platforms are impacting structural governance issues and the data culture at the local level. Our research objectively interrogates "how India's aspiring smart cities can leverage Open Data to shape more open and innovative societies". This introductory piece is followed by a statement of the methods and analytical processes in Sect. 2. We then discuss the various Indian national Open Data initiatives, their effectiveness and limitations. In Sect. 4, we elaborate on the city Open Data platforms developed by select municipal authorities. In the last segment, we offer valuable recommendations for linking the Open Data strategy with smart city movement in India and beyond for a wider audience.

Methods and Analytical Framework

We use a two-layered framework to assess the different Open Data case studies in India. We first analyse the national policies and platforms and, later, assess the local level Open Data initiatives. We review the national-level case studies based on a chronological development method. Through this, our research delves into the process of evolution of the projects, the linkage and continuity of interventions and above all their current state and impacts on the city data landscape. We examine the local Open Data projects by a framework that is inspired by an earlier "City Indicator Data Openness Measure" (CIDOM) developed by Fox and Pettit (Fox & Pettit, 2015). Our method measures each of the municipal data platforms based on a set of Key Performance Indicators (KPI), including objectivity and appeal, the design of the interface, the availability and quality of data and the sustainability of the initiative. The objectivity criteria reveal the focus of the platform and its value for a large number of city stakeholders. We have looked into the usability of the portal and how easy it is to understand the presentation of the data through a critique of the design of the interface. More significantly, the quantity and quality of data available over the case study sites are mapped out as part of this study, and we compare them with globally successful open city datastores. The final KPI assesses the continuity and trends of periodic data updates on the platform, as well as the governance and ownership of the projects.

Urban Open Data in India: National Initiatives

India, with an over 377 million urban population, spread over 7935 towns across the country, is on the forefront of the challenges of rapid urbanisation. The High-Powered Expert Committee (HPEC, 2011) estimated that the urban population in India would increase up to 600 million by 2031, and there will be about 87 million-plus cities in the country. The Census of India 2011 brought out the fact that the rate of urbanisation in India has been faster than that was expected (Bhagat, 2011). This fast growth, coupled with weak managerial capacities of urban local bodies (ULBs), has resulted in a colossal deficit in the delivery of essential services (Ahluwalia, 2017). As its urban population contributes two-thirds of India's gross domestic product (GDP), managing the sustainability of cities is critical to the country's future growth story.

While India is struggling to meet the basic infrastructure demand in its cities, the Government of India has unleashed several aspirational urban development missions, including the much talked about "Smart Cities Mission". The programme aims to drive local development and create smart solutions for citizens by harnessing the power of information and communication technology (GoI, 2015). As the cities in India are beginning to digitise infrastructure and adapt to smart technologies, greater generation of real-time data is expected to flow out of the city systems (Praharaj, Han, & Hawken, 2018a). This provides a unique opportunity for Indian cities to enhance urban intelligence by capturing the data and sharing them with the public for the development of collaborative responses to crooked problems. This new paradigm could not only support Indian cities in managing rapid urbanisation but also help them leapfrog in the emerging digital information economy.

Urban development is constitutionally regarded as a state subject in India. However, the national government primarily sets the policy agenda and rolls out various schemes and initiatives for promoting planned urban infrastructure development. Data is a contested subject and hence requires clear policy directions from the top if at all data be made public. Like urban management, the national government in India has led the development activities around the Open Data architecture over the last decade. A critical chronological review of those initiatives is summarised in the sub-sections below (Fig 16.1):

National Spatial Data Infrastructure (NSDI)

The Indian National Spatial Data Infrastructure (NSDI) was launched on February 5, 2001, under the aegis of the Department of Science and Technology (DST) and the Department of Space, Government of India.

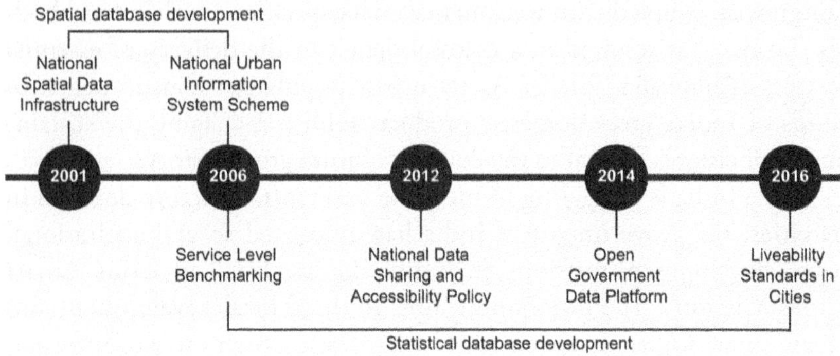

Fig. 16.1 Initiatives to develop open spatial and statistical data repositories in sequential periods

The vision of NSDI was to develop a national infrastructure for ensuring availability and access to organised spatial data and promote the use of geospatial data for planning (GoI, 2011). NSDI was mandated to develop standards for spatial data and services, metadata definitions, network and access protocols as well as facilitate technical agreements between agencies to follow standard conventions. When NSDI was envisioned in 2001, it was planned that over of a period of five years, the agency would be able to launch a spatial data portal to reshape the spatial data landscape in the country (Rao, 2007). However, the lack of machinery for coordination challenged the establishment to achieve those goals. In December 2008, when the NSDI finally introduced the India Geoportal, it was found that only limited spatial metadata of the country was made available. More importantly, it was observed that although hundreds of government and private sector organisations across the country were involved in creating spatial data, only 17 government organisations got affiliated with NSDI by March 2009 and just a few of them have uploaded their metadata to NSDI servers (Singh, 2009). While the initiative began with a much broader scope, the agency's operation today is limited to the formulation of spatial data standards by the various working groups set up under NSDI.

National Urban Information Systems (NUIS) Scheme

With an aim to develop an integrated urban database system, Town & Country Planning Organisation (TCPO) in the year 2006 initiated the National Urban Information System (NUIS) scheme. The goal of NUIS was to: (a) develop a city-level Urban Spatial Information System (USIS) and (b) create National Urban Databank and Indicators (NUDBI) for enhanced integration between socio-economic and spatial planning (GoI, 2006a). The scheme aimed at covering all the 4378 urban agglomerations and towns in the country enumerated in the Census 2001. But, resource and capacity constraints forced the authority to scale down the number of cities to 152. The urban spatial data openly made available on the Bhuvan Portal (available at https://bhuvan.nrsc.gov.in/) is mapped at the 1:10000 scale, which has significant implications for master planning. However, the TCPO was advised by local planning authorities to initiate the mapping of pilot cities at the 1:1000 scale for utility planning and plot-level space planning, which could not be undertaken due to technical constraints of the central agency (GoI, 2006b). Even the mapping system at the 1:10000 scale lacked a framework for integration with the cadastral maps and attribute data of the ULBs, making it inappropriate for detailed master planning purposes for which it was originally conceived. Moreover, the published statistical profile indicators under NUDBI were found to be aggregated city-scale data, mostly repetitive of what was already available from the Census of India publications. These databases, although having immense value, lacked public traction due to their execution over a handful of cities and lack of utility of the data for strategic urban planning.

Service-Level Benchmarking (SLB)

The Ministry of Housing and Urban Affairs, formerly known as the Ministry of Urban Development (MoUD), in 2006 launched the Service Level Benchmarking (SLB) programme covering water supply, wastewater, solid waste management, stormwater drainage and transport sectors. The initiative encompassed: (a) collation of performance data on selected indicators, (b) implementation of improved information systems at the city and state levels to support the provision of this data on an ongoing basis and (c) development of performance improvement plans based on the benchmarking data (GoI, 2010). The initiative was first of its kind, targeting accumulation of fine-grained urban data in Indian cities and therefore received strong global support with agencies such as Water and Sanitation Program–South Asia (WSP–SA) and Japan International Cooperation Agency (JICA) lending technical and financial support. In December 2009, MoUD released the SLB performance data of 28 pilot cities that kicked off an intense debate around comparative advantages and disadvantages of cities emerging from the Open Data. While the first objective of collation of data was well achieved, the initiative lacked focus on the other two. Thus, the promise of an improved periodic information collection system at the city level and urban performance improvement plans based on the outcome of benchmark data remained on the paper. The World Bank in its project synthesis report (World Bank, 2016) branded SLB as a centrally mobilised initiative with little engagement by the cities. Which meant, neither the cities took it up as a recurring exercise nor it was replicated to other cities.

NDSAP and Open Government Data (OGD) Platform

The National Data Sharing and Accessibility Policy (NDSAP) published on March 17, 2012, by the Government of India's Ministry of Science and Technology stands as a landmark towards a new Open Data paradigm in India. The policy offers an enabling provision and platform for public access to the data generated by various Government of India entities. The policy calls for streaming of machine-readable data through a wide area network all over the country in a periodically updatable manner, permitting wider accessibility and usage of Open Data by the public (GoI, 2012).

The policy came into real effect with the National Informatics Centre (NIC) publishing the implementation guidelines of this policy in November 2015 (Open Government Data Division, 2015). NIC, which is working towards the goal of opening up the information out of the government system, envisioned a single Open Data platform for the country for data sharing. Although such a platform was existing since 2012, the site has gone through a sea of change with this current version launched on December 11, 2014, known as "Open Government Data (OGD) Platform India" (Fig. 16.2). This platform received high visibility and became popular among the community once it was marketed under the "Digital India Mission" (see at https://digitalindia.gov.in) aggressively pitched by the current Prime Minister of India. As of date, the platform

Fig. 16.2 Open Government Data (OGD) platform in India. Source: https://data.gov.in/

hosts 4237 catalogues, covering 139,052 resources contributed by 113 Chief Data Officers located over across 107 departments in various central, state and local public agencies. The portal has also published 1235 visualisations and has recorded a total of 5.1 million downloads of different datasets till October 24, 2017.

The platform with an aim to engage citizens and various communities has integrated a community portal (available at https://community.data.gov.in) where multiple stakeholders can contribute through blogs, infographics, visualisations and so on using the data available on the platform. Various hackathons and challenges were organised to encourage app developers' community with technical patronage from global software giants such as Microsoft.

The OGD platform is indeed a game-changer and aberrates from the contested history of data sharing and access in India. For the first time, through NDSAP, the issues of data standardisation, metadata specification and Open Data formats are addressed comprehensively and systematically. However, the platform has not been able to resolve the scarcity of city data with just around 22 catalogues of municipal data available on the platform, most of which are macro census data and information on public investments by central ministries. The portal hosts infrastructure and community assets data from only two urban areas out of 7935 towns/cities identified by the Census of India 2011, exposing the grave challenges of data architecture in India.

Liveability Standards in Cities

The Ministry of Housing and Urban Affairs, Government of India, has developed a set of "Liveability Standards in Cities" to generate a Liveability Index and rate cities. The framework launched in May 2017 identifies a total of 79 indicators (57 core indicators and 22 supporting indicators) for measuring liveability standards in 116 pilot cities (GoI, 2016). The indicators were grouped into 15 distinct categories, such as education, health, water supply, pollution and so on. One-fourth of the indicators fall under two categories: governance and transportation and mobility. These various categories will form sub-indexes, ultimately feeding into the City Liveability Index. The index is pitched as a performance measure of the nominated smart cities.

A more in-depth look at the guideline reveals that there is very little clarity on the collection process and whether the data will be made openly available. Doubts are arising about the transparency of such ventures and how the data reliability will be ensured. The Centre for Internet & Society (CIS) questioned the predominance of the static data proposed to be used in the exercise and a lack of intention to utilise more of smart data from digital sources, to better reflect the performance of cities (Hickok, Lakshané, & Tiwari, 2016). Such technology-enabled data could also address the concerns of transparent and non-political data-generation processes. Also, there is confusion regarding multiple standards with the Bureau of Indian Standards (BIS) launching "Smart Cities—Indicators" (BIS, 2016); a few months later the Liveability Index was announced. It is not clear why two distinct sets of indicator systems were published for measuring the performance of the same set of cities identified under smart cities programme. Amidst these concerns, urban experts called for a re-look into previous experiences such as the SLB, so that the same mistakes are not repeated (Patro, 2017), and a more systemic, bottom-up data culture is promoted.

Urban Open Data in India: Municipal Local Initiatives

Data is generated in our homes, offices and public spaces, and therefore the framework at the bottom level is critical for capturing those data. City governments are in an ideal position to collect that information and create value from those assets. But, the process of data capturing, standardising and streaming is a complex process and sometimes challenges the capacity of the urban local bodies. In the following sections, let us critically examine the three-unique local urban Open Data initiatives in India.

Performance Assessment Systems (PAS)

The Performance Assessment System (PAS) was a five-year Action Research Project conceived at CEPT University, Ahmedabad, to develop appropriate tools and methods to measure, monitor and improve the delivery of water and sanitation services in urban India. The framework adopted by PAS identifies a set of Key Performance Indicators for reforms by the states, as well as Local Urban Action Indicators that are analysed to achieve city-specific improvements in the sanitation sector. PAS is recognised as the most structured attempt in India's urban history to develop a framework for the annual collection of service-level data in cities. Funded by Bill and Melinda Gates Foundation, the project primarily emphasises on urban local bodies in the state of Gujarat and Maharashtra. Presently the database hosted at PAS covers 900 towns and cities. The PAS web portal (hosted at https://www.pas.org.in) is the largest Open Data repository on urban water and sanitation in India. The platform (Fig. 16.3) offers visualisations of the data at both state and city level and gives unrestricted access to each city's database. Various tools have also been developed by the PAS team to enable interactive analysis of data and use the information for urban performance improvement (Mehta & Mehta, 2013).

PAS was indeed a successful venture for gathering, analysing and performing the open publication of data on urban sanitation-related services. The success of the project suggests that the development and sustenance of city datastores in India are much more feasible if they are conceived for a specific region that shows significant similarities in governance and political environment—as the case with Gujarat and Maharashtra. However, the limited scope of the project on a specific sector meant that it could not be referred to as a model for a comprehensive city Open Data mechanism. Moreover, the project was privately owned and funded primarily by international donor organisations. It was yet to be seen how Indian cities considering their limited resources and capacity can themselves develop infrastructure to provide Open Data access and come up with an investment model to sustain such Open Data ventures.

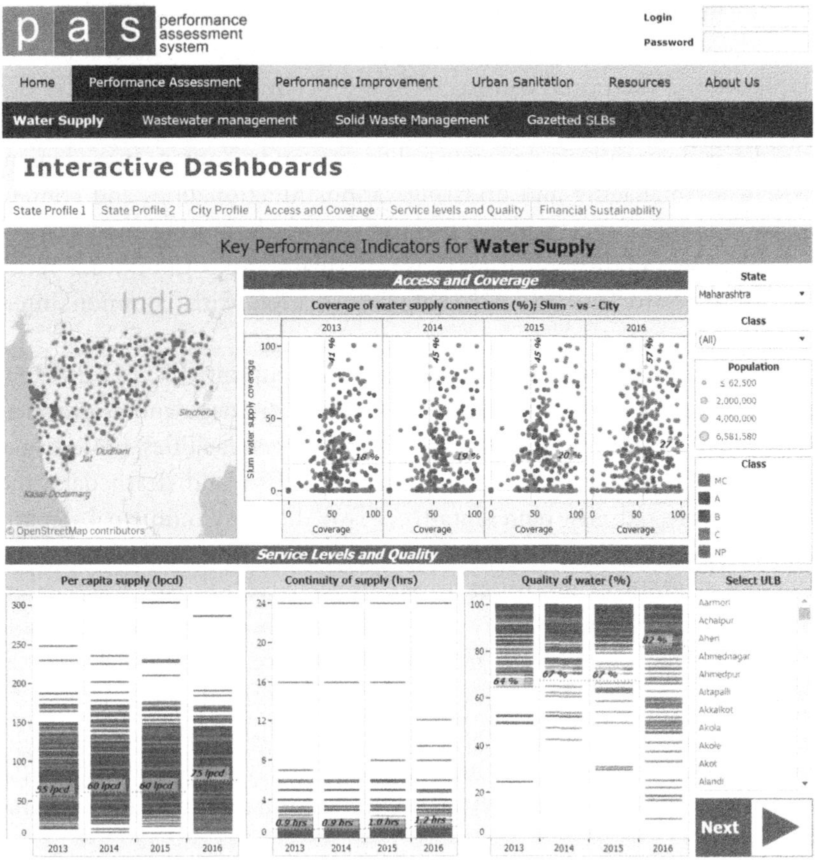

Fig. 16.3 Performance Assessment Systems' (PAS) interactive portal. Source: http://www.pas.org.in

Open Government Data Portal of Surat City

The Surat Open Government Data Portal (available at https://surat.data.gov.in/) is designed and developed by the Open Government Data Division of National Informatics Centre (NIC). The portal has been created under the Software as a Service (SaaS) model prescribed in the National Data Sharing and Accessibility Policy. The portal is intended to be used by the Municipal Corporation of Surat to publish datasets and documents collected by them for public use and open avenues for innovative uses of city government data for creating digital services, tools and applications. The portal consists of four different modules. The first two are a data management system for contributing data catalogues by various departments after a due approval process and a content management system for managing and updating various functionalities and content types. The other two modules include visitor relationship management for collating and disseminating viewer feedback and communities' module for users to interact and share information with common interest groups.

We found that the Surat Open Data portal currently lists 11 different data catalogues, including datasets related to drainage and stormwater infrastructure, weather station information, citizen facilities, professional tax, property tax, vehicle registration data, birth and death data and details of elected city representatives. Two different municipal departments have contributed 119 resources that are made available through the portal. From its inception in November 2016 till date, the datasets have been downloaded 3876 times. The number of datasets available under different catalogues and their respective download counts as reflected on the portal is diagrammatically represented in Fig. 16.4.

We find that a large number of datasets published are weather information and census data, which are not collected by the Municipal Corporation. On the contrary, very little information is made available

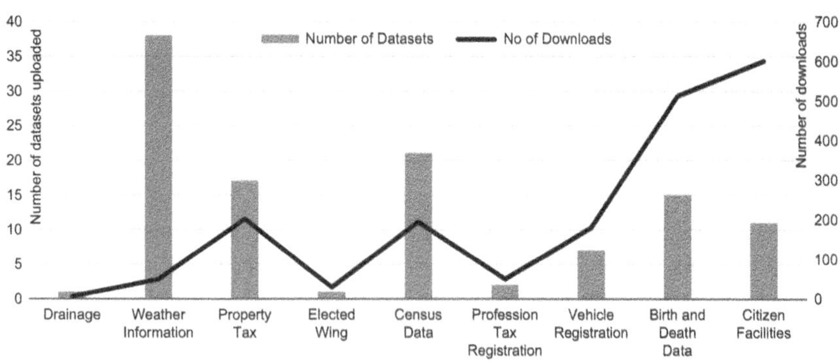

Fig. 16.4 Number of datasets available and download counts under different catalogues. Source: https://surat.data.gov.in/

on city-specific infrastructure, such as transport services and environment. Although there is a lot of interest for data on citizen facilities as reflected from the high number of downloads, there are very few datasets provided for public use or analysis. The Surat Open Data portal neither offers longitudinal information nor hosts any fine-grained dynamic datasets. More significantly, the occurrences of periodic data updates over the platform are scarce.

Pune Open DataStore

The Pune DataStore (set up at http://opendata.punecorporation.org) aims to provide collated access to datasets and apps published in an open format under the various departments of the Municipal Corporation. The portal (Fig. 16.6) is designed, developed and maintained by the Information Technology (IT) Department of the Pune Municipal Corporation (PMC) with data feeds from 19 departments from within the corporation itself and other public service providing agencies at the city level. The Pune DataStore is developed entirely using open Source Stack along with standard metadata-controlled vocabularies on government sectors, jurisdictions, dataset types, access mode and so on. PMC has appointed 13 Data Officers (DOs) from municipal departments, who are responsible for managing spreadsheets, standardising datasets and data validation before the data is updated on the portal. The platform has a robust backend data management system with dashboards available for DOs to see the current status of the datasets, usage analytics as well as queries put forward by citizens in real time. The Pune DataStore also features a community component and social media linkage to form interest groups around the datasets. Feedbacks and inputs from these groups support the PMC in prioritising the release of datasets that are valued by these communities.

As of October 25, 2017, a total of 150 datasets has been published through the portal, which was downloaded 12,472 times, at a frequency of nearly 32 downloads per day. Just over one-third of the available datasets are on various utilities such as water supply, sanitation and so on. Electoral statistics and data on the local economy and informal businesses are too abundantly found on the portal. However, little Open Data is provided on critical sectors including environment, education, health and transport.

To examine the success of the Pune DataStore, we have benchmarked it with the London DataStore—a global best practice on Open Data storage (visit at https://data.london.gov.uk/). As the Pune data portal provides classified data on urban sectors, like that of London, it was practically possible to compare the two. Our analysis in Fig. 16.5 compares the datasets published through the two datastores and highlights that Pune consistently lags behind in the number of datasets released across the urban sectors. The difference is especially staggering in the areas of demographics, environment, jobs and employment and health data availability. Whereas London did not shy away from extensively sharing data on crime statistics, road accidents, jobs growth and so on that may project the city negatively, Pune has mostly avoided the publication of such sensitive data.

The Pune DataStore is India's first Open Data platform that is entirely hosted by the City Corporation. It has challenged the general conventions that Indian cities lack the capacity and technological know-how to collect and manage data in a systemic way. However, in the limited time of the portal's existence, concerns were raised by civic activists as the cor-

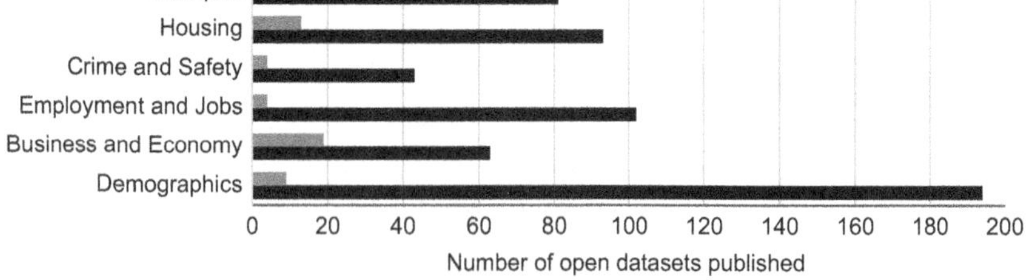

Fig. 16.5 Sector-wise datasets available in the city datastores of Pune and London

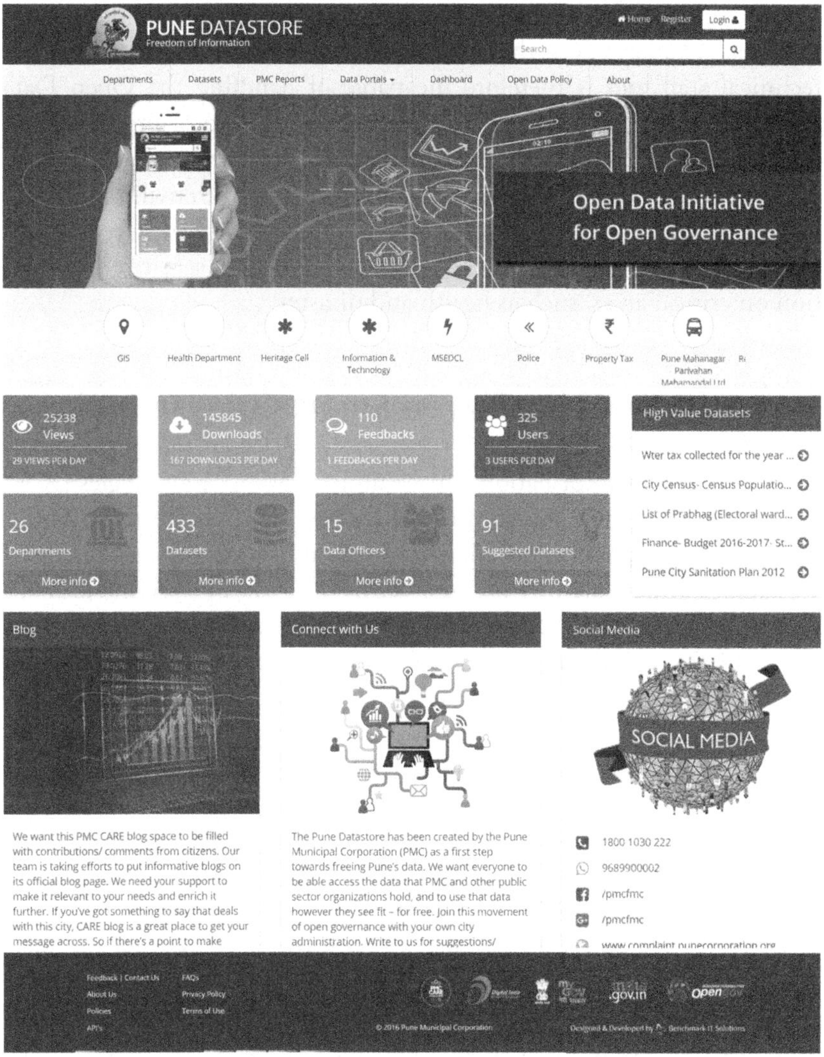

Fig. 16.6 Pune DataStore. Source: http://opendata.punecorporation.org/

poration has failed to update the data periodically. Renowned Pune activist Vivek Velankar asserts that the Municipal Corporation has inadequate technical staff and is therefore incapable of handling the Open Data challenge (Khape, 2016). While the issues around local organisational capacity are worth examining (Praharaj, Han, & Hawken, 2018b), the need for greater real-time data generation points for the environment and infrastructure-related sectors requires urgent emphasis. The city authorities must also demonstrate more willingness to share sensitive information on critical areas, such as health and disaster.

Towards the Data-Driven Smart Cities in India

Advancements in Information and Communication Technologies (ICT) have created massive opportunities for shaping more connected communities. Rapid leaps in digitisation of urban infrastructure and real-time availability of data are offering new forms of urban operation and management. Amidst this global trend, our research examines the capacities and potential of Indian national and local governments in developing an Open Data culture to realise their aspirations of becoming smart cities. Our analysis reveals that a number of national policies and projects are laid out over the last two decades to develop an architecture for data collection and sharing. However, the initiatives show a lack of continuity and convergence. Federal infrastructure, such as the NSDI and NUIS though started with a substantial promise, could create limited spatial metadata of a handful of cities. More significantly, the failure of the SLB programme to upscale the framework to urban areas beyond the pilot cases and enforce a change in the local data management system indicates a lack of organisational ability within the central and state organisations dealing with data infrastructure. The recent upcoming of the NDSAP and the Open Government Data (OGD) platform does provide a consistent policy framework and technical guideline. But, the contribution of data to the OGD by a mere two cities exposes the severe data deficiency and a lack of confidence among Indian municipalities to embrace an Open Data culture.

While the national initiatives could have a negligible impact on the city data landscape, a regional project in the form of the Performance Assessment System (PAS) came out as an unexpected winner. The success of the PAS is profoundly contributed by its focus on a specific region covering the states of Gujarat and Maharashtra. These two states have significant similarities in their urban governance and political cultures. They are also among the fewer states in India that have empowered municipal corporations with most of the planning functions transferred by the states to ULBs (Praharaj, Han & Hawken, 2018b). The success of the PAS must also be attributed to a continuous stream of funding by global aid agencies, which has led to the creation of manpower and capacities within CEPT University to sustain the project. Our findings indicate that the development and sustenance of city data projects in India are much more feasible if they are taken up at a regional scale in comparison to the national level where the sheer number of cities and organisational needs are humongous. We, therefore, recommend hosting of City Data Hubs (CDH) at targeted not-for-profit institutions and universities within different Indian regions. For example, one CDH can be established in the North-Eastern region of India to collect, maintain and share urban data periodically. Similarly, another CDH for the states in South India can substantially create and support the data needs of the cities located in the region.

The two municipal Open Data initiatives in Surat and Pune prove that cities with a forward-looking agenda and a motivation for promoting transparency can quickly develop affordable city datastores. The NDSAP guidelines serve as an enabler for cities that are looking to create such platforms. However, in the current state, the quantum of data available through both the analysed platforms are limited. More importantly, none of the Indian cities has a considerable emphasis on collecting real-time data. Even if they do gather such agile data, they are not made available to the public. A comparison of the data availability between the portals of Pune and London highlight significant deficiencies in data related to health, environment, transport and so on in the Indian urban context. The leading cities, such as Pune and Surat, must look up to the global best practices, such as London DataStore or NYC Open Data. Such benchmarking can potentially help these cities understand the spectrum of data that can be made public. It can support these progressive Indian cities to identify the areas and sectors on which more datasets can be collected and shared. In going ahead, these municipalities should promote streaming of real-time information through “Open City Dashboards” that bring together data from diverse sources, such as weather, transit, road traffic, environmental pollution, utilities, social media trends and so on. The globally successful Dublin Dashboard (McArdle & Kitchin, 2016) and City of Sydney Dashboard (Pettit, Lieske, & Jamal, 2017) can provide immense motivation and technical guidance in developing the first generation of open and data-driven smart cities in India.

This chapter contributes to the literature by measuring the challenges and constraints of building open city data infrastructure in a developing world context. We emphasise that public authorities in the global south must acknowledge data as a physical infrastructure of the city. We argue that the role of government in the digital age should be of a “platform provider” that facilitates and support public conversations for shaping healthy and democratic cities. At this stage, it is also essential to assert that merely opening up of data does not help cities to become smart. Governments and municipalities can only become smart if they can integrate and synthesise these data through analytic functions to meet some purpose, ways of improving the efficiency, transparency, sustainability and quality of life in cities. Furthermore, there are dangers that data-driven urban management can promote technological determinism (Wilson, 2015) and a concentration of power and control (Kitchin, 2014). Aspiring smart cities in India and elsewhere need to shape a culture of openness and consider the broader effects of values, politics, policy, governance and people, alongside data, while framing urban development strategies. Open Data in smart cities must not centralise power and decision making into a select set of offices armed with analytic software and giant screens. Instead, in the era of collaboration, they must empower societies and innovative businesses to work in partnership with the government in enhancing public awareness and solving the pressing urbanisation challenges.

References

Ahluwalia, I. J. (2017). Urban governance in India. *Journal of Urban Affairs, 41*(1), 1–20.

Barns, S. (2016). Mine your data: Open Data, digital strategies and entrepreneurial governance by code governance by code. *Urban Geography, 37*(4), 554–571. https://doi.org/10.1080/02723638.2016.1139876

Batty, M. (2012). Smart cities, big data. *Environment and Planning B: Planning and Design, 39*(2), 191–193. https://doi.org/10.1068/b3902ed

Bhagat, R. B. (2011). Emerging pattern of urbanisation in India. *Economic and Political Weekly, 46*(34), 10–12. https://doi.org/10.2307/3517022

BIS. (2016). *Smart cities—Indicators lCS 13.020.20 (Doc No. CED 59 (10000) WC)*. New Delhi: Bureau of Indian Standards. Retrieved from http://www.bis.org.in/sf/ced/CED59(10000)_30092016.pdf

Chakraborty, A., Wilson, B., Sarraf, S., & Jana, A. (2015). Open data for informal settlements: Toward a user's guide for urban managers and planners. *Journal of Urban Management, 4*(2), 74–91. https://doi.org/10.1016/j.jum.2015.12.001

Elwood, S. (2008). Grassroots groups as stakeholders in spatial data infrastructures: Challenges and opportunities for local data development and sharing. *International Journal of Geographical Information Science, 22*(1), 71–90. https://doi.org/10.1080/13658810701348971

Fox, M. S., & Pettit, C. J. (2015). On the completeness of open city data for measuring city indicators. In *Smart Cities Conference (ISC2), 2015 IEEE First International* (pp. 1–6). IEEE.

GoI. (2006a). *National Urban Information System (NUIS) Scheme: Guidelines for implementation*. New Delhi: Ministry of Urban Development, Government of India.

GoI. (2006b). *National Urban Information System (NUIS) Scheme: Memorandum of Understanding (MOU)*. New Delhi: Town & Country Planning Organisation, Government of India.

GoI. (2010). *Improving urban services through Service Level Benchmarking*. New Delhi: Ministry of Urban Development, Government of India. Retrieved from http://www.indiaenvironmentportal.org.in/files/GOI_Benchmarking.pdf

GoI. (2011). *NSDI Working Groups Report*. New Delhi: National Spatial data Infrastructure (NSDI), Government of India.

GoI. (2012). *National Data Sharing and Accessibility Policy*. New Delhi: Ministry of Science and Technology, Government of India. Retrieved from http://www.dst.gov.in/sites/default/files/nsdi_gazette_0.pdf

GoI. (2015). *Mission statement and guidelines. Ministry of urban development, government of India*. Retrieved from http://smartcities.gov.in/content/innerpage/guidelines.php

GoI. (2016). *Liveability Standards in Cities*. New Delhi: Ministry of Urban Development, Government of India. Retrieved from http://smartcities.gov.in/upload/uploadfiles/files/LiveabilityStandards.pdf

Hickok, E., Lakshané, R., & Tiwari, U. (2016). Comments to the BIS on smart cities indicators. Retrieved October 22, 2017, from https://cis-india.org/internet-governance/blog/comments-to-bis-on-smart-cities-indicators

HPEC. (2011). *Report on Indian Urban Infrastructure and Services*. High Powered Expert Committee, Government of India.

Khape, A. (2016, September 30). Pune Municipal Corporation launches Open Data portal with free real-time data access. *Indian Express*. Pune. Retrieved from http://indianexpress.com/article/cities/pune/pune-municipal-corporation-launches-open-data-portal-with-free-real-time-data-access/

Kitchin, R. (2014). The real-time city? Big data and smart urbanism. *GeoJournal, 79*(1), 1–14. https://doi.org/10.1007/s10708-013-9516-8

Kloeckl, K., Senn, O., & Ratti, C. (2012). Enabling the real-time city: LIVE Singapore! *Journal of Urban Technology, 19*(2), 89–112. https://doi.org/10.1080/10630732.2012.698068

McArdle, G., & Kitchin, R. (2016). The Dublin dashboard: Design and development of a real-time analytical urban dashboard. In *ISPRS Annals of Photogrammetry, Remote Sensing and Spatial Information Sciences* (Vol. IV-4/W1, pp. 19–25). Retrieved from http://www.isprs-ann-phot. https://doi.org/10.5194/isprs-annals-IV-4-W1-19-2016.

Mehta, M., & Mehta, D. (2013). City sanitation ladder: Moving from household to citywide sanitation assessment. *Journal of Water, Sanitation and Hygiene for Development, 3*(4), 481. https://doi.org/10.2166/washdev.2013.134

Mellouli, S., Luna-Reyes, L. F., & Zhang, J. (2014). Smart government, citizen participation and open data. *Information Polity*. https://doi.org/10.3233/IP-140334

Mitchell, W. (1995). City of bits. *City of bits*, 159–173. Retrieved from http://journals.cambridge.org/production/action/cjoGetFulltext?fulltextid=2129608

O'Hara, K. (2012). Transparency, open data and trust in government: Shaping the infosphere. In *WebSci '12 Proceedings of the 4th Annual ACM Web Science Conference* (pp. 223–232). https://doi.org/10.1145/2380718.2380747

Open Government Data Division. (2015). *Implementation Guidelines for National Data Sharing and Accessibility Policy (NDSAP)*. New Delhi: Ministry of Communications and Information Technology, Government of India. Retrieved from https://data.gov.in/sites/default/files/NDSAP_Implementation_Guidelines_2.2.pdf

Patro, S. (2017). Smart City indicators by BIS. Retrieved March 24, 2018, from http://www.cstep.in/uploads/default/files/pressroom/stuff/CSTEP_Smart_City_indicators_by_BIS_Feb_13_BA.pdf

Peled, A. (2011). When transparency and collaboration collide: The USA Open Data program. *Journal of the American Society for Information Science and Technology, 62*(11), 2085–2094. https://doi.org/10.1002/asi.21622

Pettit, C., Lieske, S. N., & Jamal, M. (2017). CityDash: Visualising a changing city using Open Data. In *International Conference on Computers in Urban Planning and Urban Management* (pp. 337–353). Cham: Springer. https://doi.org/10.1007/978-3-319-57819-4

Praharaj, S., Han, J. H., & Hawken, S. (2017). Innovative civic engagement and digital urban infrastructure: Lessons from 100 Smart Cities Mission in India. *Procedia Engineering, 180,* 1423–1432. https://doi.org/10.1016/j.proeng.2017.04.305

Praharaj, S., Han, J. H., & Hawken, S. (2018a). Urban innovation through policy integration: Critical perspectives from 100 smart cities mission in India. *City, Culture and Society, 12,* 35–43. https://doi.org/10.1016/j.ccs.2017.06.004

Praharaj, S., Han, J. H., & Hawken, S. (2018b). Towards the right model of Smart City governance in India. *International Journal of Sustainable Development and Planning, 13*(2), 171–186.

Rambaldi, G., Kyem, P. a. K., McCall, M., & Weiner, D. (2006). Participatory spatial information management and communication in developing countries. *The Electronic Journal of Information Systems in Developing Countries, 25*(1), 1–9. Retrieved from http://www.ejisdc.org/ojs2/index.php/ejisdc/article/view/237

Rao, M. (2007). NSDI–Then, now and whenever. *Coordinates, 3*(8), 10–13.

Rial, N. (2013, May 24). Free access to data can improve sectors like health care. Retrieved October 19, 2017, from https://www.neweurope.eu/article/power-big-data-europe/

Singh, P. (2009). Spatial data infrastructure in India: Status, governance challenges, and strategies for effective functioning. *International Journal of Spatial Data Infrastructures Research, 4*, 359–388. https://doi.org/10.2902/1725-0463.2009.04.art17

Wilson, M. W. (2015). Flashing lights in the quantified self-city-nation. *Regional Studies, Regional Science*. https://doi.org/10.1080/21681376.2014.987542

World Bank. (2016). *Republic of India: Service level benchmarking, citizen voice and performance improvement strategies in urban water supply and sanitation (Report No: ACS18249)*. World Bank. Retrieved from https://books.google.com.au/books?id=Sh9atAEACAAJ

17

Resilient Cities, User-Driven Planning, and Open Data Policy

Paul Burton, Anne Tiernan, Malcolm Wolski, Lex Drennan, and Lochlan Morrissey

Abbreviation

UDP User Driven Planning

P. Burton (✉)
Cities Research Institute, Griffith University, Gold Coast, QLD, Australia
e-mail: p.burton@griffith.edu.au

A. Tiernan
Griffith Business School, Griffith University, Nathan, QLD, Australia
e-mail: a.tiernan@griffith.edu.au

M. Wolski
eResearch Services, Griffith University, Nathan, QLD, Australia
e-mail: m.wolski@griffith.edu.au

L. Drennan • L. Morrissey
Policy Innovation Hub, Griffith University, Nathan, QLD, Australia
e-mail: l.drennan@griffith.edu.au; l.morrissey@griffith.edu.au

S. Hawken et al. (eds.), *Open Cities | Open Data*,
https://doi.org/10.1007/978-981-13-6605-5_17

Highlights

- A user-driven model for disaster planning
- Proposal for an amalgamated approach to Open Data sources
- Narrative-based communication of disaster risk
- Alignment of subjective and objective risk through Open Data platform design

Introduction

The population of Australia is predominantly urban and coastal. Two-thirds of Australians live in capital city statistical areas, with growth rates expected to continue to rise (Australian Bureau of Statistics, 2017). As more people are concentrated into a smaller geographic area, the effects of disaster events on these areas are likely to become more extreme and disruptive (Deloitte Access Economics, 2013, 2016). Furthermore, the number of disaster events and the costs associated with disaster events are expected to increase into the twenty-first century (Cowan et al., 2014; Purich et al., 2014; Zheng, Westra, & Leonard, 2015). Successful adaptive cities will be those that support the resilience of their residents. Resilience is understood in this chapter as the capacity of a system to adapt to the new situation that is ushered in by a disaster event (Tierney, 2012).

Resilience is enhanced not only through the risk-sensitive construction of the built environment but also importantly through the promotion of social capital and cohesion (Aida, Kawachi, Subramanian, & Kondo, 2013; Aldrich & Meyer, 2015; Henly-Shepard et al., 2015; Kapucu, Hawkins, & Rivera, 2013; Shaw, 2013). The resilience of cities is significantly influenced by the strength of their economy. Small to medium businesses, in particular, play a critical role in building resilience and applying these resources in the aftermath of disasters (Drennan, McGowan, & Tiernan, 2016; Tiernan, McGowan, & Drennan, 2013).

In this chapter, we present a framework for user-driven planning (UDP) that links disaster management policy and the legislation of various levels of government, Open Data produced by various agencies, and businesses. Disasters are defined with reference to government policy, using the definition provided by the National Disaster Relief and Recovery Arrangements (Attorney-General's Department, 2012, pp. 5–6). Accordingly, a disaster is "a natural disaster or terrorist event" where a natural disaster is defined as:

> a serious disruption to a community or region caused by the impact of a naturally occurring rapid onset event that threatens or causes death, injury or damage to property or the environment and that requires significant and coordinated multi-agency and community response, and is one, or a combination of, the following: a) bushfire; b) earthquake; c) flood; d) storm; e) cyclone; f) storm surge; g) landslide; h) tsunami; i) meteorite strike; j) tornado. (Attorney-General's Department, 2012, pp. 5–6)

By applying the principles of user-driven planning (UDP), Open Data on disaster risk can be contextualised to the local level. This allows objective measures of risk to be communicated meaningfully to local business owners. Such an approach can enable businesses to apply data to understand their risk environment, assess their individual exposure, and build adaptive resilience capabilities without top-down command and control from the government. Open Data has the potential to bridge governments' objective risk perspectives with local, subjective interpretations.

Building on this model of user-driven planning, we explore policy implications in the areas of disaster management and Open Data specifically with respect to Queensland and Queensland government-administered data. The Queensland government is very proactive in their approach to Open Data, with each department having its own Open Data strategy. We address some ways that UDP might be improved through policy initiatives, including more consistent, usable datasets; formulating standard measures that target specific hazards and specific properties or geographic areas; and committing funds to building a platform for residents to easily assess their own risk and for UDP at the household level. These measures provide a means of increasing social capital by using residents' and organisations' local knowledge in the disaster management process, leading to more adaptive and resilient cities than is possible under current, top-down approaches to disaster management.

Perceptions of Disaster Risk

Broadly speaking, risk can either be considered objectively or subjectively (Hansson, 2010). In the objective approach, taken by the International Standard on Risk Management and embedded in Australian government policy via the National Emergency Risk Assessment Guidelines, risks are external, knowable, and discretely quantifiable (Emergency Management Australia, 2015; ISO, 2009). From this perspective, government risk assessments, at their most basic, involve estimating the likelihood of an event, estimating consequences of that event, and implementing appropriate mitigation measures on that basis (Emergency Management Australia, 2004). Unstated in this approach is the presumption that the communities impacted by these risks, and attendant policy prescriptions, apply the same analytic frame to understanding their risk environment (Drennan, 2017).

However, extensive research has shown that individuals and by extension communities view risk through a subjective lens (Henwood, Pidgeon, Parkhill, & Simmons, 2010; Ho, Shaw, Lin, & Chiu, 2008; Paton, 2005; Paton, Smith, Daly, & Johnston, 2008). In this subjective perspective, the individual considers her perception of likelihood and balances it against estimates of her ability to cope with this event. Furthermore, factors such as risk sensitivity and beliefs about the role of government impact how individuals or communities may act to mitigate risks (Alkon, 2004; Lindell & Whitney, 2000). These factors comprise the risk narrative of individuals, businesses, and communities and often result in very different beliefs about risk levels and appropriate mitigation measures to that recommended in government policy (Corvellec, 2011; Drennan, 2017; Menapace, Colson, & Raffaelli, 2012; Sjöberg, 2000).

Disaster Management Planning and the User-Driven Turn

In this section, we discuss Disaster Management (DM) planning and its paradigm of top-down management. A model of user-driven planning is proposed to bridge the top-down, objective risk assessment paradigm, with a local and individualised subjective interpretation of risk. We argue that Open Data sets are the mechanism by which these two perspectives can be bridged as a point of common understanding.

DM evolved from civil defence and it consequently operates in a paradigm of top-down management (Clarke, 1999; W. Waugh, 2006). Yet, "the response to natural disasters is, in large measure, an ad hoc affair involving organized nongovernmental actors, governmental actors, and emergent groups that often become well organized and long lived. No one can ever have complete control; it is not possible to fully command attention or to compel compliance" (William Waugh & Streib, 2006, p. 138). This approach has received sustained criticism from scholars as it does not provide the emphasis on coordination needed to address the complexity and multi-actor nature of crises (Alpaslan, Green, & Mitroff, 2009; Boin, 't Hart, Stern, & Sundelius, 2005; Goldsmith & Eggers, 2003; William Waugh, 2009; Wise & Nader, 2002). Instead, they call for approaches that emphasise relationships, decentralisation, collaborative networks, and a participatory approach to decision-making (Alpaslan et al., 2009; Boin & 't Hart, 2010; Comfort & Kapucu, 2006; Kapucu & Garayev, 2011; Wise & Nader, 2002).

The standard approach to building disaster management capability is to document the system and then practice increasingly complex elements of the system (Clarke, 1999; ISO, 2014). This generally progresses through training individuals and training teams, to exercising one team and then multiple teams simultaneously. The Plan-Do-Check-Act cycle that underpins the Business Continuity Management Systems standard embodies this approach (ISO, 2014) (Fig. 17.1).

This approach follows logically from the expert-driven approach to disaster management planning, which sees appointed experts extracting knowledge from organisational members, developing products, and then teaching the organisation how to use these outputs. The top-down philosophy of disaster response extends into the approach to planning that

Fig. 17.1 Steps in the traditional planning process

is largely characterised by an expert-driven style of planning, which privileges technical expertise and tends to discount local knowledge (Clarke, 1999).

Debate over the role of experts versus participants arises commonly in the urban planning sector because "Planning is a social process as well as a technical one" (Larkham & Lilley, 2012, p. 648). Numerous authors argue that expert-driven, top-down planning frequently fails to engage communities or produce plans that engender widespread community support (Arnstein, 1969; Fiskaa, 2005; Jacobs, 1961; Larkham & Lilley, 2012). In her seminal work, 'The Death and Life of Great American Cities' (Jacobs, 1961), Jacobs takes explicit aim at the expert-driven planning approach as it fails to understand the lived experience of communities and their interactions with their physical environment.

This perspective is similarly embraced in the field of participatory governance, where responsibility and authority for problem-solving is devolved from central governments to local communities. Participatory governance enables "those most directly affected by targeted problems … to apply their knowledge, intelligence and interest to the formulation of solutions" (Fung & Wright, 2001, p. 18). This approach leverages local knowledge and intentionally makes the subjective perspective part of the policy solution. In this approach, the role of experts in this approach becomes to "facilitate popular deliberative decision making" and "to leverage synergies between professional and citizen insights rather than preempt citizen input" (Fung & Wright, 2001, p. 19).

The User-Driven Planning Approach

Building on the traditions of urban planning and participatory governance, we propose a user-driven approach to disaster management that seeks to emphasise the role of the participant in planning whilst retaining the expertise of the disaster management professional. By integrating local, subjective understandings of risk with the government's objective assessments, UDP has the potential to bridge these perspectives and enable businesses to build their disaster resilience capabilities.

UDP makes two critical assumptions:

1. the user knows itself and its business best; and
2. the user's disaster management process is intuitively clear to that user, but not necessarily to outside observers.

In essence, these assumptions privilege the knowledge held by everyday employees of an organisation. They also reflect the understanding that under pressure people and organisations default to what they know. This approach reverses the first two steps in the planning process on the basis of the assumptions outlined (Fig. 17.2).

UDP is a 'bottom-up' planning approach that captures the 'as-is' in the organisational response to major business interruption. Contrasting with the traditional disaster management process, in UDP, the planning process commences by conducting a disaster management exercise without training or preparing participants. This allows the observation and documentation of the 'as-is' organisational approach to disaster management. The observed process is then shaped by best practice planning approaches that seek to enhance the effectiveness of the organisation's instinctive response and embed local knowledge, rather than dictate an entirely new system. The benefits of this approach are that it leverages day-to-day organisational practice, lessens the training liability required to maintain organisational capability, increases the utility of disaster management plans, and provides a framework that is instinctively understood by its users.

Throughout the process of the exploratory exercise, debriefing, followed by a plan review exercise, participants have numerous opportunities to identify where the organisation needs to build capability. This feedback forms the basis of a capability development and maintenance programme that reflects participant understanding of their needs. Making these activities short, regular, and targeted moves the programme away from repetitive training of the plan to building the complex skill sets necessary to effectively manage an incident. The exercising component is

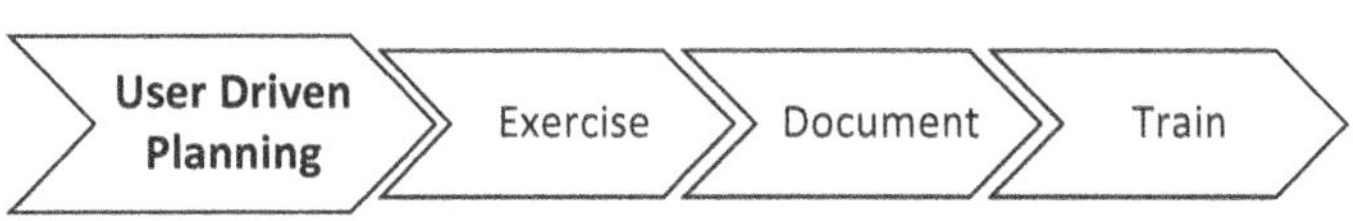

Fig. 17.2 Steps in a user-driven planning process

essential: "Regular simulation exercises nurture awareness of crisis management complexities, hone decision-making skills, and allow members of the response network to get to know, understand and respect each other. Not having this is one of the principal causes of crisis response failures" (Boin & 't Hart, 2010, p. 368).

Open Data Sources for UDP

In this section, we identify specific open datasets that may be used to improve the quality of plans devised using the UDP approach, and suggest ways that datasets might be extended to provide more functionality for UDP. In particular, we focus on datasets that are produced by the Queensland government. Queensland has become a leader in Australian Open Data policy, with each department and statutory body of the state government required to produce an Open Data strategy.[1] Accessing Queensland's Open Data is simple for even inexperienced users: the Queensland government provides a centralised search for its datasets that spans the departments and statutory bodies at https://data.qld.gov.au.

There are a number of open datasets that may be used to facilitate UDP. Data of historical disaster events is useful for UDP in that it provides a point of comparison for potential disaster events with those that are in the organisation's memory. The following Table 17.1 summarises some datasets that we propose are useful for UDP.

The resources identified above represent potential datasets that can be used for UDP to bridge the objective risk perspective of governments and the subjective application by the user. Although these datasets are easily accessible to the non-expert, two problems are readily apparent in their application. First, the data themselves are fragmented and held by a number of different custodians. While there is a centralised search function, users are required to download each dataset individually. This fact means that the data are semantically incompatible and therefore collation of the data becomes more challenging. For instance, tide-level data and flood

[1] As of the time of writing, the full list of these strategies is available at https://data.qld.gov.au/article/department-strategies.

Table 17.1 Some suggested existing open datasets and their potential use for UDP

Title	Description	Use for UDP	URL
Flood extent maps 1974	Spatial data of the extent of the Brisbane and Ipswich floods of 1974.	Afford organisations a means of observing how their assets might be affected in a similar event and to plan accordingly. A potential drawback of these datasets is that the degree of similarity of disaster events is highly uncertain.	https://data.qld.gov.au/dataset/flood-extent-series/resource/a94a60b0-76ed-424b-a52b-b307e20af224
Flood extent maps 2011	Spatial data of the extent of the Queensland floods of 2011.		https://data.qld.gov.au/dataset/flood-extent-series/resource/4c687b12-2666-4293-94c8-0fbf71346fc1
Gold Coast City Plan interactive map	An amalgamation of a number of spatial datasets that present generalised risk profiles *based on* prior disasters.	Provides organisations with spatial representations of multiple different hazards, which allow them to account for these hazards in the one plan. However, these lack the narrative cogency of a representation of a past disaster event.	http://cityplanmaps.goldcoast.qld.gov.au/CityPlan/

(*continued*)

Table 17.1 (continued)

Title	Description	Use for UDP	URL
Interactive road conditions map	A near real-time map that shows road closures, reported traffic accidents, and so on. It is confined to the Moreton Bay LGA.	Provide real-time data that can be incorporated as part of the UDP process, and that can be used to inform the enactment of a user's disaster management plan in the event of a disaster.	https://www.moretonbay.qld.gov.au/maps-disaster.aspx
QLDTraffic	A near real-time map that shows road closures, reported traffic accidents, and so on, across Queensland		https://qldtraffic.qld.gov.au/
Near real-time tide levels	Numeric data of tide levels taken in near real time at Gladstone, Queensland.		https://data.qld.gov.au/dataset/gladstone-tide-gauge-near-real-time-data

extent data represent a phenomenon that is interrelated and mutually reinforce a user's risk; yet these are available only in separate datasets. Second, many of the datasets are presented in files that require some degree of familiarity with and expertise in working with data of this kind. For instance, the flood extent maps must be downloaded and opened on a user's machine. Furthermore, each flood dataset must be opened separately, leading to a lack of continuity between the datasets.

We argue that these two problems combine to weaken the usefulness of the available data for the purposes of UDP. Both these subtract from the user understanding the relevance of the data to her own situation. Instead of presenting a unified, coherent picture of the user's risk with respect to a number of hazards, the user is only able to view data with respect to specific risks and with specific instances of these risks. Furthermore, the ability to do so assumes the user can locate the data and then apply that data in the manner intended by the data provider. Consequently, despite the potential of Open Data to support building business resilience, and the potential of its application through UDP, the fragmentation of datasets impedes the process of bridging the objective/subjective risk perception divide.

Policy Considerations

Government entities play a leading role in designing and delivering open platforms that ingest, collate, and combine a number of potentially disparate datasets. Since these datasets belong largely to government entities, public policy considerations are centrally important to successfully enabling and promoting user-driven disaster management planning. In this section we outline how governments can improve the utility of open datasets for UDP, thereby enhancing the preparedness of businesses to face a broad spectrum of natural hazard risks.

In essence, government policy should seek to ensure that combined data platforms are designed to provide a user with a more complete picture of their disaster risk and the quality of their resilience. For instance, *Disaster Hub*—a website[2] prepared by the Local Government Association

[2] See http://www.disasterhub.com.au/.

Queensland—collects data from a number of sources and provides interactive maps on which various risks are represented geographically. It provides real-time data on rainfall at rain gauges across the Local Government Area, up-to-date locations of car accidents, power outages, and strong weather warnings issued by the Bureau of Meteorology; it also provides lists of school closures, social media posts, evacuation centres, and so on. These kinds of platforms provide data that allow for plans to be devised at the individual, household, or organisation level.

To enhance the usefulness of these data platforms, careful consideration should be given to two aspects of their design. First, the data should be semantically commensurate. That is, the data should, as much as possible, represent kinds of measurement that can be understood. For instance, representing tide levels in a river system and in a bay may differ greatly if they are represented purely in numbers. This difference reduces the intelligibility of the data. If this same data were presented, for instance, pictorially, so that the extent of the tide change was represented on a map, then a commonality in the measures emerges for the non-expert user. This consideration may also render data that are represented by different measures intelligible to non-expert users. For instance, interactions between river-level measurements and rain gauge measurements by themselves are not necessarily meaningful when only the raw measurements are viewed by a non-expert. But if the combined effect of the rain measurement and the river measurement is modelled and presented pictorially, then a more complete picture of a potential hazard emerges.

As we have noted, UDP is undertaken by users who are not necessarily domain experts. Therefore, the data on the platforms should be presented in a way that does not assume such expertise. In particular, the data should be presented in a way that is sensitive to the subjective risk assessment of the users and that do not assume only objective risk measurements.

A successful amalgamated data platform will attempt to align the user's subjective risk and its objective risk as closely as possible. We argue that this may be achieved by leveraging findings of narrative policy analysis (see, for instance, Hampton, 2009; Jones & McBeth, 2010; Roe, 1994). Narrative policy analysts examine the narrative strategies that policymakers use to communicate their policy ideas and that activists use to shift the policy agenda. The effects of a narrative on the perception of a particular event and its relation to policymaking can be considerable. As Fischer and Forester (1993, p. 44) point out:

> whether or not a situation is perceived as a political problem depends on the narrative in which it is discussed. To be sure, large groups of dead trees as such are not a social construct; the point is how one makes sense of dead trees. … One may see dead trees as the product of natural stress caused by drought, cold, or wind, or one may see them as victims of pollution … [thereby constituting] a political problem.

McComas and Shanahan (1999) document the use of narratives in US climate policy and the effects of these narratives in reaching the attention of policymakers. McBeth, Shanahan, Arnell, and Hathaway (2007) illustrate the ways in which environmental activists alter their narrative strategies to suit the political and policy environment, and Hampton (2005, 2009) argues that policy narratives may be used to increase public participation in policymaking.

An example of a concrete narrative strategy for a data platform is to include a user's risk and resilience in a historical perspective. For instance, geographically representing the extent of a previous disaster allows a user to connect their memory of the event to potential future events. Furthermore, coordinating a risk in a chain of causal events—for instance, the effects of the tides on the flooding of tributaries that run through communities that are further away from the coast—will serve to allow the user to position themselves in a larger risk and resilience narrative. Government policy that recognises the power of subjective risk narratives and is mindful of how data is presented within this narrative frame will allow for greater public participation by increasing the legibility and comprehensibility of the data for citizens.

Conclusion

Small to medium businesses play a critical role in building and sustaining the resilience of cities; their contribution to strong economies is vital. Similarly, their ability to withstand disasters and recover quickly is essential to broader community resilience. We have outlined how businesses can use Open Data to build their resilience. The model of user-driven planning leverages an extensive tradition of participant leadership and problem-solving in the domains of urban planning and participatory governance. In this chapter we have sought to integrate the approach of subjective user-led planning with hazard risk assessments using Open Data as the bridge to connect these world views. We have provided a novel approach to the challenge of building resilient cities that will benefit from further research to understand how this approach can work in practice.

In more immediate terms, our analysis of the user experience interacting with current datasets points to where governments can enhance Open Data policy to benefit users through the enhanced accessibility and utility of these data. The Queensland government has made significant progress in making such data available to the public. Policy and platforms that consider usability from a risk narrative perspective will enhance the government's efforts to promote disaster-resilient communities. Queensland is exposed to numerous, and increasingly severe, natural hazards; leveraging Open Data to build the resilience of cities provides an important pathway to face the challenges of its rapidly changing climate.

References

Aida, J., Kawachi, I., Subramanian, S. V., & Kondo, K. (2013). Disaster, social capital and health. In I. Kawachi, S. Takao, & S. V. Subramanian (Eds.), *Global perspectives on social capital and health* (pp. 167–188).

Aldrich, D. P., & Meyer, M. A. (2015). Social capital and community resilience. *American Behavioral Scientist, 59*(2), 254–269. https://doi.org/10.1177/0002764214550299

Alkon, A. H. (2004). Place stories and consequences: Heritage narratives and the control of erosion on Lake County, California, Vineyards. *Organization & Environment, 17*(2), 145–169.

Alpaslan, C. M., Green, S. E., & Mitroff, I. I. (2009). Corporate governance in the context of crises: Towards a stakeholder theory of crisis management. *Journal of Contingencies and Crisis Management, 17*(1), 38–49. https://doi.org/10.1111/j.1468-5973.2009.00555.x

Arnstein, S. R. (1969). A ladder of citizen participation. *Journal of the American Institute of Planners, 35*(4), 216–224. https://doi.org/10.1080/01944366908977225

Attorney-General's Department. (2012). National disaster relief and recovery arrangements: Determination 2012, version 2.0. Canberra: Attorney General's Department.

Australian Bureau of Statistics. (2017). *Regional population growth, Australia, 2015–16*.

Boin, A., & 't Hart, P. (2010). Organising for effective emergency management: Lessons from research. *The Australian Journal of Public Administration, 69*(4), 357–371.

Boin, A., 't Hart, P., Stern, E., & Sundelius, B. (2005). *The politics of crisis management: Public leadership under pressure*. Cambridge: Cambridge University Press.

Clarke, L. (1999). *Mission improbable: Using fantasy documents to tame disaster*. Chicago and London: University of Chicago Press.

Comfort, L. K., & Kapucu, N. (2006). Inter-organizational coordination in extreme events: The World Trade Center attacks, September 11, 2001. *Natural Hazards, 39*(2), 309–327. https://doi.org/10.1007/s11069-006-0030-x

Corvellec, H. (2011). The narrative structure of risk accounts. *Risk Management, 13*(3), 101–121. https://doi.org/10.1057/rm.2011.5

Cowan, T., Purich, A., Perkins, S., Pezza, A., Boschat, G., & Sadler, K. (2014). More frequent, longer, and hotter heat waves for Australia in the twenty-first century. *Journal of Climate, 27*(15), 5851–5871.

Deloitte Access Economics. (2013). *Building our nation's resilience to natural disasters*. Sydney: Deloitte Access Economics.

Deloitte Access Economics. (2016). *The economic cost of the social impact of natural disasters*. Sydney: Deloitte Access Economics.

Drennan, L. (2017). Community narratives of disaster risk and resilience: Implications for government policy. *Australian Journal of Public Administration, 77*(3), 456–467.

Drennan, L., McGowan, J., & Tiernan, A. (2016). Integrating recovery within a resilience framework: Empirical insights and policy implications from regional Australia. *Politics and Governance, 4*(4), 74. https://doi.org/10.17645/pag.v4i4.741

Emergency Management Australia. (2004). *Emergency management in Australia: Concepts and principles*. Canberra: Commonwealth of Australia.

Emergency Management Australia. (2015). *National Emergency Risk Assessment Guidelines (NERAG) Handbook*. Canberra: Commonwealth of Australia. Retrieved from https://knowledge.aidr.org.au/resources/handbook-10-national-emergency-risk-assessment-guidelines/

Fischer, F., & Forester, J. (1993). *The argumentative turn in policy analysis and planning*. Duke University Press.

Fiskaa, H. (2005). Past and future for public participation in Norwegian physical planning. *European planning studies, 13*(1), 157–174.

Fung, A., & Wright, E. O. (2001). Deepening democracy: Innovations in empowered participatory governance. *Politics & Society, 29*(1), 5–41. https://doi.org/10.1177/0032329201029001002

Goldsmith, S., & Eggers, W. D. (2003). *Governing by network: The new shape of the public sector*. New York, NY: Brookings Institution Press.

Hampton, G. (2005). Enhancing public participation through narrative analysis. *Policy Sciences, 37*(3–4), 261–276. https://doi.org/10.1007/s11077-005-1763-1

Hampton, G. (2009). Narrative policy analysis and the integration of public involvement in decision making. *Policy Sciences, 42*(3), 227–242. https://doi.org/10.1007/s11077-009-9087-1

Hansson, S. O. (2010). Risk: Objective or subjective, facts or values. *Journal of Risk Research, 13*(2), 231–238.

Henly-Shepard, S., Anderson, C., Burnett, K., Cox, L. J., Kittinger, J. N., & Ka'aumoana, M. (2015). Quantifying household social resilience: A place-based approach in a rapidly transforming community. *Natural Hazards, 75*, 343–363. https://doi.org/10.1007/s11069-014-1328-8

Henwood, K., Pidgeon, N., Parkhill, K., & Simmons, P. (2010). Researching risk: Narrative, biography, subjectivity. *Forum: Qualitative Social Research, 11*(1). Retrieved from http://www.qualitative-research.net/index.php/fqs/article/view/1438/2925

Ho, M.-C., Shaw, D., Lin, S., & Chiu, Y.-C. (2008). How do disaster characteristics influence risk perception. *Risk Analysis, 28*(3), 635–643. https://doi.org/10.1111/j.1539-6924.2008.01040.x

ISO. (2009). *Risk management—Principles and guidelines (Vol. ISO 31000)*. Geneva: International Standards Organization.

ISO. (2014). *Societal security—Business continuity management systems—Requirements (Vol. ISO22301:2014)*. Geneva: International Standards Organization.

Jacobs, J. (1961). *The death and life of great American cities*. New York, NY: Random House.

Jones, M. D., & McBeth, M. K. (2010). A narrative policy framework: Clear enough to be wrong? *Policy Studies Journal, 38*(2), 329–353.

Kapucu, N., & Garayev, V. (2011). Collaborative decision-making in emergency and disaster management. *International Journal of Public Administration, 34*, 366–375.

Kapucu, N., Hawkins, C. V., & Rivera, F. I. (2013). Emerging research in disaster resiliency and sustainability: Implications for policy and practice. In N. Kapucu, C. V. Hawkins, & F. I. Rivera (Eds.), *Disaster resiliency: Interdisciplinary perspectives*. New York and London: Routledge.

Larkham, P. J., & Lilley, K. D. (2012). Exhibiting the city: Planning ideas and public involvement in wartime and early post-war Britain. *The Town Planning Review, 83*(6), 647–668.

Lindell, M. K., & Whitney, D. J. (2000). Household seismic hazard adjustment adoption. *Risk Analysis, 20*(1), 13–25.

McBeth, M. K., Shanahan, E. A., Arnell, R. J., & Hathaway, P. L. (2007). The intersection of narrative policy analysis and policy change theory. *The Policy Studies Journal, 35*(1), 87–108.

McComas, K., & Shanahan, J. (1999). Telling stories about global climate change: Measuring the impact of narratives on issue cycles. *Communication Research, 26*(1), 30–57.

Menapace, L., Colson, G., & Raffaelli, R. (2012). Risk aversion, subjective beliefs, and farmer risk management strategies. *American Journal of Agricultural Economics, 95*(2), 384–389. https://doi.org/10.1093/ajae/aas107

Paton, D. (2005). *Community resilience: Integrating hazard management and community engagement*. Paper presented at the International Conference on Engaging Communities, Brisbane.

Paton, D., Smith, L., Daly, M., & Johnston, D. (2008). Risk perception and volcanic hazard mitigation: Individual and social perspectives. *Journal of Volcanology and Geothermal Research, 172*(3), 179–188. https://doi.org/10.1016/j.jvolgeores.2007.12.026

Purich, A., Cowan, T., Cai, W., van Rensch, P., Uotila, P., Pezza, A., …, Perkins, S. (2014). Atmospheric and oceanic conditions associated with southern Australian heat waves: A CMIP5 analysis. *Journal of Climate, 27*(20), 7807–7829.

Roe, E. (1994). *Narrative policy analysis: Theory and practice*. Maine: Duke University Press Books.

Shaw, R. (2013). Incorporating resilience of rural communities for proactive risk reduction in Shikoku, Japan. In *Disaster resiliency: Interdisciplinary perspectives* (pp. 207–226).

Sjöberg, L. (2000). Factors in risk perception. *Risk Analysis, 20*(1), 1–11.

Tiernan, A., McGowan, J., & Drennan, L. (2013). *From disaster to renewal: The centrality of business recovery to community resilience.* Canberra: Regional Australia Institute.

Tierney, K. (2012). Foreword. In N. Kapucu, C. V. Hawkins, & F. I. Rivera (Eds.), *Disaster resiliency: Interdisciplinary perspectives* (pp. xiii–xxvi). New York, NY: Routledge.

Waugh, W. (2006). Collaboration and leadership for effective emergency management. *Public Administration Review, 66*(2), 132–140.

Waugh, W. (2009). Mechanisms for collaboration in emergency management: ICS, NIMS, and the problem with command and control. In R. O'Leary & L. B. Bingham (Eds.), *The collaborative public manager: New ideas for the twenty-first century* (pp. 157–176). Washington, DC: Georgetown University Press.

Waugh, W., & Streib, G. (2006). Collaboration and leadership for effective emergency management. *Public Administration Review, 66*(2), 131–140.

Wise, C. R., & Nader, R. (2002). Organizing the federal system for homeland security: Problems, issues, and dilemmas. *Public Administration Review, 62*, 44–57.

Zheng, F., Westra, S., & Leonard, M. (2015). Opposing local precipitation extremes. *Nature Climate Change, 5*(5), 389–390.

Correction to: Open Cities | Open Data

Scott Hawken, Hoon Han, and Christopher Pettit

Correction to:

S. Hawken et al. (eds.), *Open Cities | Open Data*,
https://doi.org/10.1007/978-981-13-6605-5

An error in the editing process by Springer unfortunately led to a misspelling of a surname for a co-author of this book. Pettit was misspelt as Petit. It has been corrected as Pettit.

The updated original online version can be found at
https://doi.org/10.1007/978-981-13-6605-5

S. Hawken et al. (eds.), *Open Cities | Open Data*,
https://doi.org/10.1007/978-981-13-6605-5_18

Index[1]

[1] Note: page numbers followed by 'n' refer to notes.

S. Hawken et al. (eds.), *Open Cities | Open Data*,
https://doi.org/10.1007/978-981-13-6605-5

The manufacturer's authorised representative in the EU is Springer Nature Customer Service Centre GmbH, Europaplatz 3, 69115 Heidelberg, Germany. If you have any concerns regarding our products, please contact ProductSafety@springernature.com

Printed and bound by CPI Group (UK) Ltd, Croydon, CR0 4YY
30/09/2026
02225235-0005